The Decentralization of Forest Governance

The Earthscan Forest Library

Series Editor: Jeffrey A. Sayer

The Decentralization of Forest Governance: Politics, Economics and the Fight for Control of Forests in Indonesian Borneo
Moira Moeliono, Eva Wollenberg and Godwin Limberg

The Forest Certification Handbook 2nd edition
Ruth Nussbaum and Markku Simula

The Forest Landscape Restoration Handbook
Jennifer Rietbergen-McCracken, Stewart Maginnis and Alastair Sarre (eds)

Forest Quality: Assessing Forests at a Landscape Scale
Nigel Dudley, Rodolphe Schlaepfer, Jean-Paul Jeanrenaud, William Jackson and Sue Stolton

Forests in Landscapes: Ecosystem Approaches to Sustainability
Jeffrey A. Sayer and Stewart Maginnis (eds)

Forests, People and Power: The Political Ecology of Reform in South Asia
Oliver Springate-Baginski and Piers Blaikie (eds)

Illegal Logging: Law Enforcement, Livelihood and the Timber Trade
Luca Tacconi (ed)

Lessons from Forest Decentralization: Money, Justice and the Quest for Good Governance in Asia-Pacific
Carol J. Pierce Colfer, Ganga Ram Dahal and Doris Capistrano (eds)

Logjam: Deforestation and the Crisis of Global Governance
David Humphreys

Plantations, Privatization, Poverty and Power: Changing Ownership and Management of State Forests
Mike Garforth and James Mayers (eds)

Policy that Works for Forests and People: Real Prospects for Governance and Livelihoods
James Mayers and Stephen Bass

The Politics of Decentralization: Forests, Power and People
Carol J. Pierce Colfer and Doris Capistrano

The Sustainable Forestry Handbook 2nd edition
Sophie Higman, James Mayers, Stephen Bass, Neil Judd and Ruth Nussbaum

The Decentralization of Forest Governance

Politics, Economics and the Fight for Control of Forests in Indonesian Borneo

Edited by
Moira Moeliono, Eva Wollenberg and Godwin Limberg

earthscan
from Routledge

First published by Earthscan in the UK and USA in 2009

ISBN-13: 978-1-84407- 586-7 (hbk)
ISBN-13: 978-0-41585- 303-3 (pbk)

For a full list of publications please contact:

Earthscan

2 Park Square, Milton Park, Abingdon, Oxon OX14 4RN

711 Third Avenue, New York, NY 10017

Earthscan in an imprint of the Taylor & Francis Group, an informa business

First issued in paperback 2013

A catalogue record for this book is available from the British Library

Library of Congress Cataloging-in-Publication Data

The decentralization of forest governance : politics, economics and the fight for control of forests in Indonesian Borneo / edited by Moira Moeliono, Eva Wollenberg and Godwin Limberg.
p. cm.
ISBN 978-1-84407-586-7 (hardback)
1. Forest policy–Indonesia–Malinau. 2. Forest management–Indonesia–Malinau. 3. Logging–Indonesia–Malinau. I. Moeliono, Moira M. M. II. Wollenberg, Eva. III. Limberg, Godwin. IV. Title: Politics, economics and the fight for control of forests in Indonesian Borneo.
SD657.I5D435 2008
333.7509598'3–dc22

2008017439

Contents

Part II Small-scale Timber Harvesting in Malinau

List of Figures, Tables and Boxes

Figures

Tables

Boxes

List of Contributors

Imam Basuki is a soil specialist working with the Center for International Forestry Research (CIFOR) in Indonesia. He has worked mostly in soil and socio-economic aspects of forested landscape, together with local institutions and communities in Borneo, Papua and Vietnam. He is interested in resource management, land evaluation and GIS.

Haris Iskandar's professional experiences range from field work as field assistant and field coordinator, to collaborative research and studies conducted jointly and/or funded by a number of international organizations and civil societies. He has experience with GIS, carbon sequestration, natural regeneration of mahogany, Reduced Impact Logging (RIL) techniques, utilization of wood wastes for the community, project scenarios of small-scale A/R CDM Project in Indonesia, and several studies related to REDD mechanisms. Throughout his professional experience with CIFOR and Winrock International he has participated in numerous seminars and workshops and has authored/co-authored a number of publications.

Ramses Iwan has been a field researcher at CIFOR since 2001. He is a community member of Setulang Village in Malinau, East Kalimantan.

Kewin Kamelarczyk holds an MSc in forestry and land use in developing countries from the Faculty of Life Science at Copenhagen University. He has been involved in work and research within several forest related topics; from community forestry and forest dependency to forest resources assessments and information management. Most recently he has worked for FAO's Global Forest Resources Assessment and FAO's country support to national forest assessments in Rome. Currently he is pursuing a PhD studying the interaction between national forest assessments processes and national policy dialogues.

Godwin Limberg has lived and worked in East Kalimantan, Indonesia, since 1990. He has been a CIFOR field researcher, focusing on governance issues and community rights, since 2000. Prior to joining CIFOR, he worked with the Rio Tinto Foundation, World Wildlife Fund in Kayan Mentarang National Park, Indonesia, and as a volunteer with a local community development organization. He holds a Master's degree in tropical crop science from Wageningen Agricultural University, The Netherlands.

Nining Liswanti is a biodiversity researcher working with CIFOR in Indonesia. She has worked on developing rapid ecological assessment methods in Indonesia, Thailand and Vietnam.

Moira Moeliono has been a researcher at CIFOR since 2001, looking at national-local policy linkages, policy learning at local level and issues of community involvement in natural resource management.

Charles Palmer is an environmental economist currently working as a postdoctoral researcher at the Swiss Federal Institute of Technology (ETH) in Zurich, Switzerland. He has previously worked with CIFOR in Indonesia, along with the United Nations Development Program (UNDP) and the International Institute for Environment and Development (IIED). Current interests include forestry carbon contracts, biofuels and land use, and the evaluation of protected area policy.

Steve Rhee received his doctorate from Yale, and has worked over the past 15 years as a researcher and practitioner of international development to improve evidence-based policy and practice, particularly as it relates to natural resource management. He has worked with a range of stakeholders, from forest-dependent villagers in Borneo to international donors, and has worked closely with CIFOR since 1999. He is currently an American Association for the Advancement of Science (AAAS) Science and Technology Policy Fellow at the US Department of State.

Douglas Sheil conducted this work with CIFOR but is now a joint director of the Institute for Tropical Forest Conservation (Mbarara University for Science and Technology), PO Box 44, Kabale, Uganda. Tel: +256 (0)39 2709753; Fax: +256 (0)39 2251753.

Uffe Strandby holds an MSc in forestry and a PhD in socio-economics and conservation ecology related to an endangered conifer in Central America. He has worked in research about timber species' regeneration ecology, stakeholder conflicts, marketing of forest products and community forestry in Indonesia, Thailand, Peru, Ecuador, Guatemala and Mexico.

Made Sudana worked for CIFOR from 1999 to 2004, based in Malinau Subprovince of East Kalimantan. He participated in the ACM project and has also worked in research about conflict and power among stakeholders related to community forestry, timber and mining companies, and local government in Malinau and Manggarai Subprovince of East Nusa Tenggara. He currently works for the Indonesian-German Development Cooperation on Rural Water Supply in Sumba.

Miriam van Heist conducted this work with CIFOR but is now a joint director of the Institute for Tropical Forest Conservation (Mbarara University for Science and Technology), PO Box 44, Kabale, Uganda. Tel: +256 (0)39 2709753; Fax: +256 (0)39 2251753.

Meilinda Wan is an agronomist with CIFOR. Her main research interest is in biodiversity and people in changing landscapes.

Eva (Lini) Wollenberg is a social scientist with a longstanding commitment to the empowerment of local communities. Lini received her PhD from the University of California, Berkeley, and has subsequently worked with the Ford Foundation and as a senior scientist with CIFOR, leading research in the tropical forests of Asia, especially in Kalimantan (Borneo). She is a senior associate of CIFOR and currently director of sustainable agriculture at the University of Vermont, United States.

Introduction

Eva Wollenberg, Moira Moeliono and Godwin Limberg

Amidst global concern for forests – whether deforestation, the protection of species-rich ecosystems or climate change – the stories of local people's struggles over forests are often lost or invisible. Yet, local people's daily struggles directly affect forests as well as the hundreds of thousands of people who depend upon them for their livelihood.

This book is about local people's decisions about valuable forest resources in the context of changing opportunities in a globally critical forest region: Indonesian Borneo. We explore what happened after decentralization and democracy reforms in Indonesia, when district governments gained new authority and local communities gained new rights, and both led small-scale forest exploitation from 1998 to 2003 in the district of Malinau, East Kalimantan. The chapters examine the resulting changes and relationships among communities, local government and the forest.

Our story is about an experiment in democratic control of a natural resource, as well as of what happened to the local use of this resource during a time of rapid policy reform. It is a story about everyday people trying to make sense of the opportunities available to them.

Malinau is not unique. What has happened in Malinau is part of a larger global trend towards local political orders. As a result of decentralization and democratization movements in many countries, for the first time in the most remote forest areas, local people have representatives from their own communities who have a say in formal government decision-making. People in these areas no longer rely only on timber companies or limited forestry sector programmes for public goods and services. Local governments are now providing health services, schools, administration and access to natural resources to people in even the remotest areas.

At the same time, many of these local arrangements are fragile and in flux. The new orders are not as democratic as their architects had intended for them to be (Larson and Ribot, 2005; Ribot, 2006). The uncertainty accompanying the new policies has often created a window of opportunity for rapid resource exploitation. As in many localities, the balance of control between the state and local people is still developing and different groups continue to seek opportunities to exploit new benefits from forests. National and international players are trying to find the balance of control that will also meet their needs.

In focusing on Malinau, we aim not only to understand the transition of these local political orders, but also to document a special place and time in the history of the world's forest and forest peoples. Borneo has undergone remarkable transformation in the last generation (Padoch and Peluso, 1996), but during the period covered in this book, experienced even more striking changes, especially in its remotest regions. Logging, both legal and illegal, skyrocketed. Between 1980 and 2003, Kalimantan's protected lowland forest declined by more than 56 per cent, or 29,000 square kilometres (Curran et al, 2004). Districts learned to have autonomy. New administrative boundaries were established. In East Kalimantan alone, four new districts were created, doubling the number of administrative centres in the province and bringing funding for new roads, schools, health services and village projects, but also competition for budget allocations from the central government. The claims of ethnic-based customary groups to state forest land were legally recognized and received 'fees' and compensation payments, sometimes in the tens of thousands of dollars. Overt conflict increased in frequency and intensity (Wulan et al, 2004) to levels not experienced since the headhunting days of the early 1900s.

Although these changes were spurred by decentralization and accompanying reforms, they reflect a much broader set of trends affecting people living in Indonesia's forest areas, including the weakening and disorganization of state institutions; the efforts of civil society to organize itself after three decades of authoritarian state control; increasing linkages between local forest communities and the international networks (Colchester et al, 2003); the decline of large-scale state-controlled forestry; and pressures to manage resources to alleviate poverty and social injustice. They also reflect a period of transition and immense uncertainty, as Indonesia boldly embarked on the path to becoming a democratic state during a period of economic, climatic and political crisis (Sunderlin, 1999) before the functional institutions for democracy existed. This book seeks to document, if only partially, this turbulent and complex period.

A final theme of the book is the role of the authors themselves in Malinau. The authors conducted research in Malinau for the Center for International Forestry Research (CIFOR). Most of this research shared the development objective to support local communities in better access to, and management of, their forests, and was conducted in partnership with local communities or local government. Our research assumed a final goal of adaptive collaborative management through multi-stakeholder processes. However, these approaches assume communications and trust or that they can be easily built, and that clear institutions for deliberations, agreements, checks and balances, and conflict management are in place. In Malinau, at the start of decentralization, these conditions did not exist. We therefore learned to be highly flexible and adaptive by being embedded in society, maintaining close relations with the people who support gatekeepers and developing programmes for multiple-interest groups. We were thus able to capture opportunities whenever they arose and adjust to local circumstances. In this way, information is more accurate, in-depth and comprehensive, especially in terms of listening to different perspectives from different groups (see Wollenberg et al, 2006, on spontaneous cooperation).

Organization of the book

The first part of the book provides the theoretical and contextual backdrop for Malinau's story. In Chapter 1, we examine the theory of 'state in society' and the implications of this theory for decentralization's impact upon power struggles between local peoples and the state. The chapter also summarizes Indonesia's development from an autocratic state to a local political order, with specific attention to the forestry sector.

The remainder of the book focuses on Malinau. Part I (Chapters 1 to 4) provides a history of how society in Malinau has organized itself, especially with respect to access to forests. After a brief geographical introduction, Moeliono and Limberg describe the local policy context of decentralization (Chapter 2). In Chapter 3, Rhee introduces the different actors in Malinau and the ways in which decentralization has affected their power relations, while in Chapter 4, Sheil et al describe local villagers' values concerning the forest and some of the trends in how these values are changing. Part II of the book offers a set of analyses on small-scale timber harvesting (*Izin Pemungutan dan Pemanfaatan Kayu*, or IPPK) impacts and their causes. In Chapter 5, Limberg examines the impact of the IPPK operations on seven villages in the Malinau watershed, while in Chapter 6, Palmer documents the conditions surrounding negotiations between communities and companies. Palmer explains how Malinau's communities overall benefited less than those in other districts such as Kutai Barat in East Kalimantan, where communities initiated the negotiation processes and were better informed about fee options. Kamelarczyk and Strandby look at the influence of IPPKs on households in three selected communities in Chapter 7. In Chapter 8, Iskandar et al provide an analysis of the impact of the operations upon the forests, while in Chapter 9, the last chapter of Part II, Iwan and Limberg describe a community effort to protect valuable forests through a local initiative.

Part III of the book examines the broader issues of property relations, conflict and political participation associated with decentralization. In Chapter 10, Sudana analyses the increased incidence of conflict after decentralization, and in Chapter 11, Moeliono and Limberg describe how tenure and *adat* emerged as political issues. In Chapter 12, Wollenberg describes how decentralization and democratization processes play out in the villages of Malinau. Moeliono and Limberg, in Chapter 13, then show how the government has tried to adjust to its new role in a decentralized state through several development experiments. The final chapter, Chapter14, offers a synthesis and conclusion, describing the rise of local political orders largely driven by ethnic alliances. In the process, marginalized groups might become even more marginalized, as shown in the case of the Punan of Malinau described in Chapters 12 and 13. In the mutual adjustments of state and society, however, forest governance has become a source of conflict. It remains unclear who, in fact, is in charge. In the same way as land tenure, forest governance is regulated by a mix of traditional non-formal and legal-formal rules and institutions.

References

Colchester, M., Apte, T., Laforge, M., Mandondo, A. and Pathak, N. (2003) *Bridging the Gap: Communities, Forests and International Networks*, CIFOR Occasional Paper 41, CIFOR, Bogor, Indonesia

Curran, L. M., Trigg, S. N., McDonald, A. K., Astiani, D., Hardiono, Y. M., Siregar, P., Caniago, I. and Kasischke, E. (2004) 'Lowland forest loss in protected areas of Indonesian Borneo', *Science*, vol 303, pp1000–1003

Larson, A. and Ribot, J. (2005) 'Democratic decentralisation through a natural resource lens: An introduction', in J. Ribot and A. Larson (eds) *Democratic Decentralization through a Natural Resource Lens: Experience from Africa, Asia and Latin America*, Routledge, London

Padoch, C. and Peluso, N. L. (eds) (1996) *Borneo in Transition: People, Forests, Conservation, and Development*, Oxford University Press, Kuala Lumpur

Peluso, N. L. and Padoch, C. (1996, and 2nd edition, 2003) 'Changing resource rights in managed forests of West Kalimantan', in C. Padoch and N. L. Peluso (eds) in *Borneo in Transition: People, Forests, Conservation, and Development*, Oxford University Press, Kuala Lumpur

Ribot, J. C. (2006) 'Representation, citizenship and the public domain: Framing the local democracy effects of institutional choice and recognition', Paper presented at the 11th Biennial Conference of the IASCP, Bali, Indonesia, March 2006

Sunderlin, W.D. (1999) 'Between danger and opportunity: Indonesian forests in an era of economic crisis and political change', *Society and Natural Resources*, vol 12, no 6, pp559–570

Wulan, Y. C., Yasmi, Y., Purba, C. and Wollenberg, E. (2004) *Analisa Konflik: Sektor Kehutanan di Indonesia 1997–2003,* CIFOR, Bogor, Indonesia

List of Acronyms and Abbreviations

AAC	annual allowable cut
ADB	Asian Development Bank
AMAN	Alliansi Masyarakat Adat Nusantara (alliance of *adat* groups)
ANOVA	analysis of variance
APBD	*Anggaran pendapatan dan Belanja Daerah* (provincial budget)
APBN	*Anggaran pendapatan dan Belanja Negara* (national budget)
asl	above sea level
BAPPEDA	Badan Perencanaan Pembangunan Daerah (district/regional planning agency)
BBP	Bakti Bumi Perdana
BDMS	Bara Dinamika Mudah Sukses (mining company)
BLU	Badan Layanan Umum (public service agency)
BPD	*Badan Perwakilan Desa* (village assembly)
BPN	*Badan Pertanahan Nasional* (national land agency)
BTRF	Borneo Tropical Rainforest Foundation
CEC	cation exchange capacity
° C	degrees Celsius
CERD	Community Empowerment for Rural Development
CIFOR	Center for International Forestry Research
cm	centimetre
CV	*Commanditaire Vennootschap* (limited partnership)
dbh	diameter at breast height
DFID	UK Department for International Development
DPRD	*Dewan Perwakilan Rakyat Daerah* (local consultative assembly/district house of representatives)
dsh	diameter at the estimated stump height
FFPRI	Forestry and Forest Product Research Institute
GER	Global Eco Rescue
GIS	geographic information systems
GPS	global positioning system
ha	hectare
HPH	*Hak Pengusahaan Hutan* (logging concession/concessionaire)
HPHH	*Hak Pemungutan Hasil Hutan* (timber permit)
IED	Institute for Environmental Decisions, Zurich, Switzerland
IFAD	International Fund for Agricultural Development
IPGRI	International Plant Genetic Resources Institute

IPPK	*Izin Pemungutan dan Pemanfaatan Kayu* (small-scale timber harvesting)
IUPHHK	small-scale timber concession of a maximum of 50,000ha (small-scale timber concession of a maximum of 50,000ha)
kg	kilogram
KKN	*korupsi, kolusi, dan nepotisme* (collusion, corruption and nepotism)
km	kilometre
KSN	*Kawasan Strategis Negara* (state area of strategic importance)
LADAS	Lembaga Adat Dayak Abai Sembuak (non-profit organization)
LATIN	Lembaga Alam Tropika Indonesia (the Indonesian Tropical Institute, an NGO based in Bogor, West Java)
LIPI	Indonesian Academy of Sciences
LTI	Lestari Timur Indonusa (a limited liability company that operated in Setulang, Malinau)
m	metre
mm	millimetre
MoF	Ministry of Forestry
MPR	Majelis Permusyawaratan Rakyat (People's Consultative Assembly)
n	total sample population size
NGO	non-governmental organization
NRM	natural resource management
NTFP	non-timber forest product
P5D	*Pedoman Penyusunan Perencanaan dan Pengendalian Pembangunan Daerah* (Guidelines for Planning and Monitoring Regional Development)
PAD	*pendapatan asli daerah* (locally generated revenue)
PADUS	Persatuan Dayak Sewilayah Semendurut (united working group for forest rights)
PDM	Pebble Distribution Method
PKK	*Pembinaan Kesejahteraan Keluarga* (Family Welfare Movement)
PMD	*Pemberdayaan Masyarakat Desa* (Village Empowerment Agency)
PT WYKI	PT Wana Yasa Kahuripan Indonesia
RTRWP	*Rencana Tata Ruang Wilayah Propinsi*
SE	standard error
SHK	*Sistem Hutan Kerakyatan* (an NGO based in Samarinda, East Kalimantan)
SK	*Surat Keputusan*
SRL	sustainable rural livelihood framework
TPTI	*Tebang Pilih Tanam Indonesia* (guidelines for Indonesia's selective cutting and replanting system)
UK	United Kingdom
US	United States
VOC	Verenigde Oost-Indische Compagnie (Dutch East India Trading Company)
WALHI	Wahana Lingkungan Hidup Indonesia (Indonesian Forum for Environment, an NGO based in Jakarta)
WWF	World Wide Fund for Nature (*formerly* World Wildlife Fund)

Part I

Theoretical and Contextual Background

1

Between State and Society: Decentralization in Indonesia

Eva Wollenberg, Moira Moeliono and Godwin Limberg

In order to understand local people's changing opportunities in Indonesia, an explanation is needed of how the spectacular upheavals of Indonesia's decentralization and democracy reforms enabled local society to assert increased influence over the previously highly autocratic, coercive state and its iron grip over forests. The work of Joel Migdal, Atul Kohli, Vivienne Shue and others (1994) provides a framework for the analysis of these changing relationships. This chapter will take an in-depth look at Indonesia's history of centralization and decentralization, and the legal framework defining decentralization during the reform period between 1998 and 2003.

States and societies

First, definitions and assumptions will be provided. The state is defined here as a set of organizations that use rule-making and coercion to exert control in a defined territory (Hall and Ickenberry, 1989). The organizations that comprise a state – departments of forestry, indigenous peoples, parks, legislatures – compete and contradict each other to meet their own interests. Society is a set of people who share an identity as members of a group due to shared norms (Migdal, 1994). As with the state, elements of society can compete or oppose one another.

States and societies shape each other through attempts at domination and resistance in multiple arenas over time (Migdal et al, 1994). Forest-related arenas include property rights, distribution of benefits from the forests, the use and protection of forest resources, opportunities for labour or employment, political security, infrastructure, knowledge and access to information. The channels of influence of the state and society differ.

The state's influence over forests is typically based on state control of policy, forestry operations, or ownership of forest and forest land (Finger-Stich and Finger, 2003). In countries with valuable forest resources, such as Indonesia, state control is often centralized in national forestry departments. Elements of society might exercise influence through informal networks, social movements, or formal organizations such as businesses, religious institutions, donors and advocacy groups; but, in practice, local societies have had little formal influence over valuable forests.

Local societies have had enormous informal influence, however. Where state presence and authority in forest areas are weak, local groups often have their own principles, norms, rules and practices, and selectively implement or ignore laws imposed by the state. These groups act as 'semi-autonomous social fields' (Moore 1973, pp720–722). In Chaggaland, Tanzania, villagers ignored land reform measures intended to convert private land to government leaseholds and continued to allocate land rights according to local lineages (Moore, 1973). Local political bosses in Bosawas National Park, Nicaragua (Kaimowitz et al, 2003), and northern Sumatra (McCarthy, 2000b) and East Kalimantan (Obidzinski, 2004), Indonesia, openly organize logging considered illegal. While such autonomy enables forest communities to influence their immediate environment, it does not enable them to engage in the larger political framework.

Where state presence and authority are strong, local people have sought to influence local officials through the usual weapons of the weak (Scott, 1998): co-opting officials (Lipsky, 1980); developing personalized patron–client networks with officials (Shue, 1994); ignoring, resisting or poorly implementing policies (Manor, 1999); or, in extreme cases, protesting through fires, destroying property, poaching, uprooting seedlings and seizing equipment (Peluso, 1992). In Indonesia, as elsewhere, parastatal concession holders informally allowed people to continue swidden farming and hunting despite clear national regulations outlawing both activities. In China, local governments in Jinping, east Guizhou, and Jinggu, south Yunnan, developed shareholding systems for community members to allocate forest access, responsibilities and benefits (Liu and Edmunds, 2003).

Decentralization

The global trend towards decentralization is increasing the state's influence in forest areas, while giving local society a vastly greater role in that influence. Governance is now placed more firmly between local society and the state. Eighty per cent of all developing countries or countries in transition are now engaged in some form of decentralization (Gregersen et al, 2005, p13, citing Manor, 1999). More than 60 countries have decentralized some aspect of natural resource management (Larson 2005, p32, citing Agrawal, 2001).

Decentralization has taken two forms that have distinctive implications for the influence and well-being of people living in or near forests. The first is *centrally driven community forestry programmes*: the granting of formal rights to conduct forestry from central or meso-level forestry agencies to community-level groups (or agroforestry). Examples include joint forest management in India, integrated social forestry in the Philippines, community forestry in Cameroon, indigenous forestry concessions in Bolivia and forestry *ejidos* in Mexico.

The second form is *local governance*, the granting of formal control from the central government to local governments for administration and decisions in multiple sectors. Local governments are given powers and authority previously held by the centre, as well as financial allocations from the centre or from locally generated taxes, fees and royalties. The degree of power, responsibility and finances in each sector varies by country, and forest management has not always been fully decentralized to local government units (Ribot and Larson, 2005). The administrative level to which decentralization occurs is usually one to two levels above the community. Examples include decentralization to *panchayats* in India, to *municipios* in Bolivia, and to districts in Uganda. This book focuses on decentralization in Indonesia to local districts, or *kabupaten*.

Despite their similar origins, centrally driven community forestry programmes and local governance differ strongly in their implications for local people. In contrast to community forest management, co-management or devolution policies, which attempt to retain control over critical decisions and benefits, and are limited to the forest sector (Sarin et al, 2003), decentralization creates opportunities for forest users to engage in a broader realm of politics and influence the state itself, including its historical domination over timber and forest land.

Local governance, by its very nature, enables local groups to exercise more influence informally *and* formally. It sets up new intermediate arenas of struggle that encourage social organizations and political engagement at the intermediate level (Shue, 1994). Decentralization makes the boundaries between state and society more indistinct and fluid. It can thereby serve as an integrating force between the state and society while fostering struggles between them.

These interdependencies create several special challenges and opportunities for people living in and near forest areas.

First, the creation of local governments *increases the presence of the state locally*. Where such local presence is also associated with strong authority, *forest communities' capacity for behaving semi-autonomously is reduced*. Yet, communities need to *engage more directly with the state* in order to maintain influence over forest management. They also *have more influence in matters of the state*. Local people have more opportunities to engage local officials and more personalized relationships with local officials. Officials with significant authority find themselves absorbed in local social relations, with the attendant opportunities and obligations. Local government may be composed of local people, who must become adept at juggling their roles. The boundary between society and the state becomes highly fluid under these circumstances. Social forces are not necessarily organized for a common purpose, however, and can have highly disparate interests.

Enhanced engagement is a double-edged sword. Local elites may co-opt or collude with the state to achieve their own personal interests, including appropriating government resources intended for the poor (Etchevery-Gent, 1993). To promote desirable outcomes from engagement, the state needs to be strong enough to guard against elite greed or inequities caused by markets, and society needs to be strong enough to exert its 'civil authority over public matters' (Antlöv, 2003, p73). Enhanced state presence and engagement will be most intense for those in closest physical proximity to the local government. For those in remote locations, there may be little change.

Second, decentralization introduces divisions that can *fragment the state* and weaken it (Kohli, 1994), especially during policy transitions. Effective horizontal and vertical relationships among local governments are necessary for the state to achieve dominance. Where these links do not exist, local governments are likely to act more like the semi-autonomous fields described above. In Indonesia, local governments, the district forest services and the national Ministry of Forestry often act almost as if the other does not exist. A fragmented state also makes it easier for local government officials and the private sector to act corruptly since local officials can overlook national laws with little fear of sanctions (Resosudarmo, 2003; Smith et al, 2003).

Third, the impersonal nature of the centralized state has been one of its defining sources of power (Kohli, 1994). Decentralization *makes the state more personal* to the extent that officials share kinship and personal histories, friendship, economic interdependence, shared cultural norms and local power relations with local society. While this reduces the state's power and makes co-optation of local officials more likely, personalization can also *make the state and society more accessible to each other*, facilitating communication, understanding and mutual engagement.

Fourth, taken together, decentralization, weaker states and the revival of traditional values and indigenous identity have allowed *local traditional or cultural groups to become more politically relevant*, especially where their scales of influence coincide with that of local governments. This tendency reinforces the personalized links between local society and officials. It suggests that local governments have to build alliances with these groups, particularly the most influential ones. Where the power balance is in flux or where conflict is high between groups in a jurisdiction, local officials may not be able to maintain stable political control.

Fifth, decentralization *limits society's possibilities for organizing large-scale political initiatives, while increasing its possibilities for organizing locally.* The resulting emerging *local political orders seek their own autonomy and capacity.* Both local governments and social forces seek to organize and mobilize people and resources to increase their influence. Local arenas and means of engagement become more important than national ones. In forest areas, issues related to economic and social development take precedence over national concerns of biological diversity, intellectual property rights and foreign currency earnings.

Sixth, emerging *local political orders have characteristic unresolved tensions that contribute to persistent conflicts and instability, and drive inevitable adaptations and shifts in power*. The tensions arise from local government's intermediate position between the centre and local society. Policies from the central government and from local government are prone to conflict. Stronger local political orders make it more difficult for the centre to achieve domination. Yet, the centre and the periphery remain strongly interdependent; both have resources that the other wants. In forest areas, the centre is reluctant to give up income from valuable timber and forest products, while local authorities need the markets, skills and capital available from the centre.

Together, these conditions create distinct opportunities for local forest communities to influence the state in new ways that blend formal and informal channels of influence. Such channels include more opportunities for direct engagement, personal relationships, traditional or customary authorities, and new local political organizations, as well as formal mechanisms such as voting or participating in public hearings. Local political orders make the classic branching hierarchy of Weberian governance of forest areas obsolete. Control is not absolute, nor is the monopoly of the state. No one element of the state or society can dominate. Instead, governance is multi-polar, multileveled and characterized by multiple struggles. Decentralization enables local society and the state to share power better, but in ways that are also vulnerable to vacillations in power. As shown in the next section, such vacillations are part of the very nature of local people's relationship with the state.

Indonesia's transformation

Decentralization is not new in Indonesia. Although the pace of reforms was most dramatic from 1998 to 2003, decentralization and democratization reforms had their origins much earlier. Indeed, decentralization has waxed and waned throughout Indonesia's history since the Dutch colonial period. Such waves of decentralization are not unusual. There have been at least four waves in Francophone West Africa since 1917 (Ribot, 1999) and three in South Asia since the mid 1800s (Agrawal and Ribot, 1999).

This section briefly reviews the history of Indonesia's efforts at decentralization, showing how both the balance between local and central power and the meanings of decentralization and autonomy have shifted regularly. As the colonial period played a significant role in the way in which the Indonesian state evolved, some early history is outlined below. For an overview of the different legal products structuring the progress of decentralization, see Table 1.1.

An earlier effort

A first experiment in decentralization was the Decentralization Law of 1903 issued by the Dutch colonial government (Matsui, 2003).

When the Dutch took over control of the archipelago in 1800, they inherited a system of indirect rule started by the Dutch East India Trading Company

Table 1.1 *Chronology of laws on decentralization and forestry*

Year	Regional Administration	Forestry
1903	Dutch Colonial Decentralization Law	
1922	Revised Decentralization Law; formation of provinces and districts	
1942	Japanese occupation	
1945	Indonesian Independence Law 1, 1945, regulating regional administration including establishment of the local national committee; the elected chief executive of the committee is both a central government officer and a leader of local government; emphasis is put on de-concentration	
1948	Law 22, 1948, establishing legislative and executive functions of the government and three autonomous levels (the province, the district/municipality and the village); emphasis is put on decentralization	
1957	Law 1, 1957, recognizing extensive autonomy to the local government concerning all powers that the central government does not have (Matsui, 2003, p7); elected local assembly	
1957		GR 64/1957 (Hutabarat, 2001), granting provinces the authority to issue logging permits for areas up to 10,000ha and district heads for areas of 5000ha
1959	Presidential Decree 6, 1959, on reverting to the 1945 constitution; return to the centralized system	
1960		Basic Agrarian Law, considered by some to be the overarching law, although more recently disputed as being merely a sectoral law equal to other sectors
1965	Law 18, 1965, the basic law on local administration; government head tasked with dual function of heading local autonomy as well as being a proxy for the central government	
1967		Law 5, 1967, the basic forestry law
1970		GR 21, 1970, on Rights of Exploiting Forest negating GR 64, 1957
1974	Law 5, 1974, on regional administration, regulating the central–regional relationships	
1979	Law 5, 1979, on village administration, introducing a uniform village structure	
1985		The Directorate General of Forestry becomes the Ministry of Forestry
1995	GR 8, 1995 – a two-year trial on decentralization to 26 districts	GR 8, 1995, providing for the partial transfer of authority over forests to 26 districts on a trial basis
1995		Ministerial Decree on Community Forestry (SK 662)

Year	*Regional Administration*	*Forestry*
1998		GR 62, 1998, transferring of partial authority within the forestry sector (rehabilitation; community forest, beekeeping, silkworm farming) to districts
1998		Revised Community Forestry Decree
1999		GR 6, 1999, granting districts the right to issue permits for small-scale logging (100ha) and provinces (10,000ha)
1999		Law 41, 1999, on forestry
1999	Law 22, 1999, on regional administration, setting the framework for general decentralization	
2000	Tap MPR III on the Hierarchy of Laws; ministerial decrees not mentioned	GR 6, 1999 repealed by a ministerial decree; but districts ignored the decree and continued to issue small-scale logging permits
2001		Second revision of the Community Forestry Decree
		Consultative Assembly Decree IX on Agrarian Reform
2002		GR 34, 2002, on Forest Use and Management
2004	Revised Decentralization Law (Law 32, 2004)	
2007		GR 6, 2007, revision of GR 34

(Verenigde Oost-Indische Compagnie, or VOC). The Dutch government started a period of hierarchic empire building, which produced a centralized state overlaying diverse traditional structures. By the beginning of the ethical policy period in the mid 1800s, the Dutch had established a centralized state with most of Java under direct rule, while most of the outer islands, including Borneo, remained under indirect rule.

At the end of the 1800s, the Dutch started to incorporate the outer islands. The Decentralization Law of 1903 intended both to lighten the burden of the government in administering remote areas and to increase control over those areas (Furnivall, 1956). With this law, the central government transferred authority to *keresidenan*,[1] or residencies, and established regional councils (Pide, 1999; Engelfriet, 2000; Ismanto, 2003). The councils had the authority to make financial decisions; but as they consisted mostly of government officials with the residents or assistant residents appointed, rather than elected, as chair, the councils generally functioned merely as advisory bodies. This first attempt to decentralize set a pattern for later efforts: rather than devolving the central administration, it supported the emergence of local administration by local organizations (Pide, 1999; Matsui, 2003) at the district level.

The 1903 law was amended in 1922 to create new provinces and provincial councils (Ismanto, 2003), where autonomy was transferred from the *keresidenan* to the province (Suharyo, 2000). The councils, however, remained an advisory body led by government officials (Pide, 1999).

When Indonesia declared its independence in 1945 and became the Republik Indonesia under President Soekarno, there were eight autonomous provinces each consisting of several *keresidenan* areas. With Law 1, 1945, regional autonomy, understood to be based on the sovereignty of the people and the freedom to self-govern, was set at both the provincial and *keresidenan* levels, with emphasis on the latter.

This right to self-governance under regional autonomy has been a matter of debate since those early days (Rinakit, 2002; Mokhsen, 2003). The Indonesian government changed the degree of autonomy granted under the regional autonomy several times, each time changing the level of accountability as well (Suharyo, 2000). Both Law 1, 1945, and Law 22, 1948, the basic law on local administration, were designed for full autonomy at the regional level and more democratic measures of governance. Unfortunately, these laws, which were intended to counter re-colonization by the Dutch, were applicable only in Java and Madura, then under the republic's direct control (Matsui, 2003).

The Dutch, meanwhile, tried to regain control over Indonesia and started several offensives between 1946 and 1949. This struggle ended in 1949 with international recognition of the Indonesian Federal Republic. However, in 1950, after less than a year as a federal state, the young state decided to revert to a unitary republic based on the provisional constitution of 1950.

In the absence of new legislation on local administration, the Administrative Law of 1948 was temporarily adopted. However, since it was written for Java and Madura, other regions soon started to resent the imposition of Javanese ways of life. As a result, the Law of 1948 was repealed and changed to Law 1, 1957. This law recognized 'as extensive an autonomy as possible' for regions concerning all powers that the central government did not have; for the first time, the law also made the head of a local government responsible to a local assembly (Matsui, 2003, p7). An important feature of this law is that autonomy was set at the provincial level (MacAndrews, 1986), rather than the district level. Unfortunately, repeated outbreaks of unrest between 1948 to 1962 in Sumatra, South Sulawesi and West Java thwarted the law's aim to revitalize regional autonomy (Brillantes and Cuachon, 2002).

Centralization

While the aim for decentralization originated from those early years, implementation, especially in administrative and fiscal aspects, continued to prove difficult. The new republic had to cope with fiscal difficulties and lack of human resources, funding and experience of the regional governments. The 1950 to 1957 period was characterized by considerable discontent, lack of agreement on social and economic growth, and regional rebellions (MacAndrews, 1986; Lloyd, 2000). As mentioned above, the new Administrative Law of 1957 did not sufficiently reduce mistrust in the regions. In response, in 1959, President Soekarno declared a return to the 1945 constitution (Matsui, 2003), thereby effectively abolishing regional autonomy.

Although the constitution does include provisions for autonomy under Article 18, the lack of clarity on the definition of autonomy and the urge to maintain the unitary state led Soekarno to adopt the centralized power so familiar from the colonial period, starting an era of 'guided democracy' with broad presidential powers. This period, which lasted until 1965, resulted in a stronger central government, which effectively imposed control on the regions and put down rebellions. This period also saw the rise of the army as the power behind the government (MacAndrews, 1986), with the dual function of heading the local autonomy and representing central government (de-concentration).

The legal basis for the centralized system was provided by Law 18, 1965, the basic law on local administration, which put local administration in line with the centralization of administrative power (Matsui, 2003). This principle was inherited by the New Order government under the second president, Soeharto, which took power in 1965. The new government had to face three challenges: the need to establish control, restore economic stability and initiate its legitimacy (MacAndrews, 1986). To impose political stability and therefore allow the influx of foreign capital, especially in the exploitation of natural resources, the New Order continued the centralistic emphasis of the earlier period. Nonetheless, in the beginning, Soeharto tried to restore 'as extensive an autonomy as possible' to local governments. In the process, however, a shift occurred and the term became a 'realistic and responsible autonomy'.

A realistic and responsible autonomy informed the new law, Law 5, 1974, which established a comprehensive system of central–provincial relationships and of local government (MacAndrews, 1986), and emphasized the mobilization of regions in the effort of national development (Lloyd, 2000). Despite the wording, this law was the epitome of the centralistic government of Soeharto. Its main intention was to promote national stability, and, in essence, it reined in local autonomy by emphasizing obligations to the central government (Lloyd, 2000).

Law 5, 1974, also reinforced the hierarchical structure of the government introduced by the earlier law, Law 18, 1965, with the use of the terms 'region level I' (the province) and 'region level II' (the regency or district). The 1974 law only de-concentrated administrative tasks to the regions down to the district level. No real power was given.

Law 5, 1974, was followed by Law 5, 1979, on village government, which imposed a uniform local political structure throughout Indonesia in the name of efficiency. This law introduced a secular administrative structure of villages (*desa*) to replace the existing customary (*adat*) structures, although the two often co-existed. Literally, the law allowed village autonomy and promoted public participation. In reality, it established a system that ensured central government penetration to the village level (Ismanto, 2003; Matsui, 2003) and a rigidly hierarchical top-down process of decision-making. At the same time, the military established a parallel system where their presence was ensured from the central to village level (Rinakit, 2002; Matsui, 2003).

This centralistic hierarchical structure was primarily a reaction against the earlier regional rebellions and the fear of national disintegration. Soeharto used the threat of communism to strengthen his hold. The imagined 'unity' of *Bhineka Tunggal Ika* (Unity in Diversity) was translated as 'uniformity'. Laws and regulations were drafted at the national level without regional consultation and were uniformly applied throughout the country to the more than 600 ethnic groups. By the end of 1985, Soeharto's regime had suppressed all political and social dissidents, and all citizens were to recognize *Pancasila*, or the five principles, as the only principles for Indonesian citizens to live by. Centralization was therefore nearly perfected (Matsui, 2003).

During this time, national laws strengthened the position of the state as the supreme authority over natural resources, including land and forest resources. Since the 1960s, the forests and forest land have provided state revenue, as well as a resource for political patronage and a safety valve for scarcities of land and resources in densely populated Java. They have also been a vehicle for integrating remote areas in the state (Barber, 1990, 1997). Earlier on in 1957, provinces were given the right to issue permits to exploit timber on areas of up to 10,000ha, while districts or regencies were given authority to issue permits for areas of up to 5000ha (Simon, 2000).

Realizing the importance of forest resources as a 'key arena for the New Order programme of economic development, political control, and social and ideological transformation' (Barber, 1990, 1997), the government passed the 1967 Basic Forestry Law. This law reinforced Article 33 of the 1945 constitution, which provided that the state controls all natural resources – in this case, specifically forests and forest land. In 1970, one of the first implementation regulations was issued[2] to regulate the rights to forest exploitation, thereby revoking the provincial rights to issue permits. The 1970 regulation also explicitly negated the prior rights of local and *adat* communities (Article 6) (Simon, 2000). Timber concessions were granted based on political patronage to members of the military and a select group of businesses, in this way developing a network of loyal cronies to consolidate the regime's power. As was common practice among Javanese kingdoms, the power of the centre rested on these networks of state patronage and clientelism (McCarthy, 2000a; Obidzinski, 2004).

The designation of forest land generally followed the earlier delineations by the Dutch. During the 1980s, the government had started a process of revising forest boundaries in which large areas of the outer islands were declared state forest land, ignoring and dispossessing the approximately 65 million people living there (Fay and Sirait, 2002).

The management of these forest and forest lands was entrusted to the Directorate General of Forestry in the Ministry of Agriculture, which became the Ministry of Forestry in 1985. In some areas of production forest management rights were given to the state-owned forest enterprise (*Perhutani* for forests on Java and *Inhutani* for forests in the outer islands) established in 1973.

The forest service structure took form along with the increasingly centralized government. In line with the concept of de-concentration introduced by Law 5,

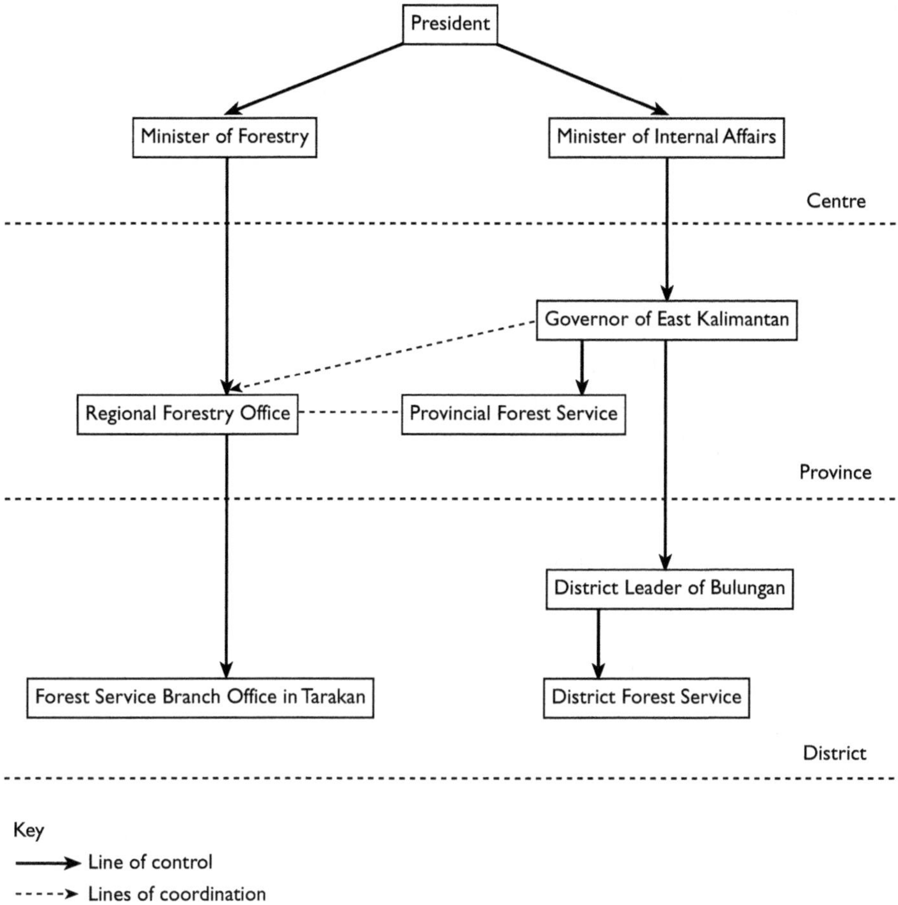

Source: Wollenberg et al, 2006, p426

Figure 1.1 *The structure of forest administration before decentralization*

1974, the Ministry of Forestry operated at the provincial level through the regional office (*Kantor Wilayah*), which was accountable to the ministry, but also supervised the provincial forest service (*Dinas Kehutanan*), which was accountable to the Ministry of Home Affairs through the governor. The rigidly hierarchical structure (see Figure 1.1) reached the district level through branches of the provincial office and implementation units of regional offices, and allowed all decisions pertaining to forest exploitation to be made at the centre.

In order to promote capital investment for the growth-oriented economic development programmes, the central government then facilitated land appropriation for industrial purposes, often using intimidation and force against strong opposition of the local land owners (Konsorsium Pembaruan Agraria, 1997).

Rising opposition

In the later years of the New Order, general resentment grew among the populace, not only over the hegemonic control of the centre, but also cronyism, the huge wealth it generated for Soeharto's inner circle, the lack of benefits to poor communities, and the environmental problems that it wrought. The provinces began noting income foregone that they paid to the centre. Environmental and agrarian reform movements emerged supported by non-governmental organizations (NGOs), labourers, the rural and urban poor, and intellectuals (Konsorsium Pembaruan Agraria, 1997). During the 1990s, these movements became more organized and challenged the government through campaigns and careful protests that began to influence the policy and implementation climate (Wrangham, 2002). In 1994, several environmental NGOs represented by the Indonesian Legal Aid Institute and including Wahana Lingkungan Hidup Indonesia (WALHI), Lembaga Alam Tropika Indonesia (LATIN), Pelangi and Sistem Hutan Kerakyatan (SHK) even went so far as to bring the first legal suit against Soeharto for the misuse of reforestation funds. They lost the case on procedural grounds and were then immediately blacklisted by the government. This made them ineligible to receive foreign funds; but they continued to operate essentially as before.

The challenges created by these social movements, together with declining financial returns in the timber industry, foreshadowed changes in the central government's willingness to share the costs of forest management.

Together with a weakening central state and strengthening social forces, the economic crisis that hit Asia in 1997 led to a tumultuous period of reforms. The national elections in 1998 that led to the re-election of Soeharto were followed by large-scale social unrest and highlighted the problems of *korupsi, kolusi, dan nepotisme* (KKN) – collusion, corruption and nepotism – that had plagued the Indonesian government since its beginning. General instability and increasing violent protests and demonstrations fuelled the toppling of the New Order in May 1998.

In the following three years, the institutions of the central government became progressively weaker and fragmented. The People's Consultative Assembly (Majelis Permusyawaratan Rakyat, or MPR) issued a decree in 1998 (Tap MPR XV, 1998) expressing the need to have regional autonomy (Zakaria and Fauzi, 2000). The government subsequently produced Law 22, 1999, and Law 25, 1999, on regional government, whereby autonomy was established at the district and municipality level (cities are outside of the district structure). The three years saw three presidents, with the two brief terms of Presidents Habibie and Abdurachman Wahid (also known as 'Gus Dur') succeeded by the ineffectual Megawati Sukarnoputri. The decentralization laws issued by Habibie were executed by the subsequent governments that had less sympathy for the full autonomy allowed by the laws.

Meanwhile, the struggles between the districts and the centre had already begun, with local people and the districts stepping into the political space afforded by the weakening centre (Rhee, 2000).

At the local level, reforms released the anger of the people who had suffered repression for almost 30 years. People vented against the symbols of centralized oppression and capitalist exploitation, such as state forests, private plantations, forest concessions and government buildings (Simarmata and Masiun, 2002). It was also during this period of reforms in 1999 that the *adat* mobilized and the alliance of *adat* groups – Alliansi Masyarakat Adat Nusantara (AMAN) – was established (*Down to Earth*, 1999).

Increasing space for society

Although the post-Soeharto era unleashed a rash of dramatic reforms, most reforms had their origins in prior years. When the Cold War ended at the end of the 1980s, communism lost its role as common enemy, and political fragmentation became a threat again. During the 1990s, domestic and external pressure for more political openness and economic deregulation increased, supported by more opportunities for society groups to disseminate information.

Thus, in 1995, an experiment with local autonomy was launched. Decree 8, 1995, of the minister of the interior designated one district or municipality in each of the 26 provinces as the site for a two-year pilot project on decentralization (Suharyo, 2000; Matsui, 2003). It involved the devolution of 19 government functions (Suharyo, 2000), including forestry. The project resulted mostly in the transfer of powers from the province to the district. Not much was ceded by the central government. As a result, the experiment undermined the powers of the province and reinforced the links between the central government and the districts. Some believed that the experiment ironically served to strengthen central powers (Matsui, 2003).

Policy changes related to public participation increased in the official discourse. During the mid 1980s, the regional planning agencies took on 'bottom-up-planning'. The government introduced a new national planning process, known as P5D (*Pedoman Penyusunan Perencanaan dan Pengendalian Pembangunan Daerah*, or Guidelines for Planning and Monitoring Regional Development). Development planning started at the village level through village meetings (*Musyawarah Pembangunan Desa*). Village plans were discussed at the sub-district (*kecamatan*) and district level. Acceptance for funding was subject to the plan's conformity to the district- and provincial-level plans. Although essentially lacking in participation, the policy established a framework that allowed for more inclusive and informed planning.

Between the end of the 1980s and the early 1990s, the forestry sector also became more receptive to local demands, with logging concessions required to implement village development programmes (Wrangham, 2002) and increasing interest in community forest management. The Ministry of Forestry developed a programme on community forestry in 1995 (Ministerial Decree 662); but the decree was amended and replaced in 1998 and 2001, progressively decreasing the rights of the communities. Participation was constrained to forest rehabilitation, replanting and gathering of non-timber products, while decision-making power remained with the Department of Forestry. Around the same time, based on the experience of the Krui damar (*Shorea javanica*) agroforests in Krui, Sumatra, and related policy discussions that started during the New Order in 1995, the ministry issued a Decree on Special Purpose Areas, Ministerial Decree 47, which legally recognized the legitimacy of community management for the first time.

Public participation became an element of other legislation in the mid 1990s (e.g. the Law on Spatial Planning of 1992 and the Environmental Law of 1997), which explicitly stated the need for public participation.

In 1998, the Ministry of Forestry began to decentralize by transferring some authority to the districts (Government Regulation 62). The authorities transferred, however, were those of a more financial drain than gain (Syaukani, 2001) – that is, rehabilitation and reforestation, soil and water conservation, management of protected forest, extension, and smaller-scale economic activities regarded as community forestry: beekeeping and silkworm farming. In 1999, however, the government also decentralized the issuing of harvesting permits to provinces for harvesting areas less than 10,000ha and districts for one-year concessions on 100ha (Government Regulation 6). Unfortunately, the new permits were captured by local entrepreneurs and elite, and led to rampant logging (see Part II of this volume). The decree was repealed in 2000; but by that time, the districts had tasted the profits from forest exploitation and were geared up for autonomy. They ignored the ministerial decree. Ministerial decrees also lost their influence in the post-Soeharto era since the People's Consultative Assembly (Tap MPR III, 2000) declared ministerial decrees not part of the legal hierarchy.

Later in 1999, the new Forestry Law 41, 1999, long planned and discussed, was finally issued. This new law made it possible to engage new actors outside of the state apparatus in forestry, including *adat* communities (Lindayati, 2002, Wollenberg and Kartodihardjo, 2002). It also allowed for public participation and limited community rights. These innovations, however, were couched within the old paradigm of absolute control of the state and a centralistic system of management, rather than the new decentralization laws (Nurjaya, 2000). In addition, the districts' capacity to implement forestry was weak (Hutabarat, 2001). From the start, the new forestry law was at odds with the decentralization law, and the central government and regional governments were not able to agree on the extent of the division of authority over forests.

Law 22, 1999, on regional government established autonomy at the district and municipality levels. Unlike Law 5, 1974, which only de-concentrated administrative tasks, Law 22 also devolved real decision-making power (Mokhsen, 2003). The law, in principle, granted the freedom of regional diversity as well as the right for regions to manage natural resources in their territories. Districts were given the opportunity to manage their own affairs, including those of local communities, and develop specific policies in accordance with local needs (WALHI, 2000; Simarmata and Masiun, 2002). The expressed intention, like most decentralization policies in other countries (Ribot, 2001), was to support democracy, public participation, equality and justice, and to eliminate corruption. Law 22 was accompanied by Law 25, which specified how revenues were to be shared between the central and the regional governments, and by Law 28 on clean government.

Reforms also included attempts to reinstitute the rights of local and customary communities. Decree IX, 2001 of the People's Consultative Assembly on Agrarian Reform, and the Second Amendment to the Constitution give explicit recognition of *adat* communities and their rights, (as had Forestry Law 41, 1999). For the first time, these *adat* rights, including communal rights, were recognized and could be registered. However, recognition depends upon the discretion of the district head. With the land office only partly decentralized and the land law in revision, the situation remained unclear.

During the 2001 People's Consultative Assembly, a decision was made that, starting in 2004, the president would be directly elected by the people. In 2004, the people of Indonesia directly elected their president for the first time, choosing Susilo Bambang Yudhoyono, a respected army general.

Full cycle?

Decentralization reflected the weakening of the central state during Soeharto's third decade of control (Ascher, 1999) and the increasing demands of the wealthier provinces, such as Aceh, East Kalimantan, Riau and Papua, for more autonomy and a bigger share of benefits from their own natural resources. As a consequence, local autonomy was seized by the districts rather than being a voluntary transfer of power from the central government.

Implementation of Laws 22 and 25 was thus hampered by the reluctance of the central government to transfer real control and by the issuance of numerous conflicting and unclear laws. For example, the first implementing regulation defined the authority given to the provinces, implying that districts had authority only over the remaining domains. On the other hand, by establishing autonomy at the district level, the decentralization laws established a direct link between the district and central government, bypassing the province. As a result, districts completely disregarded the coordinating and monitoring function of the provinces.

As soon as the decentralization laws were passed, they gave rise to public criticism. The laws were considered to have been drawn in haste, lacking public consultation in drafting and sufficient preparation before implementation (Suharyo, 2000). Uncertainty over the meaning and implementation of the laws was rampant and led initially to increased conflict among the government, local communities who understand autonomy as a return of their traditional ownership rights, and local contractors eager to have their chance to exploit lucrative natural resources (Wulan, 2004).

As the chapters in this book vividly illustrate, the districts interpreted the laws to their own advantage, citing the right to harvest 100ha and using Laws 22 and 25 to justify taking control over the forests in their territory. Unfortunately, many districts saw forests as an easy source of financial revenue to be exploited rather than managed, much as the centralized regime had regarded them.

Local and traditional community groups, supported by NGOs, attempted to claim their right and ability to manage and utilize forest resources in their area (Barr et al, 2001). A national law entitling communities to compensation from timber exploitation in their territories, as well as a provincial decree setting the amount of fees to be paid, empowered local communities to demand a share of the benefits from forest exploitation from the local government and their private contracting companies. This recognition, however, had the unintended consequence of converting community rights to a cash value, and left issues of land rights or ownership unresolved.

Adjusting the balance

Since the sweeping decentralization and forest reforms in 1999, the government has attempted to reinstall some order. In 2004, a revised decentralization law, Law 32, 2004, was issued, re-establishing the hierarchy between province and district. Although considered an attempt to recentralize, the new law was accepted almost as a relief as autonomy became better understood.

The Ministry of Forestry, meanwhile, had also attempted to adjust the delegation of powers in response to the initial wave of uncontrolled timber exploitation by districts and their demand for revenues from managed forests. The first implementing regulation in 2002, Government Regulation 34, responded to the district's uncontrolled exploitation of timber (2000 to 2002). It explicitly stated that only the central government could issue permits for timber harvesting, although they would be based on the recommendations of lower levels of government. This regulation provided the minister with authority to gradually and selectively delegate the rights to issue permits for the utilization of wood products in production forest to the regions if they demonstrate readiness in terms of institutions, vision or mission; however, no specific guidelines or conditions were given (Patlis, 2004). The Ministry of Forestry also attempted to maintain its supremacy by emphasizing the strategic importance of forests and its own role in setting criteria and standards, classifying forests and, above all, setting forest tariffs and retaining the right to issue transportation permits at the provincial level.

Then, in 2007, a second overriding implementing regulation, Government Regulation 6, was passed, giving district heads the right to issue management permits, but also specifying the nature of forestry management units and outlining requirements for their management. The establishment of forest management units, in particular, was seen as another attempt to reassert authority over the forest by the central government.

The reforms and decentralization created unrealistically high expectations and hope for improving the lives of Indonesians and natural resource management. The ongoing processes have resulted in incorporating forestry in the district government system, which has given local government and local society more control over their forests. These processes, however, have disappointingly been unable to increase local society's commitment towards better governance of natural resources.

Implications

As shown, decentralization is not new, nor is the fact that it moves in waves of increasing and decreasing degrees of freedom. What is new in Indonesia in this last wave is the development of a civil society and devolution of control over a still-valuable forestry industry. There is growing awareness among the general public of their rights with respect to the government and possibilities for participation. Indeed, there is a general trend of the public demanding accountability, as indicated by the rise in court cases against government and assembly members accused of corruption.

The present wave has also particularly strengthened cultural (*adat*-based) groups. A space was created for local communities to openly contest the ownership of the state and regain their customary or *adat* rights (Simarmata and Masiun, 2002). Decentralization does not absolve the state, including its local government, of the responsibility to defend and protect the rights of all people. In many areas, including Malinau, there are many indigenous groups making claims for the same resources and lands. However, the local governments have their own strong interests and claims to resources, and are often in the middle of conflict over natural resources. There is no turning back, however. Although, as Larson (2005b) states, 'Power given by decree can be taken away', and the central government might try. Decentralization will probably continue, but with some significant changes. Although the districts might not willingly give up power, they might actually be relieved to transfer some responsibilities back

Notes

1 *Keresidenan* was the seat of the local Dutch administrator (resident), who, in Java at least, shared the rule with the native regent(s), or *Bupati*.

2 Government Regulation 21, 1970, on the Rights of Exploiting Forest.

References

Agrawal, A. (2001) 'The decentralizing state: Nature and origins of changing environmental policies in Africa and Latin America 1980–2000', Paper prepared for the 97th Annual Meeting of the American Political Science Association, 30 August–2 September, San Francisco, CA

Agrawal, A. and Ribot, J. (1999) 'Accountability in decentralization: A framework with South Asian and African cases', *Journal of Developing Areas*, vol 33, pp473–502

Antlöv, H. (2003) 'Not enough politics! Power, participation and the new democratic polity in Indonesia', in Aspinall, E. and Fealy, G. (eds) *Local Power and Politics in Indonesia: Decentralisation and Democratisation*, Institute of Southeast Asian Studies, Singapore

Ascher, W. (1999) *Why Governments Waste Natural Resources*, Johns Hopkins University Press, Baltimore and London

Barber, C.V. (1990) 'The legal and regulatory framework and forest production in Indonesia', in Zerner, C. (1990) Community Land Rights, Customary Law, and the Law of Timber Concessions in Indonesia's Forests: Legal Options and Alternatives in Designing the Commons, Draft (March 1990) of Forestry Studies report UTF/INS/065, Ministry of Forestry/FAO, Jakarta, Appendix

Barber, C.V. (1997) *Environmental Scarcities, State Capacity, and Civil Violence: The Case of Indonesia*, Occasional Paper Project on Environmental Scarcities, State Capacity, and Civil Violence Cambridge: American Academy of Arts and Sciences and the University of Toronto, www.library.utonronto.ca/pcs/state/indon/indonsum.htm

Barber, C.V. and Churchill, G. (1987) *Land Policy in Irian Jaya: Issues and Strategies, Project Report No INS/83/013*, Government of Indonesia, United Nations Development Program and International Bank for Reconstruction and Development, Jakarta

Barber, C.V. and Talbott, K. (2003) 'The chainsaw and the gun: The role of the military in deforesting Indonesia', in Price, S.V. (ed) *War and Tropical Forests: Conservation in Areas of Armed Conflict*, The Haworth Press Inc, Philadelphia

Barr, C., Wollenberg, E., Limberg, G., Anau, N., Iwan, R., Sudana, I. M., Moeliono, M. and Djogo, T. (2001) *The Impacts of Decentralization on Forests and Forest-Dependent Communities in Malinau District, East Kalimantan*, Case Studies on Decentralisation and Forests in Indonesia, CIFOR, Bogor, Indonesia

Brillantes, A. B., Jr. and Cuachon, N. G. (2002) *Sourcebook on Decentralization in Asia*, Asian Resource Centre for Decentralization, Manila, the Philippines

Down to Earth (1999) 'NGO support for the indigenous peoples' movement', Special Issue, *Down to Earth*, October 1999, http://dte.gn.apc.org/SIngo.htm

Echeverri-Gent, J. (1993) *The State and the Poor: Public Policy and Political Development in India and the United States*, University of California Press, Berkeley, CA

Engelfriet, Aad (2000) *ARCENGEL: The History of Indonesia Mirror Site*, http://home.iae.nl/users/arcengel/Indonesia/1830.htm

Fay, C. and Sirait, M.T. (2002) 'Reforming the reformists in post-Soeharto Indonesia', in Colfer, C. J. P. and Resosudarmo, I. A. P. (eds) *Which Way Forward? People, Forests and Policymaking in Indonesia*, Resources for the Future (RFF)/Center for International Forestry Research (CIFOR), Washington, DC

Finger-Stich, A. and Finger, M. (2003) *State versus Participation: Natural Resources Management in Europe*, IIED, Brighton, London

Furnivall, J. S. (1956) *Colonial Policy and Practice*, New York University Press, New York

Gregersen, H. M., Contreras-Hermosilla, A., White, A. and Phillips, L. (2005) 'Forest governance in federal systems: An overview of experiences and implications for decentralization', in Colfer, C. J. P. and Capistrano, D. (eds) *The Politics of Decentralization, Forests, Power and People*, Earthscan, London

Hall, J. A. and Ickenberry, G. J. (1989) The State, *Open University Press*, Milton Keynes

Hutabarat, S. (2001) 'Forestry developments with regard to decentralization', *The Indonesian Quarterly*, vol XXIX, no 2, second quarter, pp151–158

Ismanto, I. (2003) 'Decentralization and democratization in Indonesia's political changes', in Legowo, T. A. and Takahashi, M. (eds) *Regional Autonomy and Socio-Economic Development in Indonesia – A Multidimensional Analysis*, Institute of Developing Economies, Japan External Trade Organization, Chiba, Japan

Kaimowitz, D., Faune, A. and Mendoza, R. (2003) 'Your biosphere is my backyard: The story of Bosawas in Nicaragua', *Policy Matters*, vol 12, pp6–15

Kohli, A. (1994) 'Centralization and powerlessness: India's democracy in a comparative perspective', in Migdal, J. S., Kohli, A. and Shue, V. (eds) *State Power and Social Forces:*

Domination and Transformation in the Third World, Cambridge University Press, Cambridge

Konsorsium Pembaruan Agraria (1997) 'Suara Pembaruan Agraria', *Jurnal Konsorsium Pembaruan Agraria*, no 3, Tahun

Larson, A. M. (2004) 'Democratic decentralisation in the forestry sector: Lessons learned from Asia, Africa and Latin America', Paper presented at Workshop on Decentralization, Federal Systems in Forestry and National Forest Programmes, Interlaken, 22--30 April

Larson, A. M. (2005a) 'Democratic decentralization in the forestry sector: Lessons learned from Africa, Asia and Latin America', in Colfer, C. J. P. and Capistrano, D. (eds) *The Politics of Decentralization, Forests, Power and People*, Earthscan, London

Larson, A. M. (2005b) 'Formal decentralization and the imperative of decentralization "from below": A case study of natural resource management in Nicaragua', in Ribot, J. and Larson, A. (eds) *Democratic Decentralization through a Natural Resource Lens*, Routledge, Oxford

Lindayati, R. (2002) 'Ideas and institutions in social forestry policy', in Colfer, C. J. P. and Resosudarmo, I. A. P. (eds) *Which Way Forward? People, Forests and Policymaking in Indonesia*, Resources for the Future (RFF)/Center for International Forestry Research (CIFOR), Washington, DC

Lipsky, M. (1980) *Street-Level Bureaucracy: Dilemmas of the Individual in Public Services*, Russell Sage Foundation, New York, NY

Liu, D. and Edmunds, D. (2003) 'The promises and limitations of devolution and local forest management in China', in Edmunds, D. and Wollenberg, E. (eds) *Local Forest Management: The Impacts of Devolution Policies*, Earthscan, London

MacAndrews, C. (ed) (1986) *Central Government and Local Development in Indonesia*, Oxford University Press, Singapore

Manor, J. (1999) *The Political Economy of Democratic Decentralization*, World Bank, Washington, DC

Matsui, K. (2003) *Decentralization in Nation State Building of Indonesia*, IDE Research Paper No 2, Institute of Developing Economies (IDE-JETRO), Jakarta, Indonesia

McCarthy, J. F. (2000a) 'The changing regime: Forest property and reforms in Indonesia', *Development and Change*, vol 31, pp91–129

McCarthy, J. F. (2000b) *Wild Logging: The Rise and Fall of Logging Networks and Biodiversity Conservation Projects on Sumatra's Rainforest Frontier*, CIFOR Occasional Paper, vol 31, CIFOR, Bogor, Indonesia

Migdal, J. S. (1994) 'The state in society: an approach to struggles for domination', in Migdal, J. S., Kohli, A. and Shue, V. (eds) *State Power and Social Forces: Domination and Transformation in the Third World*, Cambridge University Press, Cambridge

Migdal, J. S., Kohli, A. and Shue, V. (eds) (1994) *State Power and Social Forces: Domination and Transformation in the Third World*, Cambridge University Press, Cambridge

Moeliono, M. (2002) '*Adat* and globalization: Living apart together', Paper submitted for the International Association for the Study of Common Property 9th Biennial Conference, Victoria Falls, Zimbabwe

Mokhsen, N. (2003) *Decentralization in the Post New Order Era of Indonesia*, Commonwealth Advanced Seminar: Leading Public Service Innovation, Victoria University of Wellington, Wellington, New Zealand

Moore, S. F. (1973) 'Law and social change: The semi-autonomous social field as an appropriate subject of study', *Law and Society Review*, vol 7, pp719–746

Nurjaya, I. N. (2000) Proses Pemiskinan di Sektor Hutan dan Sumberdaya Alam: Perspektif Politik Hukum dalam ICRAF dan JAPHAMA (2001). Masyarakat adat dalam mengelola sumber daya alam: KEDAI I diselenggarakan oleh ICRAF dan JAPHAMA, Cisarua, 26–28 Mei, 2000. Bogor, Indonesia. World Agroforestry Centre – ICRAF, SEA Regional Office.

Obidzinski, K. (2004) Illegal logging and the fate of Indonesia's forests in times of regional autonomy. Paper Presented at the Xth Meeting of the International Association for the Study of Common Property, 9–13 August 2004, Oaxaca, Mexico.

Patlis, J. M. (2004) *A Rough Guide to Developing Laws for Regional Forest Management*, CIFOR, Bogor, Indonesia

Peluso, N. L. (1992) *Rich Forests, Poor People: Resource Control and Resistance in Java*, University of California Press, Berkeley

Pide, Andi Mustari. (1999) Otonomi Daerah dan Kepala Daerah. Memasuki Abad XXI. Jakarta: Gaya Media Pratama

Resosudarmo, I. A. P. (2003) 'Shifting power to the periphery: the impact of decentralisation on forests and forest people', in Aspinall, E. and Fealy, G. (eds) *Local Power and Politics in Indonesia: Decentralisation and Democratisation*, Institute of Southeast Asian Studies, Singapore

Rhee, S. (2000) 'De facto decentralization and the management of natural resources in East Kalimantan during a period of transition', *Asia Pacific Community Forestry Newsletter*, vol 13, no 2, pp34-40

Ribot, J. (1999) 'Decentralization, participation and accountability in Sahelian forestry: Legal instruments of political-administrative control', *Africa*, vol 69, no 1, pp23-65

Ribot, J. C. (2001) 'Integral local development: "Accommodating multiple interests" through entrustment and accountable representation', *International Journal of Agriculture, Resources, Governance and Ecology (IJARGE), Special Issue on Accommodating Multiple Interests in Local Forest Management*, vol 1, no 3/4, pp327–350

Ribot, J. C. and Larson, A. M. (eds) (2005) *Democratic Decentralization through a Natural Resource Lens: Experience from Africa, Asia and Latin America*, Routledge, London

Sarin, M., Singh, N., Sundar, N. and Bhogal, R. (2003) 'Devolution as a threat to democratic decision-making in forestry? Findings from three states in India', in Edmunds, D. and Wollenberg, E. (eds) *Local Forest Management: The Impacts of Devolution Policies*, Earthscan, London

Scott, J. (1998) *Seeing Like a State*, Yale University Press, New Haven, CT

Shue, V. (1994) 'State power and social organization in China', in Migdal, J. S., Kohli, A. and Shue, V. (eds) *State Power and Social Forces: Domination and Transformation in the Third World*, Cambridge University Press, Cambridge

Simarmata, R. and Masiun, S. (2002a) 'Decentralization and the character of local government laws and regulations: New pressures on the environment and indigenous communities: A preliminary diagnosis', Paper presented at the 9th Biennial Conference of the IASCP, Victoria Falls, Zimbabwe, 17–21 June 2002

Simon, H. (2000) 'Kilas Balik Sejarah Peraturan Tentang Kehutanan', www.arupa.or.id/papers/24.htm, accessed 18 March 2003

Smith, J., Obidzinski, K., Subarudi, I. and Suramenggala, I. (2003) 'Illegal logging, collusive corruption and fragmented governments in Kalimantan, Indonesia', *International Forestry Review*, vol 5, pp293–302

Suharyo, W. (2000) *Voices from the Regions: A Participatory Assessment of the New Decentralisation Laws in Indonesia*, UNSFIR, Jakarta, Working Paper no 2

Syaukani, H. R. (2001a) 'The implementation of decentralization in forestry sector within the framework of sustainable forest management in the District Kutai Kertanegara', *The Indonesian Quarterly*, vol XXIX, no 2, second quarter, pp143–150

WALHI (2000) *Otonomi Daerah dalam konteks pengelolaan sumberdaya Alam dan Lingkungan Hidup*, Unpublished position paper

Wollenberg, E. and Kartodihardjo, H. (2002) 'Devolution and Indonesia's new forestry law', in Colfer, C. J. P. and Resosudarmo, I. A. P. (eds) *Which Way Forward? People, Forests and Policymaking in Indonesia*, Resources for the Future (RFF)/CIFOR, Washington, DC

Wollenberg, E., Moeliano, M., Limberg, G., Iwan, R., Rhee, S. and Sudana, M. (2006) 'Between state and society: Local governance of forests in Malinau, Indonesia', *Forest Policy and Economics*, vol 8, no 4, pp421–433

Wrangham, R. (2002) 'Changing policy discourses and traditional communities, 1960–1999', in Colfer, C. J. P. and Resosudarmo, I. A. P. (eds) *Which Way Forward? People, Forests and Policymaking in Indonesia*, Resources for the Future (RFF)/CIFOR, Washington, DC

Wulan, Y. C., Yasmi, Y., Purba, C. and Wollenberg, E. (2004) *Analisa Konflik: Sektor Kehutanan di Indonesia 1997–2003*, CIFOR, Bogor, Indonesia

Zakaria and Fauzi (2000) 'Men-siasat-i Politik Otonomi Daerah Demi Pembaruan Agraria: Masuk dari Pintu yang Tersedia, Keluar dari Pintu yang Baru', in *Masyarakat adat dalam mengelola sumber daya alam: KEDAI I diselenggarakan oleh ICRAF dan JAPHAMA*, Cisarua, 26–28 May 2000, World Agroforestry Centre, ICRAF, SEA Regional Office, Bogor, Indonesia

2

The Geography of Malinau

Moira Moeliono and Godwin Limberg

The district of Malinau lies landlocked along the border with Malaysia in the hinterland of East Kalimantan at about 3 degrees north of the equator (between 2° 45' to 3° 21' N and 115° 48' to 116° 34' E). The capital, Malinau Town, lies at the confluence of the Sesayap and Malinau rivers and, until very recently, was only accessible by river boat or small plane run by a Christian mission. Even today, with new roads built and subsidized air transport, the most common entry to Malinau is a three-hour boat trip along the Sesayap River from the island of Tarakan.

Before 1999, Malinau was part of the larger district of Bulungan. Bulungan was split into three, with Malinau being the largest, with 42,000 square kilometres, and having the lowest population density of 1 person per square kilometre. The district became autonomous in 2001 following the decentralization law and elected its own district head (*Bupati*) that same year. The government is therefore very young and throughout the early years of decentralization was in the process of becoming established. The new government's first priority was improving access. New roads are being constructed to connect Malinau Town with the sub-district capitals. In the meantime, more than 80 per cent of its area remains inaccessible by road.

A second priority for Malinau, as in many other new districts in Indonesia, was the establishment of a new government centre in Malinau Town, finished in 2004. Here, the *Bupati*, his administration, the 17 government agencies, the *Dewan Perwakilan Rakyat Daerah* (DPRD, the local consultative assembly) and the military have their offices. In addition, in 2002, the district reorganized its administration to include 9 sub-districts and 135 villages (see Figure 2.1).

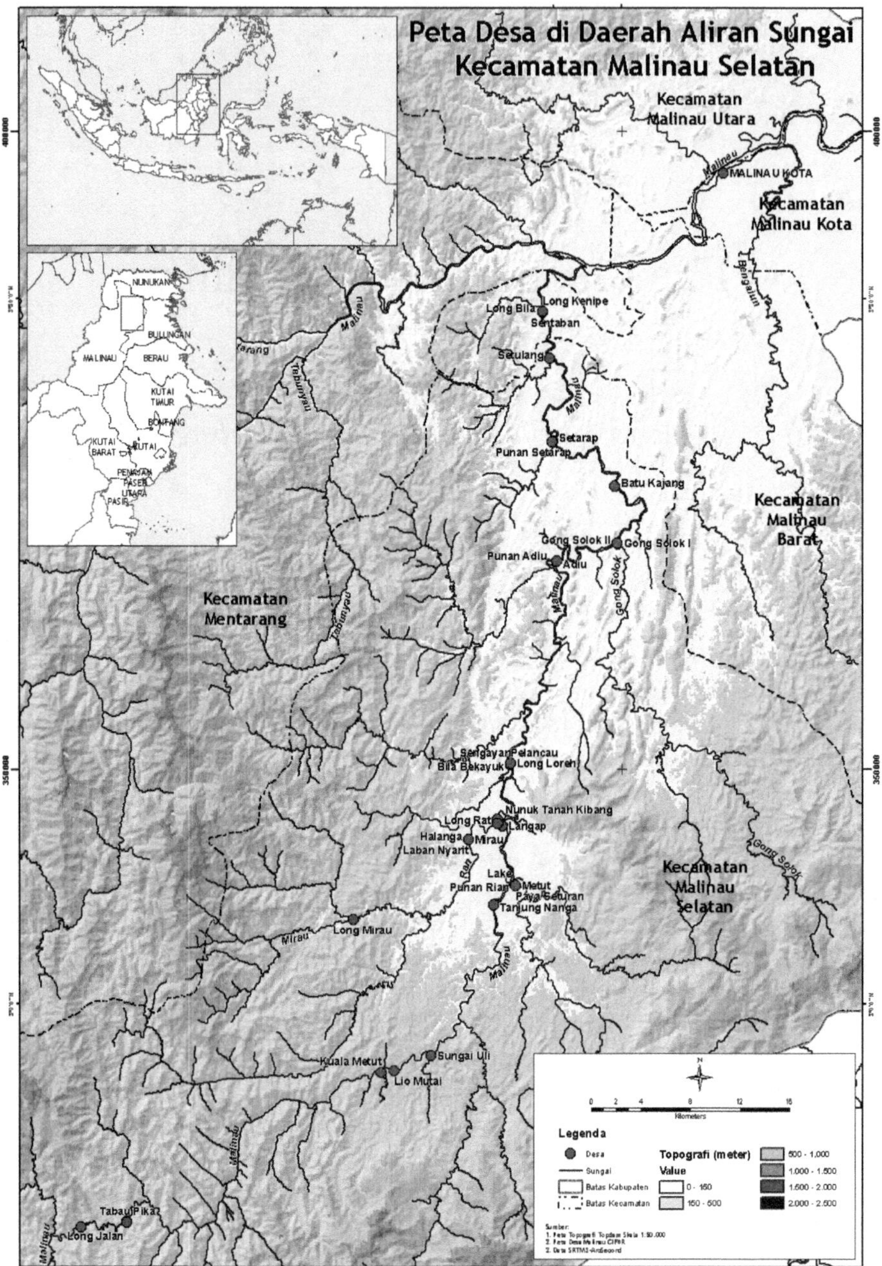

Source: CIFOR GIS Unit, 2007

Figure 2.1 *Administrative map of Malinau*

The Malinau River watershed, now comprising the sub-district of Malinau Selatan, is the most densely populated rural area of the district and the most developed. The Center for International Forestry Research (CIFOR) Adaptive

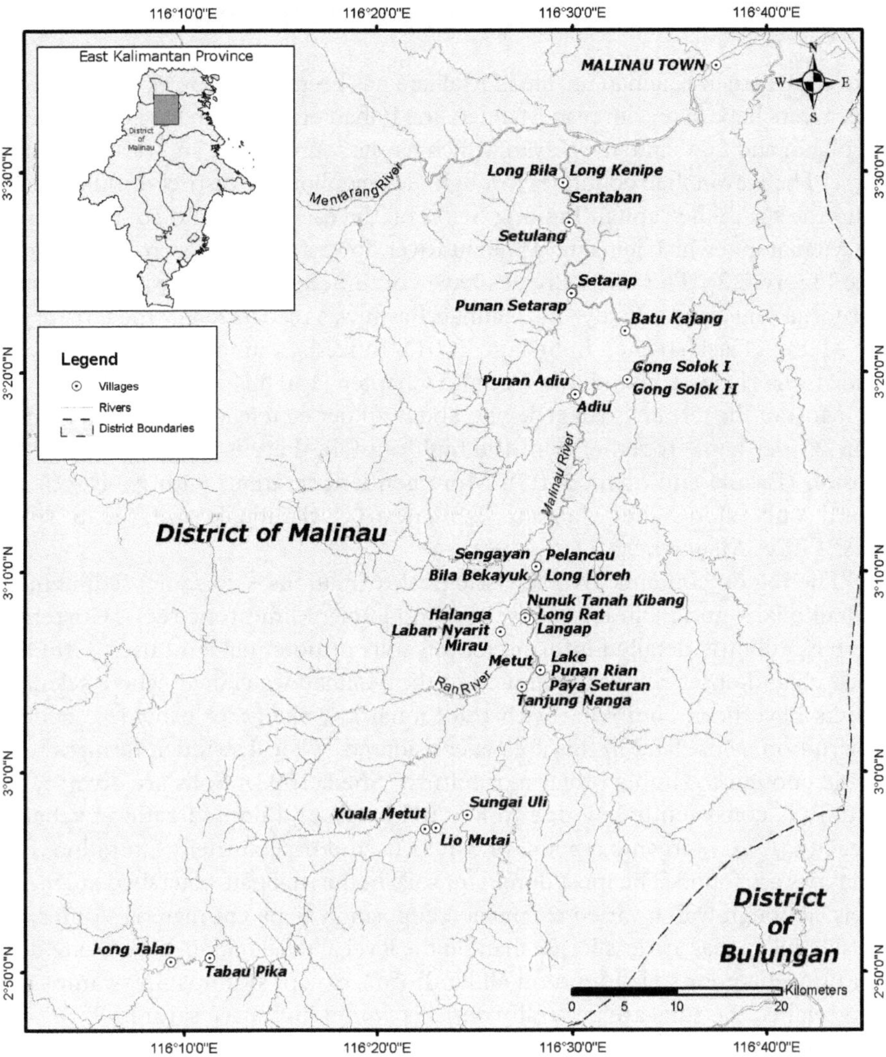

Source: CIFOR GIS Unit, 2007

Figure 2.2 *Location of the 27 villages of the upper Malinau River*

Collaborative Management Team chose to work in the 27 villages of the upper Malinau River (see Figure 2.2), from Sentaban to Long Jalan. Approximately 6673 people (Malinau Voting Census 2003, April 2003 data) reportedly live in the 500,000ha watershed, although we suspect that this figure is slightly higher than the actual resident population. The villages are distributed among 21 settlements with as little as 15 to as many as 997 individuals each. About 51 per cent of the households are considered poor according to national standards of housing, clothing and number of meals per day (BKKBN, 2001).

Topography, soils and climate

Like elsewhere in Kalimantan, life in Malinau has been shaped by its rivers. Several large rivers have their source in Malinau: the Bahau originates in the sub-district of Pujungan and flows into the Kayan, which has its source in the sub-districts of that name. The Kayan then continues through the neighbouring district of Bulungan to enter the sea at the capital, Tanjung Selor, on the east coast. The Tubu flows into the Mentarang, which joins the Malinau River downstream to become the Sesayap (see Figure 2.2). The water divides have continued to be the basis for defining territorial units; accordingly, the Malinau Basin was recognized as the territory of the Merap (Kaskija, cited in Sellato, 2001). Indeed, even today, land claims are made using rivers as boundaries (see also Chapter 11 in this volume).

Most of the terrain is rugged, with about 40 per cent featuring slopes of more than 25 per cent. Together with the high rainfall, there is a high potential for erosion (Basuki and Sheil, 2005). More gentle formations with good soils are found only along the Malinau, Semendurut, Sembuak and Salap rivers (BAPPEDA Malinau, 2001).

The region contains diverse geological formations – volcanic, sedimentary and alluvial – and includes coal seams and ancient limestone reefs. Currently, there is still little detailed information on soils or potential land uses. CIFOR's research and observations are limited to the Malinau watershed, where scientific results have been combined with traditional knowledge to provide a general description. Most land in this area is considered by local swidden farmers to be fertile enough to support their agriculture (Sheil, 2002). Soils are diverse, but fertility is consistently low due to low pH levels and limited cation exchange capacity (CEC), organic carbon, phosphorus and base nutrient saturation. Five soil types are found. The most dominant soils in the Malinau watershed are inceptisols, although with a varied chemical composition as parent materials differ; the soils on hilly areas are less fertile than on the level alluvial region. Oxisols are deep but nutrient-poor soils found on all landforms except swamps. In swamps and floodplains, the soils are generally entisols with a high base saturation (67 per cent) and sufficient fertility to offer potential for cultivation. Local people, however, considered the swampy or stony nature unfavourable. In scattered locations, heavily leached ultisols are found. These have low fertility with only 20 per cent base saturation. Alfisols were found in only two sites. These are considered good soils with a high base saturation of nutrients and good depth (Sheil, 2002; Meijaard et al, 2005).

In general, all soils have low to very low fertility with low nutrient content, relatively high acidity, low cation exchange capacity and low base saturation (Sheil, 2002). In addition, pan formation, aluminium toxicity and the steep terrain limit cultivation opportunities (Basuki and Sheil, 2005).

The climate is generally a typical humid tropical climate with a dry period of less than two months. The lowest temperature recorded in the lowlands is 23.5° C, although at the higher altitude of the Apo Kayan, approximately 400m above sea level (asl), temperatures are lower. Relative humidity is high, ranging from 75 to 98 per cent.

The forest resources of Malinau

Ninety-five per cent of the district is officially designated state forest land (Barr et al, 2001), which includes 1.03 million hectares of the Kayan Mentarang National Park. These forests comprise the largest remaining contiguous dipterocarp forest in South-East Asia and are extraordinarily rich in biodiversity. About 60 per cent of the tree families and 36 per cent of the tree genera of Kalimantan are found here (Machfudh, 2002).

The isolation of the area and difficult access have protected this richness both from exploration and exploitation. As a result, there is still limited knowledge on biodiversity. Recent surveys such as the expedition to the Krayan area of the Kayan Mentarang Park, in neighbouring Nunukan District, continue to discover new species (WWF, 2003). A CIFOR survey in 2003 to 2004 showed that the waters of the Malinau River contain at least 47 fish species, of which 15 are endemic to Kalimantan. Many of the species preferred for food by the local people are frugivorous and are believed to be associated with the presence of *Dipterocarpacea* and *Ficus* growing along riverbanks (Sheil et al, 2003).

Of the 37 mammals species identified, the bearded pig (*Sus barbatus*) has provided a source of protein for local people for generations, together with many other species (deer, porcupine, apes and monkeys), many of which are protected (Sheil et al, 2003). Also, of the more than 300 species of birds identified, at least 65 species are protected (Meijaard et al, 2005).

As elaborated upon in Chapter 4, people also obtain a large proportion of food from plants. Sago, which is usually obtained from four species, is perhaps the most important. In addition, people gather roots (*Xanthosoma* sp., *Colocasia gigantea*, *Dioscorea pentaphylla*, etc.) as well as fruits and vegetables (Uluk et al, 2001; Sheil et al, 2003).

Nonetheless, outside the national park, the forest is valued mostly for its richness in timber species such as *Agathis* spp and the various dipterocarps. Almost half of the forest is allocated as production forest logged commercially since the 1970s. During those early years, logging occurred mostly in the easily accessible areas along navigable rivers. With the timber boom of the late 1970s, timber concessions were also allocated in the remoter parts of Malinau. Although the number of logging concessions – *Hak Pengusahaan Hutan* (HPHs) – in the area varied over time, by 1999, 11 concessions were present in Malinau, covering a total area of more than 1.5 million hectares (Barr et al, 2001). By 2001, however, the days of the big timber concessions seemed to be over. Regional autonomy established by Law 22, 1999, was understood as district control over natural resources within the district. As elaborated upon in Part II, a new timber boom implemented through numerous small-scale concessions known as IPPK began and lasted until 2002. The forests then became an important source of revenue for the new district. Production increased by 10 per cent annually during those three years (1999 to 2002). In 2002, 651,000 cubic metres were produced, contributing almost 70 per cent to the local economy, although involving only 3 per cent of the local population (Anonymous, 2003). Interestingly, the official district statistics mention 570,000 cubic metres for 2002 (BAPPEDA and BPS, 2005).

Another important source of revenue from beneath the forest is the high-quality coal found in the area. Like timber, the production of coal has increased by almost 50 per cent during the early decentralization years. In 2001 alone, approximately US$14 million worth of coal from Malinau was shipped through Tarakan (Anonymous, 2003). Unfortunately, coal mining is not compatible with sustainable forest management or with the traditional agriculture practised by the local inhabitants. However, with the amount of compensation and fees promised from coal mining, many people have been tempted to hand over their land for coal mining.

As a result of logging, mining and extensive swidden cultivation, most of the accessible lowland forests, such as the area along the lower Malinau River, are now fragmented and degraded. Meanwhile, most of the district's 40,000 inhabitants[1] practise swidden agriculture, hunting and gathering from the forest. Dependence upon forests is still relatively high, with households near intact forest most reliant upon forest products for income and daily consumption (Anau et al, 2002; Levang et al, 2002). Indeed, even the most agriculturally oriented groups place a very high value on wild resources and forest. The forest may not provide everything that people want or need, but it does provide much of it and are valued accordingly (Sheil et al, 2003).

A short history

At different points in time, Malinau was considered part of the sultanates of Bulungan and Berau, while the seats of two of the Tidung kings were in the area. During these times, from the late 1800s to the early 1900s, the Tidung kings obtained ownership of the birds' nest caves, legalized by the Sultan of Bulungan (Sellato, 2001). In 1919, the Dutch military set up a post in Malinau Town. As in other areas, they imposed 'order' by defining *kampung* (village) and ownership of resources, in this way introducing and reinforcing the concept of territory. The Dutch also interfered with the *adat* structure by appointing chief adat leaders over the community *adat* leaders. In 1935, the Dutch established the *Majelis Kerapatan Besar Tanah-tanah Tidung* (High Council of the Tidung Territories), which comprised adat leaders of the Merap and Tidung and four others groups (Sellato, 2001).

Before the sultanates and petty kingdoms, people were governed through customary leadership structures, some of which were more hierarchical than others. Nevertheless, all used social controls and periodic consultation – especially with a close circle of influential villagers, or *tokoh masyarakat* – to govern the group and manage conflict. Unresolved conflict or dissatisfaction with the leaders was handled through the splitting of the group, with one faction moving to a new settlement, or (in the case of external groups) by violence. Marriage or trade alliances were used to build relationships with external groups. Rights of access to village territories were based on these ethnic and trade alliances, as well as respect for customary authority, requests for permission and verbal agreements, although people often casually entered watersheds for hunting or collecting forest products without permission. Rivers and mountain tops marked boundaries. Reflecting the economic value of forests at the time, some leaders requested fees from outside groups wishing to collect forest products in their area. Evidence suggests that in the upper Malinau, the Merap *kepala adat besar*, initially referred to as *Kepala Besar*,[2] was the reigning local power, together with the local Tidung kingdoms, for most of the 20th century (Sellato, 2001).

Although often claimed as part of Bulungan, Malinau area and other upstream groups were independent and tribute free, and traded freely with downstream settlements. The two Tidung petty kingdoms, on the other hand, established their centre in Malinau towards the end of the 1800s (Sellato, 2001). The Dutch then also established their administrative post in Malinau and from here incorporated the upstream regions such as Apo Kayan and Bahau-Pujungan in 1925.

Independence brought new laws and changes, including Law 5, 1979, on village administration. With this law, villages had to conform to a national model of '*desa*' where village leaders were accountable to the government rather than to the people. People had to be registered and settled. In Malinau, the government of Bulungan designated settlements as villages, even though they did not fulfil the criteria of '*desa*' as set by law. On the other hand, a number of Punan settlements were never formally registered (Barr et al, 2001). During this period, the government also encouraged the resettlement of small isolated communities from the upper to the lower watersheds. From the 1970s to as late as 1990, there was a continuous influx of people into the Malinau (Rhee, 2000; Anau et al, 2001), resulting in a mix of at least several ethnic groups. A curious aspect is that whole villages moved – not only the people, but also their administrative status and structure. In addition, they remained separate entities despite settling in another village territory. Thus, in the Malinau Basin, there are several settlements containing more than one village, each of which comprises a different ethnic group.

The rise of the 'New Order' was accompanied by a heavy emphasis on economic development, mainly through large-scale exploitation of the available natural resources. Resources were at the discretion of the central government and formal rights to the use of forest were given to large timber companies. In addition, Malinau was opened to logging. While local communities did not legally have formal rights, *de facto adat* rights were recognized and companies allowed communities a certain freedom. They sometimes negotiated directly with villages to determine where shifting cultivation and small strips of village forests could occur (Barr et al, 2001). This was the situation until the more recent changes, which began in 1997. The formation of the separate district of Malinau and the enactment of the regional autonomy law changed the political, economic and social landscape of Malinau dramatically.

When regional autonomy was introduced in 2000, the district government issued IPPK permits, through which companies had to cooperate with local people. Interviews indicate that conflict sharply increased in a short period. The granting of IPPKs encouraged more community claims over forest based on village and agricultural land boundaries. Many IPPKs also overlap the still-effective HPH areas, which adds to the overall complexity of the conflict.

People understand regional autonomy to mean that their *adat* rights have been returned, and, indeed, the 1998 Amendment of the Constitution explicitly recognizes customary rights (TAP MPR IX/2001). Having for years passively suffered the presence of logging companies exploiting their forests, they are now actively declaring ownership over the resources. This has been strengthened by the attitude of district governments who, for decades, have seen locally raised revenue from natural resources sent to the central government. A common comment is: 'Now it is our turn to enjoy the benefits from our resources.' As a result, people now freely and sometimes forcefully demand compensation or a share in benefits. Nowadays, it is the companies who are in need of legal protection.

People and conflict

Malinau is inhabited by a population of at least 18 ethnic groups, including the largest group of Punan hunter-gatherers in Borneo.[3] Most of the Punan are now semi-settled, although at any one time, a large proportion of people might be away in the forest. The various Kenyah groups are generally much more settled and are doing well through rice cultivation. In addition, there are the Merap, Putuk, Abai, Tidung and Berusu, as well as Javanese transmigrants, Bugis and Banjar migrants and Chinese traders (see Figure 2.3).

The history of Malinau is a history of struggles over land and forest. But while the struggles may be longstanding, the reasons and means of managing them have changed. We can trace the shifting authorities that worked to overcome these struggles and served to coordinate control over land during the last several centuries. For the earliest periods, we can only draw evidence from historical documents and oral histories, and try to extrapolate – although risky – from conditions observed in more traditional villages, (see Sellato, 2001).

Between 1750 and 1850, ethnic communities of the Ngurik or Baoe in areas of the Bahau River, Lurah and Pujungan were attacked frequently by people from Apo Kayan in alliance with the Kenyah (Kaskija, 2000). The Baoe people then moved to the Malinau River seeking protection and asked for support from the Punan and Abai to fight their enemies. In return, the Baoe allied themselves with the Punan people, vowed to become brothers, to be like 'the black and the white of the eye' that cannot be separated. Between 1830 and 1840, they lived together on the Malinau and Tubu rivers, and intermarried with other groups such as the Abai, Tebilun, Milau, Berau and Berusu. Later, the Baoe, also called Merap, fought the Berusu over ownership of birds' nest caves.

Until the early 1900s, such inter-group warfare – often over control of resources such as birds' nest caves – was common, as was migration in response to

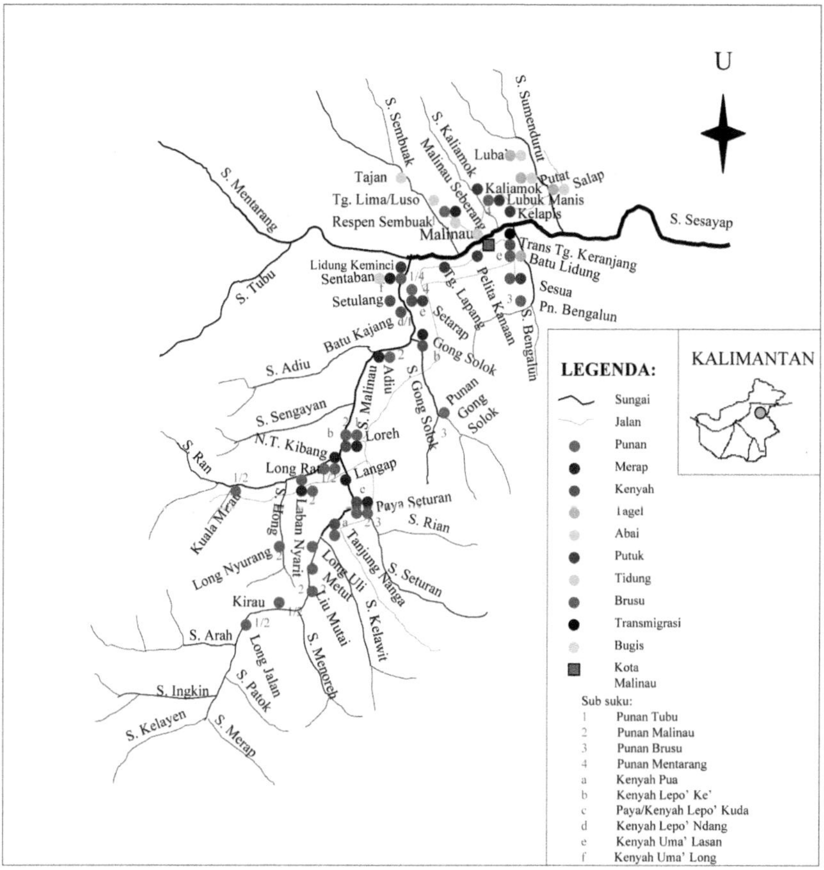

Keterangan:

1. Peta ini adalah peta sketsa untuk menjelaskan lokasi pemukiman dan kelompok suku-suku. Ada beberapa desa-desa yang terdapat di dalam satu lokasi pemukiman tidak dicantumkan dalam peta ini. Untuk mengetahui nama desa-desa dengan lengkap dapat dilihat di dalam tabel nama desa dan lokasi pemukiman.
2. Suku-suku yang dicantumkan di dalam peta ini di kelompokkan secara umum. Untuk kelompok sub suku secara detail dijelaskan di dalam tabel terlampir.
3. Sub suku Punan Malinau sering disebut Punan Bao mereka bergabung dengan Suku merap. Orang Kenyah dipanggil Paya oleh orang Merap. Lepo' Kuda dulu pecahan dari Lepo' Ke'.
4. Punan Beketan berasal dari Mahakam lewat Kayan lalu bercampur dengan Punan Malinau, sekarang lebih dikenal dengan Punan Malinau.

Source: Malinau ethnic map prepared by Asung Uluk (CIFOR consultant) in 1999

Figure 2.3 *Distribution of ethnic groups in Malinau*

the threat of war. Minor customary leaders (*kepala adat*) and their circle of close advisers helped to control access to land, manage conflicts and coordinate decisions within their ethnic group, while major customary leaders (*kepala adat besar*) helped to coordinate these matters at still larger scales. Most settled rice farming communities (e.g. the Kenyah) seemed to have sought control over their territories, while others (e.g. the Punan) seemed to have been more concerned with access to their settlements and key resources. The territories appear to have been defined in terms of a central settlement point or river and its watershed, with less emphasis on the exact location of their outer bounds. In many places, customary leaders controlled rights to valuable resources such as birds' nest caves within the territory of a sultanate.[4] The sultanate, in turn, levied taxes on the traded products.[5]

Early in the 20th century, the Dutch colonial government arrived in the region and enforced law and order to boost trade. Tribal wars were outlawed and many groups started to migrate downstream. The Putuk (or Lundaye), and the Abai moved from upper Mentarang to the village of Setarap in 1909. In 1961, a group of Punan settled in the same area and in 1972, the population increased again with the arrival of a group of Kenyah from the upper Pujungan (Anau et al, 2001; see also Chapter 9 and 10 in this volume). The Tidung, who had converted to Islam early on, claimed preferential rights based on their position as descendants from the kings of Tidung, who ruled over two petty kingdoms encompassing part of lower Malinau (Sellato, 2001), as well as on being owners of several of the birds' nest caves.

This migration was triggered by the marriage of the daughter of the chief *adat* leader of the Bahau River to the son of the *adat* leader of the Malinau River. Initially, the Kenyah were welcomed by the people already living along the Malinau River; but over the next few years, the increasing population caused conflict over boundaries of farmlands and villages.

Along the Malinau River, one can see a general pattern of claims and conflicts. In the far reaches of the upper Malinau, where mostly Punan groups lived, conflict focused on access over forest products. In the central portion of the upper Malinau, the rich coal deposits gave rise to conflicts over compensation claimed for use of cultivated and fallow fields. In the lower stretches, problems focused on access to agricultural lands and historically problematic relationships of mistrust (Anau et al, 2001).

While the source of conflict is over access to resources, it has developed a strong link to claims over territories. Coincidentally, in the period prior to the actual implementation of the regional autonomy law, the World Wide Fund for Nature (WWF) and CIFOR had started participatory mapping activities to delineate village boundaries and to help solve conflicts over natural resources. These have provided tools for making land claims, but strengthened a general impression that villages can now formerly claim rights in their territories (Barr et al, 2001).

Livelihoods

Forest people in remote areas are commonly highly dependent upon forest and forest products. Such is the case in Malinau, although the degree of dependence differs greatly among the groups and locations. The Dayak, in general, are considered agriculturalists, practising swidden cultivation of upland rice. The Punan, on the other hand, are considered predominantly nomadic hunter-gatherers. The distinction, while convenient, does not always hold, especially today. Almost all Dayak groups collect forest products in slack times and most Punan open

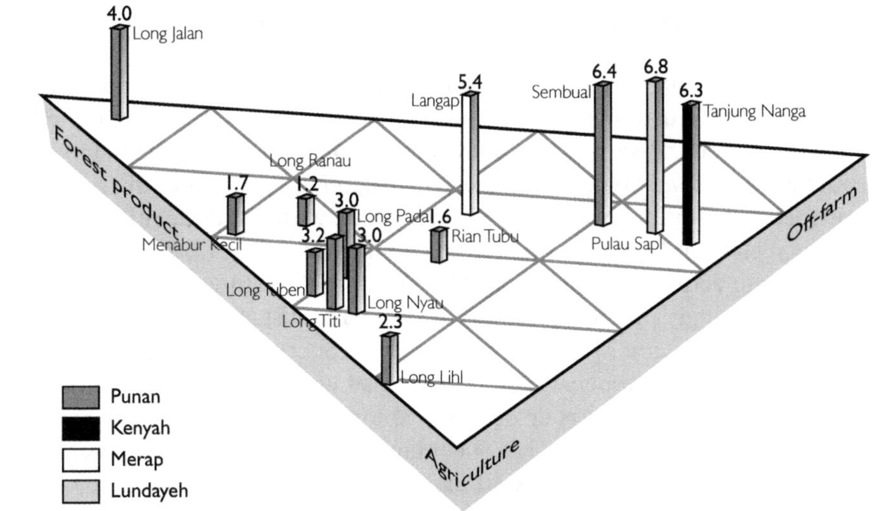

Source: Levang et al (2002)

Figure 2.4 *Average household income (million rupiah per year)*

swidden on a regular basis (Levang et al, 2002). The forest, however, remains of paramount importance to the entire local population.

Levang et al (2002) discovered a general trend where the more remote villages depend more heavily upon the gathering and selling of forest products, while downstream people make a living from agricultural activities. Here, swidden cultivation becomes essential and people have tried to diversify by planting coffee or cocoa. Further downstream, salaried activities become more important (see Figure 2.4).

During the past several years, salaried jobs have become more important, although primarily centred around Malinau Town. Nevertheless, the expansion of the government has created many new positions, including the four additional sub-district offices, established in 2002.

Local government

As warfare declined, the Dutch,[6] and later in 1950s the government of Indonesia,[7] became additional layers in the institutional hierarchy, first in the Bulungan sultanate and later in the *kabupaten*, or district of Bulungan. Bulungan, having defeated Tidung, emerged as a powerful sultanate about 1731 and was recognized as such by the Dutch colonial government. After independence, Bulungan became an autonomous region and later, in 1959, a *kabupaten*. As part of Bulungan, Malinau changed status several times in accordance with national administrative changes. In recognition of the Tidung Kingdom in the area during the Dutch colonial period, it became *Kewedanaan Tanah Tidung*, a special subdivision of the *kabupaten*, and, finally, a sub-district. An official known as *Bupati Penghubung Perwakilan Tanah Tidung* was stationed in Malinau Kota, covering Malinau, Mentarang and Lumbis. Interestingly, in 2007, the *Kabupaten Tane*

Tidung was formed, comprising sub-districts from Bulungan and Nunukan only, not from Malinau.

Meanwhile, centralized control and the Law of 1979 confused the role and accountability of customary leaders over time since many were appointed by outside officials as village government heads (*kepala desa*) and became upwardly accountable to a *camat* (sub-district leader), *bupati*, governor and the president. Local social institutions were severely weakened with the de-legitimization of customary laws and leaders. Government maps of villages bore little, if any, relationship to actual settlements and their claims. Unfortunately, local villagers became increasingly politically disenfranchised and distanced from the government under this system. Most matters of village concern continued to be settled by customary leaders, especially those who also worked for the government as village heads. These leaders made decisions among a group of ethnically homogenous people. Access to land and forest continued to be managed as before.

In addition to establishing new administrative overlays, the state staked extensive claims to forested territories starting in the 1960s as the timber boom began. Suddenly, villages found themselves sharing the forest with timber harvesters and being told that the land belonged to the government of Indonesia. The state's assertion of authority over land through the logging companies' presence openly challenged local sovereignty and claims to land in a way that the sultans and the Dutch had never done. Villagers generally accommodated the concessions, largely out of feelings of intimidation. (Military officials usually accompanied logging company staff or otherwise harassed potentially troublesome villagers). The gradual degradation of forest, loss of wildlife for hunting and lower water quality that they experienced were also partially offset by the construction of roads, provision of transportation, generation of employment, albeit limited, and occasional contribution to a village project. Local concessions did not strictly enforce hunting and burning prohibitions, and allowed swiddening in some forest areas in an effort to maintain good relations. During this time, forest-related conflicts were predominantly inter-village quarrels over access to agricultural plots and, for a few individuals, over efforts to maintain claims to birds' nest caves. The state did not allow conflicts to occur with the local government or concessions.

The authority of village customary leaders further eroded and land claims became more complex as ethnic groups began to share territories. Government resettlement programmes of the 1960s through the 1980s, and an *ad hoc* case of government-sponsored resettlement in 1999, reallocated mostly formerly Merap[8] lands along the upper Malinau River to other Dayak groups who had moved from more remote parts of the area. The newcomers did not, however, always sever ties with their former territories. In all cases, some people stayed in or close to the original village territory. As a result of these programmes, 2 to 4 ethnic groups now reside collectively in 9 out of the 16 settlements (*lokasi*) along the upper Malinau River. Population pressure has also increased substantially on local resources. These groups now make claims to multiple territories that overlap with other group's claims, calling into question who maintained authority over which land and the role of customary and government authorities in settling these claims. Because of the resettlement programmes, the upper Malinau River area suffers more from these kinds of multiple claims than any other part of the district.

The final and most recent major development occurred with three overlapping phenomena: decentralization reforms, new access for villagers to monetary payments for timber and land, and the creation of the new *Kabupaten* Malinau. With the initiation of decentralization and the associated uncertainty, local people from all sectors of society have sought to seize their share of Malinau's resources. Even before decentralization policies were formally implemented at the *kabupaten* level, *de facto* decentralization began taking place, with villages making claims to *adat* lands and negotiating directly with local investors (Rhee, 2003). Villagers made demands more freely for compensation or benefits from timber and mining companies and requested larger amounts than ever before.[9] Provincial Decree No 20/2000 added momentum by providing a legal basis for people to demand compensation.

Since former President Soeharto stepped down, villagers say they can express their discontent without fear and have been much freer about speaking out against their leaders and the government. Military officials today only rarely accompany timber companies or government entourages. New political associations have formed among different ethnic groups. Village leaders can be seen just as often in the central town of Malinau, meeting with government or company officials as in their own villages.

With the formation of the new Malinau District Government in 2001, for the first time, major positions were filled by people originating from within the *kabupaten*. *Adat* gained importance as people sought alliances and political position through ethnic groups (Anau et al, 2001), resulting in ethnic-based competition for land and power. The organization of *adat* groups based on the government hierarchy is also an inheritance from the Dutch colonial times. At that time, *adat* leaders were co-opted through formal recognition and the payment of honoraria. The Dutch, in their need for order, established a hierarchy where local *adat* groups were organized in larger associations led by a *Kepala Besar* (high *adat* leader). In Malinau, the *Majelis Kerapatan Besar Tanah-tanah Tidung* was established. Members of this council were the *Raja Tidung*, the Merap leader of Langap and four other chieftains (Césard, 2001; Sellato, 2001). The *adat* leader of the Merap was appointed high *adat* leader of Malinau in 1940. His position was legitimized through a colonial decree, and he remained in this position until his death in 1980.

Today, *adat* groups are starting to reorganize increasingly by strengthening organizations or establishing new ones, and calling on the government for formal recognition. In February 2002, the Tidung invigorated their customary institution (*lembaga adat*) by choosing new leaders in a grand ceremony. The Punan also had established the *Lembaga Adat Besar Punan Kalimantan Timur*, as well as the Adat Punan Foundation (*Yayasan Adat Punan*) in 1994.

Having prospered since their migration to the Malinau Basin, the Kenyah started to challenge the political authority of the Merap (Rhee, 2003). Since they were better organized, having a lembaga *adat* and a written *adat* law since 1968

(N. Anau, pers comm, 2001), they have become an important power in Malinau. The current *Bupati* is Dayak Kenyah. In response, the Merap reorganized and, in 1998, succeeded in filling the empty place of *Kepala Adat Besar Malinau*. Meanwhile, after a long vacuum, the Merap with support of the Kenyah tried to revive the *Lembaga Adat Besar Se-Sungai Malinau* and elected Bpk Impang Alang as *Kepala Adat Besar* in 2003. In 2006, Impang Alang resigned and was replaced by his son, who also heads the Malinau Parliament.

The Lundaye and Tidung groups have also become more politically aggressive over the past years. These groups are attempting to consolidate their claims to land, while Punan groups have little representation in the *Kabupaten* government and weak historical claims to lands. Furthermore, they are always the weaker partner in alliances with other ethnic groups (Anau et al, 2001).

The Kabupaten administration today

As mentioned earlier, the district of Malinau was established in 1999. Until 2001, it was governed by an interim district head. Thus, there was a one-and-a-half-year period of temporary leadership of the district that was not accountable to a local assembly. It was during this time that the bulk of small-scale logging concessions were issued (see Part II). In 2001, an elected *bupati* was sworn in. The same year, a government formal structure emerged with nine administrative offices (SK 117 Bupati Malinau 2001).

With the formation of the new Malinau District Government in 2001, major positions were filled by people originating from within the *kabupaten*. Previous officials were mostly from Java, Sulawesi or other parts of Kalimantan (especially Samarinda or Banjarmasin). The *dayakization* of local government has meant that authority is now rooted in the local politics of more than 18 different ethnic groups. Local relationships of power are more intertwined and complex than ever before.

Realizing these complex relationships, the *Bupati* has worked hard to achieve a balance between the different ethnic and religious groups (see Table 2.1 for a description of the current ethnic composition of the government).

Table 2.1 *Distribution of ethnic affiliations of government staff in Malinau*

Ethnicity	*Office of the Bupati Vice Bupati/ secretary/assistants*	*Head of section*	*Sectoral offices/ agencies*	*Total*
Kenyah	2	2	3	7
Lundaye	2	1	6	9
Toraja		1	4	5
Jawa			4	4
Tidung	2			2
Kutai			1	1
Banjar			1	1
Total	6	4	19	29

Source: Andrianto (2006)

While the Merap have no representative in the government, they hold the leadership of the district house of representatives – the *Dewan Perwakilan Rakyat Daerah* (DPRD) – for the second time. This position has been very powerful and has resulted in the strengthening of Merap claims over forests, including part of the new coal mine in the Langap area.

While in the beginning it seemed that decentralization and a new government had brought the state closer to society, as the new district government became well established and its authority and presence recognized, access to government officials has become formalized and distant. In defence of the local government, the accessibility of the early years meant an overwhelming number of requests and demands presented by personal contacts. By reinstalling the camat as a link between villages and district government, district officials reduced their interaction with villagers to a more manageable level.

The new district government office complex is daunting for ordinary citizens to enter due to its size and splendour and the formal procedure required for admission. As well, it is located some distance from the business centre of Malinau. Although local government is at ease, furthermore, with the rhetoric of civil society participation, it remains suspicious and unsure of exactly how to deal with it.

Modernity and political upheaval have brought many new challenges and opportunities to the people of Malinau. For the past five years, there has been considerable upheaval relating to the laws and regulations governing rights over land and natural resources. Such conditions have brought great insecurity as well as new options for developing cash incomes. Both aspects of accessibility are in flux (i.e. the physical opportunities to reach a site and extract resources, and the

Source: Photograph taken by Godwin Limberg, 2004

Figure 2.5 *In 2004, the government moved to the new office complex, depicted here*

legal and social controls that might influence choices to do so). The introduction of boat engines, new roads, increasing trade and rising fuel costs all influence physical access, as do patterns of resource depletion associated with logged areas. At the same time, diverse and overlapping claims and controls by concessions, local government and communities have led to considerable confusion. Currently, this 'tragedy of the commons' seems to favour liquidation of forests over any long-term options. The cost to biodiversity, local communities and other stakeholders seems likely to be considerable. Stability and confidence in tenure are desperately needed.

Notes

1 Estimated from 2003 election census.

2 See *Soerat Pengangkatan Alang Ampang sebagai Kepala Besar Bao Dajak di Langap oleh Pemerintah Keradjaan Boelongan*, 30-1-1940 [*Decree of the Kingdom of Boeloengan on Installing Alang Ampang as Chief Adat Leader of the Bao Dayak People in the Village of Langap*].

3 See Sellato (2001) for an historical overview of the upper Malinau River during the last 150 years.

4 Fox (2002) characterizes overlapping sovereignties during the late 1800s in Thailand, writing that they were 'neither single nor exclusive' (Fox, 2002, p2), but rather (citing Winichakul, 1994, p88) 'capable of being shared – one for its own ruler and one for its overlord – not in terms of a divided sovereignty, but rather a sovereignty of hierarchical layers'.

5 Interestingly, the letters of tax payment have been used in more modern times to establish ownership over the caves.

6 According to Sellato (2001), Dutch control in the Bulungan Sultanate began in 1850 with a *Politiek* contract, which was extended in 1877 with an agreement whereby the Dutch would handle some of the Sultanate's affairs; it was formalized in the late 1880s as part of the Dutch colony. During the early 1900s, they forced the Sultan to turn over control of the remoter regions of the Bahau River, Pujungan River and Apo Kayan. They also worked with the Sultanate, for example, to put down a Dayak rebellion in 1909 in the Tidung lands, which include the current Malinau River area.

7 According to Sellato (2001), in 1950, Bulungan became a *Wilayah Swapraja* (autonomous territory) of Indonesia after the Japanese occupation and then, in 1955, a *Wilayah Istimewa* (special territory). In 1959, after the death of the last sultan, Jalaluddin, in 1958, the sultanates were abolished and Bulungan became an ordinary regency (Daerah Tingkat II or kabupaten).

8 Prior to the Merap, it is believed that the Berusu and Punan occupied the area (Kaskija, 2000; Sellato 2001).

9 Even though demands for compensation had been made previously, villagers received few, if any, benefits in response.

References

Anau, N., van Heist, M., Iwan, R., Limberg, G., Sudana, M. and Wollenberg, E. (2001) *Pemetaan Desa Partisipatif dan Penyelesaian Konflik Batas: Studi Kasus di Desa-Desa*

Daerah Aliran Sungai Malinau, Januari s/d Juli (2000) Laporan Pengelolaan Hutan Bersama, CIFOR, Bogor, Indonesia

Anau, N., Iwan, R., van Heist, M., Limberg, G., Sudana, M. and Wollenberg, E. (2002) 'Negotiating more than boundaries: Conflict, power and agreement building in the demarcation of village borders in Malinau', in CIFOR (ed) *Forest, Science and Sustainability: The Bulungan Model Forest. Technical Report Phase I 1997-2001, ITTO Project PD 12/97 Rev 1 (F)*, CIFOR, Bogor, Indonesia

Andrianto, A. (2006) *The Role of District Government in Poverty Alleviation: Case Studies in Malinau and West Kutai Districts, East Kalimantan, Indonesia*, Governance Programme Report CIFOR, Bogor, Indonesia

Anonymous (2003) 'Kabupaten Malinau', Kompas, 17 June, http://kompas.com/kompas-cetak/0306/17/otonomi/373649.htm, accessed 14 January 2008, 23 October 2003

BAPPEDA (Badan Perencanaan Pembangunan Daerah) and BPS (Badan Pusat Statistik) Kabupaten Malinau (2005) *Kabupaten Malinau Dalam Angka 2004/2005*, BAPPEDA Malinau, Malinau, Indonesia

BAPPEDA Malinau (2001) *Kabupaten Malinau Dalam Angka 2000*, BAPPEDA Malinau, Malinau, Indonesia

Barr, C., Wollenberg, E., Limberg, G., Anau, N., Iwan, R., Sudana, I. M., Moeliono, M. and Djogo, T. (2001) *The Impacts of Decentralization on Forests and Forest-Dependent Communities in Malinau District, East Kalimantan: Case Studies on Decentralisation and Forests in Indonesia*, CIFOR, Bogor, Indonesia

Basuki, I. and Sheil, D. (2005) *Local Perspectives of Forest Landscapes: A Preliminary Evaluation of Land and Soils, and Their Importance in Malinau, East Kalimantan, Indonesia*, CIFOR, Bogor, Indonesia

BKKBN (Badan Koordinasi Keluarga Berencana Nasional) (2001) *Daftar Desa Miskin di Kabupaten Malinau, Kalimantan Timur*, BKKBN, Malinau, Indonesia

Césard, N. (2001) *Four Ethnic Groups (Punan, Kenyah, Merap, Lun Dayeh) Faced with Changes along the Malinau River (Kalimantan Timur)*, Forest Product and People Programme, CIFOR, Bogor, Indonesia

Fox, J. (2002) 'Siam mapped and mapping in Cambodia: Boundaries, sovereignty, and indigenous conceptions of space', *Society and Natural Resources*, vol 15, no 1, pp65–78

Kaskija, L. (2000) *Punan Malinau and the Bulungan Research Forest*, Report submitted to CIFOR, Bogor, Indonesia

Levang, P. and the FPP-Bulungan Team (2002) 'Peoples dependencies on forests', in CIFOR (ed) *Forest, Science and Sustainability: The Bulungan Model Forest. Technical Report Phase I 1997-2001, ITTO Project PD 12/97 Rev 1 (F)*, CIFOR, Bogor, Indonesia

Machfudh (2002) 'General description of the Bulungan Research Forest', in CIFOR (ed) *Forest, Science* and *Sustainability: The Bulungan Model Forest. Technical Report Phase I 1997-2001,* ITTO *Project PD 12/97 Rev 1 (F)*, CIFOR, Bogor, Indonesia

Malinau Voting Census (2003) *Malinau Voting Census Data*, Regional Government of Malinau District, East Kalimantan Province, Indonesia

Meijaard, E., Sheil, D., Nasi, R., Augeri, D., Rosenbaum, B., Iskandar, D., Setyawati, T., Lammertink, A., Rachmatika, I., Wong, A., Soehartono, T., Stanley, S. and O'Brien, T. (2005) *Life after Logging: Reconciling Wildlife Conservation and Production Forestry in Indonesian Borneo*, CIFOR, ITTO and UNESCO, Bogor, Indonesia

Rhee, S. (2000) '*De facto* decentralization and the management of natural resources in East Kalimantan during a period of transition', *Asia Pacific Community Forestry Newsletter*, vol 13, no 2, pp34–40

Rhee, S. (2003) '*De facto* decentralization and community conflicts in East Kalimantan, Indonesia: Explanations from local history and implications for community forestry', in Tuk-Po, L., de Jong, W. and Ken-ichi, A. (eds) *The Political Ecology of Tropical Forests in Southeast Asia: Historical Perspectives*, Kyoto Area Studies on Asia 6, Kyoto University Press, Kyoto, Japan, pp152–176

Sellato, B. (2001) *Forest, resources and People in Bulungan: Elements for a History of Settlement, Trade and Social Dynamics in Borneo, 1880–2000*, CIFOR, Bogor, Indonesia

Sheil, D. (2002) 'Biodiversity research in Malinau', in CIFOR (ed) *Forest, Science and Sustainability: The Bulungan Model Forest. Technical Report Phase I 1997-2001, ITTO Project PD 12/97 Rev 1 (F)*, CIFOR, Bogor, Indonesia

Sheil, D., Puri, R., Basuki, I., van Heist, M., Wan, M., Liswanti, N., Rukmiyati, Sardjono, M. A., Samsoedin, I., Sidiyasa, K., Chrisandini, Permana, E., Angi, E. M., Gatzweiler, F., Johnson, B. and Wijaya, A. (2003) *Exploring Biological Diversity, Environment and Local People's Perspectives in Forest Landscapes*, 2nd edition, CIFOR, Ministry of Forestry and International Tropical Timber Organization, Bogor, Indonesia

SK 117 Bupati Malinau (2001) *Tentang Susunan Organisasi Dan Tata Kerja Sekretariat Daerah Kabupaten Malinau*, Malinau, Indonesia

Soerat Pengangkatan Alang Ampang sebagai Kepala Besar Bao Dajak di Langap oleh Pemerintah Keradjaan Boelongan, 30-1-1940

TAP MPR (Ketetapan Majelis Permusyawaratan Rakyat Republik Indonesia) (2001) *Nomor IX/MPR/2001 Tentang Pembaruan Agraria dan Pengelolaan Sumberdaya Alam*, Jakarta, Indonesia

Uluk, A., Sudana, M. and Wollenberg, E. (2001) *Ketergantungan Masyarakat Dayak Terhadap Hutan di Sekitar Taman Nasional Kayan Mentarang*, CIFOR, Bogor, Indonesia

Widyastuti, R. S. (2003) 'Otonomi: Kabupaten Malinau', *Kompas*, 17 June 2003, http://www2.kompas.com/kompas-cetak/0306/17/otonomi/373649.htm

WWF (World Wide Fund for Nature) (2003) *Management of Kayan Mentarang National Park to Promote Trans-boundary Conservation along the Border between Indonesia and Malaysian States of Sabah and Sarawak*, ITTO Project, PD 38/00 REV I (F) WWF, Jakarta, Indonesia

3

The Cultural Politics of Collaboration to Control and Access Forest Resources in Malinau, East Kalimantan

Steve Rhee

Introduction

Decentralization, good governance, local forest management – these are concepts that natural resource management (NRM) practitioners often consider as integral components in strengthening the voice of disadvantaged forest-dependent groups. Within a normative framework, these concepts complement each other; yet field realities demonstrate that the institutional mechanisms that constitute these concepts are embedded in a history of ethnic relations and state-making, which are played out in the process of social change. Understanding how ethnicity and state-making are intertwined in decentralization processes is critical for equitable and sustainable forest management. Indeed, as the introduction of the book discusses, the mutual co-production of the state and society, and the reproduction of that relationship strongly affect control and access to forest resources.

This chapter discusses the relationships between social groups in the district of Malinau and explains the struggles to consolidate and access power in the district in the broader context of an ambiguous and ambitious decentralization process. In essence, it is an analysis of a 'culturally oriented political economy', which is useful in showing how power works (the processes) and the material consequences of those processes (Li, 2002, p417).[1]

This chapter first provides a brief history of interethnic relations in Malinau to contextualize current-day collaborations between actors. It then describes the context in which the social actors in Malinau operate and make use of available opportunities. I then discuss how these actors forge relationships and alliances to control resources, focusing on how these linkages are legitimized. The chapter concludes with a brief summary of the broader lessons learned from the Malinau case study with respect to stakeholder collaborations or alliances.

Brief history of interethnic relations in Malinau

The history of inter-group relations in Malinau is one marked by trade, migration and clientism. Prior to the Dutch arrival to Malinau during the early 20th century and similar to that of many parts of Kalimantan, this region was characterized by trade relationships between upland Dayak groups and coastal sultanates and petty kingdoms, who in turn traded non-timber forest products (NTFPs) with seafaring merchants, e.g., the Taosug and Bugis (Warren, 1981; Peluso, 1983).[2] For example, before the eastward migration of the Kenyah beginning in the 1960s, the Merap were the dominant political force in the upper and middle courses of the Malinau River. Originating west of the Malinau, the Merap conquered the uplands of the Malinau and adjacent Tubu River in the second half of the 19th century with the assistance of local Punan groups, and subsequently expanded their control downstream along the Malinau. An important factor in the Merap expansion downstream was the control of birds' nest caves along tributaries of the Malinau and other valuable NTFPs such as rattan (*rotan*), eaglewood (*gaharu*; *Aquilaria* spp) and resin (*damar*).[3] Due to the stratified social structure of the Merap, village chiefs and war leaders held private control of valuable resources such as birds' nest caves. The current Merap leader, a descendant of the aristocracy, continues to own the birds' nest caves to this day.

The Merap, who are traditionally swidden cultivators, controlled and exploited forest resources through their long-term relationship with the nomadic Punan, a patron–client relationship that was, and to a certain extent still is, characterized by subservience shown toward the Merap (Sellato, 2001; Kaskija, 2002). The Merap used the nomadic Punan as collectors of NTFPs, a military force and as guards of birds' nest caves, while also controlling the flow of resources between Punan collectors and downstream traders. In return for their collecting, the Punan received goods they could not obtain themselves, such as iron and tobacco (Kaskija, 2002). The patron–client relationship between agriculturalists and nomadic Punan is a common one throughout Borneo; but the equity of these relationships varies depending upon the social structure of the agricultural group with whom the Punan are associated (Rousseau, 1990; Sellato, 1994; Sellato, 2001; Kaskija, 2002).

Since the early 1900s, the Dutch interventions in this region did not dislodge Merap political dominance, although the abolition of slavery and increased Chinese and Malay competition most likely reduced the economic advantage of the Merap aristocracy (Kaskija, 2002).[4] In important ways, the Dutch legitimated and strengthened Merap political control of the uplands of Malinau, while also bringing it under formal administration.[5] For instance, in establishing a list of birds' nest caves and their owners, the Dutch sanctioned Merap control of the caves in Malinau (Sellato, 2001). The Dutch authorities and the Sultan of Bulungan later made the Merap leader the *Kepala Besar Sungai Malinau/Tubu* (chief leader of the Malinau/Tubu rivers) and this leader's son, the *Kepala Adat Besar Sugai Malinau* (chief customary or *adat* leader of the Malinau River) upon his father's death (Kaskija, 2002).

As in much of Kalimantan, for the Dayak living in the Malinau region, the post-independence period has meant adapting their traditional social structure and livelihood practices to a whole suite of challenges. Government interventions in the interior and the opening up of the frontier to extract the valuable timber and mineral resources have shifted the political–economic terrain at the local level. Most important for the Dayak in the Malinau region have been the effects of the government's resettlement initiatives, the nationalization of forests and the extension of the central government into the village political structure.

Most of the Kenyah currently settled in the Malinau Sub-district migrated east from the Bahau and Pujungan river basins in the 1960s and arrived *en masse* in the early 1970s during a general period of resettlement[6] spurred by government policy.

Until recently, the Kenyah from the Bahau remained relatively isolated and hence have maintained their highly stratified social structure, demonstrated by the presence of traditional aristocrats in powerful political positions in Kenyah villages today (Sellato, 2001). The Kenyah moved to Malinau to be closer to the market, schools and other benefits of 'development'. Their migration to the Malinau area followed traditional legal procedure. During the 1960s, the Kenyah village chiefs requested land from the Merap chief, who subsequently agreed since there was little competition for land and population densities were low (Kaskija, 2002). In return for the land transfer, the Kenyah paid a small gong as tangible proof of the land transfer and agreed to abide by Merap *adat* (customary law).[7] In this instance, the government's role was limited to sanctioning the transfer. Under the government's resettlement programmes, other groups such as the Punan and certain Merap villagers were strongly encouraged and enticed by the government to resettle into larger villages during the 1970 and 1980s (Sellato, 2001). The Kenyah have prospered since their arrival in the Malinau: 'Strong social structure, leadership, and discipline … as well as their sheer numbers and economic dynamism' have allowed them to succeed economically and challenge the political authority of the Merap (Sellato, 2001, p32).

Stakeholders or identities in the district of Malinau

With decentralization, establishing legitimacy at the district and village level has become critical to creating, accessing and consolidating power to control natural resources. The primary vehicle for articulating this legitimacy has been through identity, or, more specifically, identifying and conveying a perceived common ground or common good and, conversely, difference. Thus, in Malinau, as in other regions, it is perhaps less informative to speak about 'stakeholders' and more important to consider the 'identities that matter'. I emphasize identities because one individual may possess several identities, each of which is strategic in a given time and context and is not perceived as contradictory. In essence, the alliances in Malinau and their attendant leveraging positions hinge on social boundaries of people as 'insiders' or 'outsiders'. In Malinau, the identities that matter include the following: *putra daerah* ('children of the region'), *orang asli* (indigenous person), local ethnic groups (Kenyah, Merap, Punan, Lundaye and Tidung), district government-sanctioned companies, and local businessmen.[8] These are discussed below in the context of the opportunities, constraints and conflicts that each have incurred with the implementation of decentralization.

Prior to decentralization, politics in Malinau District, as in other regions, was characterized by a highly centralized authoritarian government in Jakarta that provided licence and legitimacy to timber concessionaires and mining companies, while de-legitimizing the culture and livelihood strategies of swidden cultivators or other rural populations (Poffenberger, 1990; Barber et al, 1994; Lynch and Talbott, 1995). Due to the oppressiveness of the state and the lack of opportunities available to Dayak groups during the Soeharto regime, disputes among Dayak ethnic groups rarely emerged as central issues that required state intervention. Decentralization, while full of uncertainty and ambiguity, allowed district- and village-level actors more opportunities to exercise power. Their voices are now heard by government and others to the extent that they have even been able to influence the formal timber sector. Actors at the district-level now attempt to leverage their positions and manoeuvre strategically by creating new, or working through existing, linkages with other district- and village-level actors. Moreover, ethnicity has come to play a more important and political role in how villagers define and align themselves vis-à-vis others.

Articulating ethnicity among Dayak groups and others who can claim historical roots to the area has become a key means of asserting or accessing power, thereby exacerbating and creating conflicts between villagers. With decentralization, villagers in the Malinau watershed use a variety of categories to define and align themselves. Perhaps one of the most divisive categories is the distinction between *orang asli* (indigenous person) and *pendatang* (migrant/newcomer). This distinction has been used by Merap, certain Punan groups, Tidung and, to a lesser extent, Lundaye to claim territory based on historical roots in the area. This is particularly threatening to the Kenyah and other groups who migrated to the Malinau region during the 1960s and 1970s since this dichotomy and being classified as *pendatang* could jeopardize their claims to territory. Indeed, the Merap, who are few in number, claim the entire middle portion of the Malinau watershed based on their claims of history.[9] Furthermore, the Tidung through their *adat* association have claimed the profits from small-scale timber harvesting – IPPKs (*Izin Pemungutan dan Pemanfaatan Kayu*) – in the territory of other villages based on claims to birds' nest caves and surrounding forest, legitimated by documents from the Dutch era.[10] In one case, the Tidung excluded the villagers in Gong Solok from obtaining profits from the IPPK operating in the Gong Solok territory based on historical claims. In another case, the village of Setarap partnered with an IPPK; but the Tidung forced the village of Setarap to give the majority of profit to the Tidung based on the same justification (see Chapters 5 and 6 in this volume for a complete discussion of IPPKs).

A partial countervailing force to claims of indigenousness is the large number of Kenyah residing in the Malinau watershed and also the political power of the Kenyah within the district government, even though the Kenyah are considered *pendatang*. There is, indeed, a tense relationship between the Kenyah, Tidung and Lundaye in the struggle for power and resources in the district as a whole. This uneasy balancing act hinges upon several factors: claims to indigenousness, population numbers and the history of interethnic relations.

The 'indigenous' Merap do not possess the influence of these other groups because the Merap population is small, as is the number of Merap elite. The Punan, recognized as indigenous[11] by other groups, have also been unable to mobilize as a collective because they historically lack a collective power base and solidarity. Thus, their claims of indigenousness have not had the same impact as that of the Merap or Tidung, although 12 per cent of Malinau's population is Punan compared to 2 per cent Merap.

Another politically powerful category is *putra daerah* ('children of the region' or 'local sons'), which refers to any group that can claim historical roots in the region. District officials who are in the most influential positions are *putra daerah*. This category, however, has fuzzy boundaries, lacking precision in the meanings of 'region' and/or 'local', as well as 'historical'. For example, the *Bupati* is considered *putra daerah* since he is ethnically Kenyah Uma' Kulit, who originated from the Apo Kayan but moved to the middle Kayan in Bulungan District. Being Kenyah[12] provides him with the status of *putra daerah*, although the sub-ethnic group could possibly exclude him from the category. Another example is the district secretary – also considered *putra daerah* – who is ethnically Lundaye from the Kerayan, which is part of Nunukan District, north of Malinau, but who has relatives in Malinau District. In fact, Malinau, Bulungan and Nunukan were all part of one district until late 1999; therefore, the original Bulungan area perhaps best represents the geographic region associated with *putra daerah*. It should also be noted that *putra daerah*[13] is a broader category than *asli* (indigenous) – one can be *putra daerah* without being asli (e.g. the Kenyah in the Malinau watershed). Furthermore, other ethnic groups such as the Chinese, Javanese and Bugis are not considered *putra daerah* regardless of how many generations they have been in Malinau. However, if they have spouses who can claim *putra daerah* status, they can, in part, claim that status as well. For example, one local Chinese businessman claims *putra daerah* status based on his mixed heritage of Chinese and Berusu, a Dayak ethnic group local to Malinau, which greatly facilitates his business dealings with certain Dayak villages.

Local entrepreneurs, regardless of origin, also play an important role in district politics, particularly as middlemen in the IPPK timber trade. In this case, 'local' is broadly defined geographically to include Tarakan, the major port city in northern East Kalimantan, and refers to being well connected to decision-makers and local communities, and established prior to decentralization. In the IPPK timber trade, the role of the local entrepreneur is twofold: on the one hand, they play a critical function in identifying accessible forests areas that are well stocked with commercial timber species and in brokering harvest agreements with nearby communities that may have tenure claims to these areas. On the other hand, the local businessmen are generally responsible for arranging the IPPK permit and all supporting documentation with the district government. These individuals are key interlocutors and have strong political connections to key government officials, which, in some cases, are reinforced by family ties and/or shared ethnic bonds (Barr et al, 2001). Furthermore, in several cases, the timber brokers are also traders in *rotan* (rattan), *gaharu* (eaglewood), *damar* (resin) and other NTFPs. Many of the individuals have longstanding ties to forest-based communities engaged in collecting these products. It is often the case that households within such communities are heavily indebted to NTFP traders, and there is a strong likelihood that debt relations of this sort have shaped the agreements made between the brokers and communities regarding IPPK timber extraction. For example, the local timber broker who arranged C. V. Hanura's IPPK permits within Inhutani II's logging concession boundaries served as the local treasurer for the government-controlled political party Golkar from 1993 to 1998. He also sat on the committee that oversaw the selection of district house of representatives members – the *Dewan Perwakilan Rakyat Daerah* (DPRD) (Barr et al, 2001). Local businessmen also use their connections with local communities to indirectly pressure the district government to issue permits.

Actors with only central government power bases, such as Inhutani II and the mining company Bara Dinamika Mudah Sukses (BDMS), have found that they can no longer operate with the certainty and stability of the New Order. Indeed, their bargaining positions have become much weaker, and they must now lobby and appease district-level actors to continue to operate, although even these strategies do not guarantee their positions. For example, Inhutani II, which has a timber concession permit issued by the central government (*Hak Pengusahaan Hutan*, or HPH), stopped its operations due to conflicts with local communities backed by DPRD,[14] but leases their equipment for other purposes, such as IPPKs. Furthermore, the operations of BDMS have been significantly interrupted by villagers' demonstrations and road blocks. BDMS has also been hampered by the district government's successful lobbying for a new coal mining company with local backing that has obtained a mining permit for part of an area previously surveyed and targeted by BDMS. However, Meranti Sakti, a large-scale timber concessionaire (HPH) that began operations prior to decentralization, has been able to continue operations because the company provided benefits during the Soeharto period to local communities and individuals who are now in the district government, thus having established a local base of support. In addition, Meranti Sakti is an East Kalimantan-based conglomerate, which connects it to provincial decision-makers, and is associated with the army, providing it with central government support as well.

In addition to these identities and/or actors who have provided vehicles for alliances, international organizations concerned with the region's forests and livelihoods of forest-dependent people and political parties have also, at times, proved to be fruitful mediums in discovering, asserting and enacting collaborations and alliances to leverage power. International organizations worth mentioning are the Center for International Forestry Research (CIFOR) and the World Wide Fund for Nature (WWF), which are concerned with livelihoods and forest. Since 1996, CIFOR has engaged in applied research focusing on improved livelihoods and sustainable forest management. CIFOR's various research activities have provided a number of forest-dependent Dayak groups in Malinau with potential 'room to manoeuvre' (Tsing, 1999). This has allowed these local groups to link with the global concern for tropical forests and forest-dependent people, and raise international leverage to support their claims (Li, 1999). Both CIFOR and WWF were involved in community mapping of traditional village territories, using state idioms, modern cartography and geographic information systems (GIS) technologies to lay the foundation for advocating recognition of ancestral lands (Harwell, 2000). While the material consequences of these connections through these two organizations are difficult to specify as direct effects, evidence indicates that they have provided vehicles for villagers to speak and think in certain ways to claim certain rights (Eghenter, 2000).

The emergence of a number of political parties in the post-Soeharto era has also added another element to fashioning an identity that matters, particularly since political parties possess and promise both symbolic and material resources. Indeed, political parties and the promises to both their candidates and constituencies have increased both the possibilities of new alliances and the disruption of old ones. For example, during the Soeharto regime, the Kenyah voted uniformly for the ruling political party Golkar, whereas the Lundaye were more divided in their vote. Although voting had little influence in government at the time, the different patterns of voting between the Kenyah and Lundaye have had significant repercussions in the post-Soeharto era. Indeed, the Lundaye have found themselves well positioned with the emergence of political parties in the post-New Order period since they never voted in a unified fashion. The learning curve for the Kenyah, however, has been much steeper with respect to dealing with the fracturing of their ethnic vote, learning how to 'read' politicians from their ethnic group who represent different political parties, and understanding the consequences of the fractured Kenyah vote for the Kenyah ethnic group as a collective. This is further complicated by the fact that although constituents vote for a particular individual, it is the political party that obtains the vote, determines the ranking of candidates and decides who will win the seat in the legislature should the party win. The Punan are in a particularly difficult situation because they are often listed as candidates for political parties, but always on the bottom of the ranking of candidates, and their votes, in effect, go to someone else in the party.

District government as a microcosm of broader ethnic tensions and power struggles

This section discusses how the district government was established and developed, highlighting that this process was indicative of broader ethnic tensions touched upon earlier. Furthermore, although ethnicity has been taken into consideration, the formation of the district government was strongly influenced by the legacy of the New Order.

Critical to and illustrative of the shifting identities and alliances of actors in Malinau are the establishment of the district government and the struggle for power within it. When Malinau District was formed in 1999 by dividing Bulungan District, the Department of Home Affairs appointed a civil servant living in Samarinda, the capital of East Kalimantan Province, as temporary regent. A year later, in 2000, the members of the district assembly were appointed based on election results in Bulungan District in 1999 (prior to the district's division). In 2001, these individuals then elected the current *Bupati*, Marthin Billa (a career civil servant), from among the three top candidates. The members of the DPRD were ostensibly influenced in their decision by community groups. Marthin Billa had the support of his own ethnic group, the Kenyah, who voted as one block, while the other two candidates were Lundaye, which split that ethnic group's vote. Moreover, Marthin Billa's running mate was Tidung, thereby safeguarding against the Tidung voting against him. The individual who was appointed the head of the district legislature also supported Marthin Billa's candidacy since it was considered a repayment to the Kenyah for backing his appointment as the head of the local legislature. Although there is no concrete evidence of 'money politics' in the election of the Malinau District head,[15] 'purchasing' the position of district head through money and favours is public knowledge throughout the country.

Being perceived as 'local' and linked to other influential ethnic factions played critical roles in the selection of DPRD members and the election of the *Bupati*. Indeed, even the administrative boundaries of Malinau District, which are irrational from both a biophysical and administrative perspective, were established to ensure that one ethnic group did not dominate the district (Barr et al, 2001). The administrative boundaries cut across watersheds, and some communities must travel three days and through another district's capital to arrive at the town of Malinau. The current boundaries reportedly were established so that neither the Kenyah nor the Lundaye would dominate the district.

Since his first election, in 2001, Marthin Billa has astutely appointed government officials to safeguard against accusations of ethnic favouritism, while also positioning allies strategically. There is representation of all major ethnic groups in ostensibly senior and influential positions in the district government: the regent is Kenyah; the vice-regent, Tidung; the district secretary, Lundaye; and the head of the legislature, Merap; and the vice-head of the legislature is Lundaye. However, the *Bupati* has placed allies in key political and technical gatekeeper positions. Thus, for example, the *Bupati*'s assistant handling development issues (*Assistan*

II) and the head of facilities who makes purchases such as computers, vehicles, etc. (*Kepala Bagian Umum dan Perlengkapan*) are both Kenyah. These two positions are not influential per se; but due to these individuals' closeness to the *Bupati* and the personalized nature of the politics of government and administration, they have become key positions. While from an official reading of the bureaucratic map of the roles and responsibilities of various district government positions, these 'gatekeeper' positions would presumably be kept in check by more senior positions in the hierarchy, the persistent patron–client networks maintain these individuals in influential positions.

District government initiatives to consolidate control and power: Bringing government closer to keep people farther away

Several factors drive how the district government operates. First, 'local' influential government officials were trained as civil servants under the New Order system and, to a large extent, still promote New Order values of 'development/progress', upward accountability and patron–client relationships. Second, the struggles between ethnic groups in Malinau are played out within the district government. Being 'local' disguises the tense, delicate balance of 'local' powers and constrains further consolidation within the district government. Finally, the process of decentralization has been paralleled by the growth and filling out of the district government bureaucracy. These factors strongly influence the quality of the relationship between district government and their ostensible constituents.

One dramatic change with decentralization has been the inclusion of local ethnic groups (collectively known as Dayaks) in positions of authority in the district government; yet these Dayaks, such as the *Bupati*, embrace several identities that are often conflicting. For example, when government officials speak casually or outside of their official capacity, their rationale and statements are not different from the villagers themselves (e.g. they note that swidden cultivation is appropriate for the area); yet once they put on their uniforms, they are transformed into bureaucrats who are still strongly influenced by the Soeharto regime's notions of 'development'. Moreover, ethnic affiliation has in no way ensured that government officials and other representatives are accountable to their constituencies.

The central organizing principle articulated by the district government is 'development' and 'economic growth' (read modernity). If we look at how the district government envisions 'development' and 'growth', we see the same focus as the New Order on infrastructure (roads, bridges and buildings), plantation development (oil palm) and projects rife with rent-seeking opportunities for government officials and their contractors. For example, there is currently a road being built to one of the more remote regions of the district ostensibly to connect those villagers to a road network; yet the difficult topography of the region suggests that building the road is illogical since the number of individuals and potential produce carried between Malinau Town and the remote areas do not justify the cost of road construction and maintenance.[16] Furthermore, even if the road were built, maintenance responsibilities and costs have yet to be addressed, which further suggests that the real purpose of the road is to extract timber.

Other district government initiatives to 'develop' the district and people are ostensibly to improve the efficiency of services and administration; yet these initiatives closely resemble New Order attempts to control and surveil its citizens through a standardization of villages. For example, the district government has a broad initiative mandated by the Ministry of Home Affairs to improve the self-sufficiency and economy of each village in the district, stipulating a minimum population number that can claim the status of village, clumping together those that do not meet this minimum population number, surveying each village for its economic 'potential', and creating institutions within the village to coordinate and regulate development.[17] From one perspective, this initiative can be seen as the district government better trying to serve its constituents; but given that the legacy of the New Order is still strongly intact, it is possible that this initiative is also a means to better regulate and control the movements and actions of villagers. For example, the district government has noted that each village will have a Village Empowerment Institution (*Lembaga Pemberdayaan Masyarakat*) that will be responsible for village development initiatives, but will have to raise its own funds and salary with little or no financial assistance from the district government. In one sense, this is a means for the district government to avoid its function and responsibility. Fortunately or not, to date, these initiatives have yet to be fully implemented by the district government.

Another expression of the further involvement of the state within villagers' lives is the division of the district into a greater number of sub-districts, again ostensibly to improve government services and administration, and settle conflicts. To date, however, the authority of the sub-district head and staff seems to be extremely limited, and it should be noted that the head of the sub-district is appointed by the *Bupati* and, hence, is upwardly accountable. Thus, creating new sub-districts and their attendant bureaucracies may have the effect of creating more difficulty and distance for villagers to influence and access decision-makers, while creating the impression of proximity and subsidiarity. An essential part of this difficulty and distance is the creation of a 'thicker' bureaucracy that villagers must navigate. When the district was first established and decentralization implemented, villagers noted their approval since they could have direct access to decision-makers (e.g. the *Bupati*). However, as the district government has filled out its bureaucracy and created more sub-districts, villagers are required to follow protocol. This means that they must first take issues and complaints to lower-level officials with little decision-making power, such as those in the sub-district government, who will attempt to resolve the situation or take it to his or her superior. Thus, by bringing the state closer to its citizens, the real effect may be to create more distance between them through the added layers of bureaucracy, while the government can more closely surveil and regulate its citizens.

Villagers' relations to district government: Access and accountability couched in networks of patronage

Prior to decentralization and, to a large extent, still today, the relationship between local government and communities can be characterized as patron–client, lacking formal functioning institutional mechanisms of downward accountability and transparency. Moreover, the authoritarian Soeharto regime successfully stunted meaningful participation of Indonesian citizens in political processes. As a result, decentralization and its attendant transfer of governing authority to district governments have taken place in a context of ill-prepared district government officials, weak civil society groups and stakeholders' extremely limited experience with participatory democratic political processes.

Villagers themselves perceive the relationship between the district government and themselves as patron–client. For example, several villagers noted that they often bring meat or forest fruit to government officials, which makes it much easier to be given an appointment with them.

This patron–client relationship is embedded in several important social structures. First, the New Order ideology that the government 'guides' or 'develops' (*membina*), and which will lead villagers on the path to 'progress' and 'modernity', remain entrenched. Thus, the idea that villagers may know better what they need than the government is not yet conventional thinking. Moreover, it is in the interest of government to maintain this relationship and ideology because 'development' and 'progress' are normally accompanied by large-scale projects that not only provide opportunities for rent-seeking, but also are easier to administer and lower transaction costs. Second, as mentioned earlier, the history of relationships among ethnic groups (between Dayak and non-Dayak, and between Dayak groups themselves) plays a critical factor in gaining access to government officials. Villagers often approach government officials based on ethnic affiliations; a common heard phrase among villagers about government officials is that they are *orang kita* ('one of us'). Thus, the Punan, who have virtually no representation within the district government, have the most difficult time obtaining an audience with government officials. Indeed, one Punan village head noted that the most difficult obstacle to alleviate poverty for the Punan is *not* having a Punan in the district legislature. Third, traditional intra-village social structures such as elders, aristocrats and kinship play an important role in decision-making, settling disputes, and whether/how government officials are approached. For example, in most Kenyah villages in the Malinau watershed, the village government is dominated by *paren*, or those with aristocratic lineage. This may also be the case for Merap, Punan and Lundaye to lesser and greater degrees (see Rousseau, 1995, Sellato, 2001, and Kaskija, 2002, regarding the extent of hierarchy in these groups). Furthermore, the Punan, who historically were clients to swidden cultivators, still remain strongly dependent upon the patronage of these groups, who now maintain important positions within local government. Moreover, government officials think of local government in terms of these traditional structures, noting that the relationship between the village head, sub-district head and regent are analogous to the relationship between child, father and grandfather.

The above institutions overlap to form the basis for how villagers' make sense of the ways in which they can access government officials and the services that government should provide, and conversely how government officials perceive villagers. The role of the district government in the broader government framework of Indonesia is uncertain, and the rights and responsibilities of the district government with respect to villagers – its ostensible constituents – are ambiguous. Accordingly, one avenue of recourse for villagers in the post-New Order period has been to protest or demonstrate as a means of making government accountable. Indeed, protests and demonstrations are a common site everywhere in Indonesia, and in Malinau, as elsewhere in the country, they are sometimes used or abused by a few individuals to demand cash compensation based on fraudulent claims (see Chapter 10 in this volume for a discussion of conflicts).

District government officials generally perceive their relationship with villagers in a reciprocal manner, as villagers' patrons. Despite earlier rhetoric, district government officials generally do not know, do not want to know and/or are not concerned with the desires and aspirations of villagers – rarely, if ever, does the district government request input from villagers.[18] Moreover, government officials tend to perceive villagers as ignorant and incapable of making 'informed' decisions. This is demonstrated by government officials not wanting to distribute legal literacy literature to villagers because villagers would 'misunderstand' or 'misinterpret' the information.[19] Conversely, government officials feel that they are more knowledgeable than villagers and know what is best for villagers, especially since they are from the region. For some government officials, this perception is based on the fact that they are in office, which means that villagers have put their faith in them and, hence, there is no need for villagers to provide input to the government. Another reason that government officials rarely meet with villagers is, of course, to avoid their complaints and demands, which have become more aggressive and common with decentralization.

The district government's attitude towards villagers is embedded in an institutional structure that does not reward downward accountability. Furthermore, district officials have little previous experience with this mechanism and other democratic processes. Most of their experience as civil servants has been about being upwardly accountable during the Soeharto regime; to a large extent, the institutional structure of district government has not changed with the ostensible transition to decentralization.

Conclusions

Establishing legitimacy at the district and village level has become critical to creating, accessing and consolidating power to control natural resources. The primary vehicle for articulating this legitimacy has been through identifying and conveying a perceived common ground or common good, as well as differences.

Thus, in Malinau, as in other regions, it is perhaps less informative to speak about 'stakeholders' and more important to consider the 'identities that matter'. In essence, the alliances in Malinau and their attendant leveraging positions hinge on social boundaries of people as 'insiders' or 'outsiders'. Indeed, actors at the district level now attempt to leverage their positions and manoeuvre strategically by creating new linkages or working through existing ones with other district- and village-level actors. While decentralization has created opportunity for the voices of local communities to be heard by government and others, intra- and inter-community conflicts, both pre-existing and new, are much more explicit and numerous due to the opportunities that decentralization has provided. Many of these conflicts are predicated on ethnicity, which has, indeed, come to play a more important and political role in how villagers define and align themselves with respect to others.

In addition, international organizations concerned with the region's forests and livelihoods of forest-dependent people and political parties have also, at times, proved to be fruitful means to finding, asserting and enacting collaborations and alliances to leverage power, and both symbolic and material resources.

Together with these means for significant alliances and identities, the district government has attempted to consolidate control and power, keeping people's demands at bay by bringing government bureaucracy closer to them; yet it has had to deal with the aforementioned ethnic tensions within the district government. Influential government officials, although local, were trained as civil servants under the New Order system and, to a large extent, still promote New Order values of 'development/progress', upward accountability and patron–client relationships. The struggles between ethnic groups in Malinau play themselves out within the district government where being local disguises the tense, delicate balance of 'local' powers and constrains further consolidation within the district government. Finally, the district government bureaucracy has grown and spread together with the decentralization process.

Patron–client relations figure importantly, particularly from the perspective of villagers on their relationship with government. Prior to decentralization and to a large extent still now, the relationship between local government and communities can be characterized as patron–client, lacking formal functioning institutional mechanisms of downward accountability and transparency. Moreover, the authoritarian Soeharto regime successfully stunted the meaningful participation of Indonesian citizens in political processes. Hence, decentralization and its attendant transfer of governing authority to district governments have taken place in a context of ill-prepared district government officials, weak civil society groups, and all stakeholders' extremely limited experience of participatory democratic political processes.

Notes

1 According to Li (2002, p417):

> *... to address the intersection between the projects that rural people pursue, and the constraints and opportunities they encounter, requires a robust concept of agency such as that supplied by the culturally oriented political economy developed by William Roseberry, among others. Roseberry (1989) argues that economism, the analysis of formal relations of production both within rural areas and between rural producers and larger social formations, is an inadequate framework for understanding agrarian change. It must be embedded in an exploration of human agency, which can be grasped* historically *as the product of transnational pressures, regional processes of class formation, and particular localized constellations of power.* Economically, *agency is manifested in people's livelihood strategies and the projects they pursue.* Politically, *agency can be understood as the forms that militancy (or quiescence) takes at certain periods.* Culturally, *it is present in people's sense of how and why they do things, and the frameworks within which they assert entitlements and make claims.*

2 Trade in NTFPs with China is recorded in Chinese archives dating to circa 1000 CE (Sellato, 2001)

3 *Damar* is a resin from trees of the *Dipterocarpaceae* family and 'is used to manufacture matches, varnish and turpentine' (Peluso, 1983).

4 Opinions differ with respect to the extent and character of Dutch involvement in the interior. Whittier (1973, p38) notes that in the Apau Kayan, the Dutch were 'relatively unobtrusive in terms of demanding changes in the Kenyah lifestyle with the exceptions of headhunting and warfare'. Sellato (2001) shares the same sentiment with respect to the Merap in Malinau, playing down the effects of the elimination of headhunting, warfare and slavery. According to King (1993, pp158–159), their elimination resulted in redirecting power and influence to the Dutch, as well as in indigenous leaders looking for 'other avenues of advancement and leadership'.

5 Rousseau (1990, p198) notes that 'colonial rule brought about a stabilization of regional leadership into formally defined political offices under administrative control'.

6 The government did not directly force upland people to resettle in this area. Villagers themselves initiated some migrations, while in other cases, the government strongly encouraged resettlements.

7 Kaskija (2002) notes that this was not a transfer of ownership, but rather a granting of use rights. Regardless, all groups moving into the upper and lower courses of the Malinau acknowledged the Merap's territorial ownership and requested land from them.

8 Each of these identities is part of a dichotomy that establishes and maintains the boundary between 'insider' and 'outsider': *putera daerah* (regional native) versus *orang luar* (person from outside); *asli* (indigenous) versus *pendatang* (migrant/newcomer); local versus elsewhere; and district government versus central government.

9 I should note here that the Merap themselves migrated to the Malinau watershed approximately 120 years ago; thus, their claims based on history are relative (Sellato, 2001).

10 The Tidung case is instructive with respect to the frequent 'misuse' of historically based claims and the deeply contested nature of these claims. While acknowledging claims to birds' nests caves, Dutch documents in no way articulate the Tidung terri-

torial claims to surrounding forests, which fits the pattern of this period of not delineating boundaries with respect to local claims since local people had no legitimate claim to vast territories, land was abundant, and specific commodities and labour were the scarce resources. 'Hard' and 'fixed' territorial boundaries did not exist for local people at that time. The problem of a lack of fixed boundaries manifests itself now when Tidung birds' nests cave owners are asked about the extent of their claim to forest areas around the caves. Some owners, in an offhand way, have noted that the extent is determined by how far one can hear a gong that is hit from the cave. This, of course, begs the questions of how large the gong is and how hard it is hit. These owners have a difficult time answering these questions.

11 The Punan arrived in the Malinau watershed either before or with the Merap (see Kaskija, 2002).

12 Having been born in the village of Naha Kramo, now part of Malinau, also strengthens his 'nativeness'.

13 Our observations show that the term *putra daerah* includes aspects of birth place, genealogy, a place where individuals have lived for a substantial period of time and recognition by other people as a *putra daerah*. For example, the district secretary of Malinau originates from Krayan and so does his family; but because people in villages in Malinau – Tanjung Lapang, Pulau Sapi and Sempayang — supported him as *putra daerah*, he was acceptable as district secretary. In contrast, a person also from the Krayan but not accepted by the Lundaye community in Malinau was rejected by them for the post of the head of the district planning agency (BAPPEDA).

14 DPRD and local government officials generally have more revenue-generating and rent-seeking opportunities with IPPKs than with HPHs. In the particular case of Inhutani II, the parastatal timber company holds a concession that largely overlaps with the territory claimed by the father of the head of the DPRD, who is also the chief customary leader.

15 One member of the Malinau legislature was offered money to vote for a certain candidate (R. Iwan, pers comm, 2003).

16 Field observations show that the number of individuals and potential produce transported between Malinau Kota and the remote areas does not justify the cost of road construction and maintenance. The route chosen also passes through areas with high timber values and stops midway due to impossible topography.

17 These administrative and regulatory changes arise out of the passing of the Law on Regional Government Administration (UU 22/1999 *Pemerintahan Daerah*). Although the law does not stipulate that villages need to be structurally uniform across the country, and should be tailored to local conditions, local government officials have infrequently taken action on this because they lack the experience and training to do so. Indeed, although this law and other regulations appear progressive and participatory, in reality, they are often not the case because implementing regulations, training of government staff and budgetary support are lacking. See Antlöv (2003) for a full analysis of the opportunities and constraints for the democratization of village-level politics with the passing of the Law on Regional Government Administration.

18 In past district elections, villagers voted for political parties and not candidates. The winning party then decided who would fill the position that the party won. Villagers voted for district legislature, which then chose the district head to be approved by the central government. This system was not conducive to downward accountability. This system was changed in 2004 to direct elections of candidates instead of parties.

19 'Misunderstand' and 'misinterpret' should mean that villagers would not interpret the law in the way in which the district government wants them to. One example of the government not wanting to distribute legal literacy literature related to briefs providing information on certain legal issues. CIFOR had drafted the original briefs and requested input from the district government. The briefs were to be distributed to villagers after correcting for the district government's comments.

References

Antlöv, H. (2003) 'Village government and rural development in Indonesia: The new democratic framework', *Bulletin of Indonesian Economic Studies*, vol 39, no 2, pp193–214

Barber, C., Johnson, N. C. and Hafild, E. (1994) *Breaking Through the Logjam: Obstacles to Forest Policy Reform in Indonesia and the United States*, World Resources Institute, Washington, DC

Barr, C., Wollenberg, E., Limberg, G., Anau, N., Iwan, R., Sudana, I. M., Moeliono, M. and Djogo, T. (2001) *The Impacts of Decentralization on Forests and Forest-Dependent Communities in Malinau District, East Kalimantan: Case Studies on Decentralisation and Forests in Indonesia*, CIFOR, Bogor, Indonesia

Eghenter, C. (2000) *Mapping Peoples' Forests: The Role of Mapping in Planning Community-Based Management of Conservation Areas in Indonesia*, Biodiversity Support Program, Washington, DC

Harwell, E. (2000) 'Remote sensibilities: Discourses of technology and the making of Indonesia's natural disaster', *Development and Change*, vol 31, no 1, pp307–340

Kaskija, L. (1999) *Stuck at the Bottom: Opportunity Structures and Punan Malinau Identity*, CIFOR, Bogor, Indonesia

Kaskija, L. (2002) *Claiming the Forest: Punan Local Histories and Recent Developments in Bulungan, East Kalimantan*, CIFOR, Bogor, Indonesia

King, V. T. (1993) *The Peoples of Borneo, Peoples of South-East Asia and the Pacific*, Blackwell Publishers, Oxford

Li, T. M. (1999) 'Compromising power: Development, culture and rule in Indonesia', *Cultural Anthropology*, vol 14, no 3, pp295–322

Li, T. M. (2002) 'Local histories, global markets: Cocoa and class in upland Sulawesi', *Development and Change*, vol 33, no 3, pp415–437

Lynch, O. J. and Talbott, K. (1995) *Balancing Acts: Community-based Forest Management and National Law in Asia and the Pacific*, World Resources Institute, Washington, DC

Peet, R. and Watts, M. (1996) 'Liberation ecology: Development, sustainability, and environment in an age of market triumphalism', in Peet, R. and Watts, M. (eds) *Liberation Ecologies: Environment, Development, Social Movements*, Routledge, London

Peluso, N. L. (1983) *Markets and Merchants: The Forest Product Trade of East Kalimantan in Historical Perspective*, MA thesis, Cornell University, Ithaca, NY

Poffenberger, M. (ed) (1990) *Keepers of the Forest: Land Management Alternatives in Southeast Asia*, Kumarian Press, Hartford

Roseberry, W. (1989) Anthropologies and Histories: Essays in Culture, History, and Political Economy, Rutgers University Press, New Brunswick and London

Rousseau, J. (1990) *Central Borneo: Ethnic Identity and Social Life in a Stratified Society*, Clarendon Press, Oxford

Sellato, B. (1994) *Nomads of the Borneo Rainforest: The Economics, Politics, and Ideology of Settling Down*, University of Hawaii Press, Honolulu, HI

Sellato, B. (2001) *Forest, Resources and People in Bulungan: Elements for a History of Settlement, Trade and Social Dynamics in Borneo, 1880–2000*, CIFOR, Bogor, Indonesia

Tsing, A. (1999) 'Becoming a tribal elder, and other green development fantasies', in Li, T. M. (ed) *Transforming the Indonesian Uplands: Marginality, Power and Production*, Harwood Academic Publishers, London

Warren, J. F. (1981) *The Sulu Zone, 1768–1898: The Dynamics of External Trade, Slavery, and Ethnicity in the Transformation of a Southeast Asian Maritime State*, Singapore University Press, Kent Ridge

Whittier, H. L. (1973) *Social Organization and Symbols of Social Differentiation: An Ethnographic Study of the Kenyah Dayak of East Kalimantan (Borneo)*, PhD thesis, Michigan State University, East Lansing

4

Biodiversity, Landscapes and Livelihoods: A Local Perspective

Douglas Sheil, Miriam van Heist, Nining Liswanti, Imam Basuki, Meilinda Wan, with the help of the people of Paya Seturan, Long Lake, Punan Rian, Langap, Laban Nyarit, Long Jalan, Liu Mutai and Gong Solok

The need for conservation in Borneo

People across the world are concerned about the loss of Borneo's forests and the unique plants and animals they contain. These rain forests are globally important for their high species richness and endemism (species that occur nowhere else) (Whitmore, 1986). Some recent global overviews suggest that Borneo's forests contain more plant species than perhaps any comparable region (Kier et al, 2005). Despite the limited data for much of Borneo, it seems that at least 37 birds, 44 land mammals and over one third of all the estimated 10,000 to 15,000 plants are found only on the island (MacKinnon et al, 1996). As the forests continue to be degraded or disappear, many of these species are increasingly threatened.

Until recently, Malinau was little explored biologically, like most of East Kalimantan. Conservationists suspected that the rugged forested landscape adjacent to the biologically rich Kayan Mentarang National Park (Sorensen and Morris, 1997; Wulffraat and Samsu, 2000) would also have a high biological conservation value – rich with significant populations of species that are threatened with extinction elsewhere,[1] such as the Malayan Sun Bear (*Ursus malayanus*) and the clouded leopard (*Neofelis nebulosa*). However, information found outside the park area was relatively scarce and localized (e.g. Puri, 1998, 2001; Fimbel and O'Brien, 1999; Gomez Gonzalez, 1999).

The Malinau region is part of the last relatively intact contiguous forest ecosystem for many of Borneo's species. Although the area is inhabited, human populations are largely small and widely scattered (densities typically less than 1 person per square kilometre). Primary forest dominates the landscape, with localized patches of swidden (shifting) cultivation and secondary re-growth. But like most of Borneo, Malinau's landscapes are undergoing rapid change. This greatly influences the value that local people and global conservationists perceive in these lands. While the views of conservationists, loggers and oil palm investors are relatively well recognized, the concerns and interests of local people are often overlooked. We sought to address that.

The recent devolution of power to the district level now gives local authorities more control over land-use decisions, and local conditions are changing rapidly (Barr et al, 2001; Sheil et al, 2006). Shame deriving from regional and national prejudice about 'primitive' forest lifestyles remains a barrier to communication on such topics (Sheil et al, 2006).

As local democracy develops, new opportunities to address how land and resources are controlled and governed are emerging. However, mindsets may not be keeping up – for example, land-use planning is not yet an effectively consultative process. Not only politicians and civil servants, but also those involved in conservation policy should be encouraged to consult more with local people.

The need to gauge local perceptions

Is there a way to achieve conservation that benefits rather than hinders the people of Malinau? How might this be possible and what do we need to know and do? What is required is an understanding of local needs and of basic conservation ecology, and a means to make such understanding more influential in any decision-making that affects Malinau's people, forest and rich biodiversity.

Revealing the services and benefits that accrue locally in tropical landscapes, as well as identifying local costs of ill-conceived policies or management strategies are critical first steps in encouraging better-informed choices. Articulating local preferences makes them harder to neglect. An underlying principle in our effort is the belief that conservation outcomes should not be sought exclusively in large protected areas or be the sole preserve of professional conservationists; rather, biodiversity should be maintained in areas also used for other purposes. This requires the cooperation of local people, timber managers and other stakeholders, but can also reflect their choices and perceptions.

This chapter is about local people's perceptions of biodiversity, and also about our approach to examining them. Our method provides a structured framework for developing mutual understanding that gives a foundation for action. Such understanding encourages a deeper discussion between development practitioners, policy-makers and forest communities. The ultimate goal is better informed decisions on land use that will improve forest conservation, protect the needs of local people and advance a wiser use and conservation of tropical forests.

We documented the landscape and biodiversity of the upper Malinau watershed to clarify local interests and priorities. The research had three components:

1 finding out which species, habitats and special sites occur where;
2 assessing who valued them and in what way; and
3 identifying how to maintain these values in the future.

When brought together, these elements clarify many local priorities regarding local biodiversity and their wider implications.

We chose to work with seven Merap and Punan communities (see Table 4.1); they represent distinct cultures in the Malinau watershed. The most obvious difference between the two groups has been the historical emphasis that the Merap place on rice farming, while the semi-nomadic Punan have specialized more in forest-based hunting and gathering. The Merap are a politically active group, whereas the Punan are less politically visible. In recent decades, efforts by the government have sought to settle the Punan and encourage their agricultural development.

Initial meetings with each community were undertaken jointly by the whole survey team; but for most tasks, the researchers divided into two main groups. The village team, together with several local assistants, held community meetings, focus group discussions, household surveys and interviews. They collected a wide range of information about the judgements, needs, culture, institutions and aspirations of the local communities, and examined their knowledge and perceptions of, and their relationship with, the landscape. Key informants were used to identify and locate local forest products and landscape units.

Table 4.1 *Communities surveyed in assessing people's perceptions about biodiversity*

Name of communities	*Traditional territory* (km^2)[b]	*Households*	*Population*	*Principal ethnicity*[d]
Gong Solok I	324	44	208	Merap
Paya Seturan[a]	22[c]	13	116	Merap
Punan Rian[a]	22[c]	9	39	Punan
Langap	469	99	415	Merap
Laban Nyarit	256	29	138	Merap and Punan
Liu Mutai	370	11	53	Punan
Long Jalan	748	31	114	Punan

Notes: a These two communities were the first examined. Due to some methodological revisions, not all of their data are comparable with those of later studies.

b These are unofficial estimates; several territorial boundaries are disputed.

c This area is shared by Paya Seturan and Punan Rian, although both these communities actually use much larger areas (land claimed by Langap, but seldom visited by them).

d In general, efforts were made to keep ethnic identities separate in the data recording, although this was not always practical in some settings, such as community meetings.

Source: data from CIFOR surveys 1999 and 2000 (cifor.cgiar.org/mla)

Table 4.2 *Value categories used to quantify the importance of the landscape*

Category	*Our explanation (based on a pilot study)*
Food	Primary and secondary foods, including famine foods
Medicine	Medicinal and health related
Light construction	Poles and cut timber for huts, forest camp structures, fences
Heavy construction	Poles and cut timber for houses
Boat construction	Timber for boats (not including oars or punting poles)
Tools	Plant parts used for tools in agriculture, hunting, boating; includes blowpipes, spears, oars, punting poles, rice pounders, tool handles
Firewood	Fuel
Basketry/cordage	String, weaving, etc.
Ornamentation/ritual	Ceremony, dress, jewellery
Marketable items	Products sold or traded
Hunting function	Poisons, bait and gums used to catch animals
Hunting place	Indirect value of a plant as improving a hunting location by attracting valued animals
Recreation, toys, fun	Area or forest products used for fun or entertainment
The future	General (not explained in detail)
Other	This allows for any values not included in the previous classes
(respondents are invited to suggest what has been omitted) (items that do not fit into categories 1 to 14)	

Source: Sheil et al, 2002

Scoring exercises, known as the Pebble Distribution Methods (PDMs), were used to quantify the importance of these products and landscape units. The various categories distinguished in these PDMs, as well as in the plant use assessment and in scoring field sites, are listed in Table 4.2. These were devised by the researchers and the first two communities working together, and proved acceptable and applicable in all the communities. Throughout this chapter we will refer to these types of values as 'value categories'.

Scoring exercises were carried out to assess the relative importance of various landscape units (land and forest types) and of specific uses. Each community was represented by four subgroups: in all but one case these were young and old, men and women. In one mixed ethnicity village, Laban Nyarit, the ethnic mix present required distinguishing between Merap and Punan, men and women, rather than age groups. The scores were based on a perception of 'overall relative importance', which people expressed by distributing a total of 100 counters (beads, buttons, maize seeds or matches). This was a subjective measurement rather than a 'value' expressed in terms of any standard economic unit or currency. More specifically, in applying these methods, we assumed that importance could be expressed as a holistic rating of relative preferences without any assumed relation to financial or easily defined economic characteristics. We sought to avoid confusing monetary value with importance. For example, we avoided words such as 'price/cost', 'fee', 'cash', 'expensive', 'cheap' (in Indonesian: *harga*, *ongkos*, *uang*, *mahal*, *murah*), but we emphasized concepts of 'general value' (*nilai*), 'usefulness' (*manfaat*) and 'importance' (*kepentingan*).

Within the context of each exercise, the scores (or mean scores) add up to 100 and can therefore be viewed as the relative percentage of overall importance contributed per item scored. For simplicity, these scores are referred to as 'importance'.

The field team visited and assessed selected sample locations recording various site properties, including plant and soil resources, both through scientific approaches and as perceived by the local people. We were seldom able to sample at large distances from the villages due to logistical constraints. Factors used in site selection included land cover, use, local topography, altitude, and presence of specific soil features and special sites (such as former village sites, salt springs and limestone outcrops). We wanted to include sites that were important to local people and that might have restricted biota. To seek these out, we used the joint mapping exercises to identify special sites and visited them. In total, 200 geo-referenced sample points were assessed, representing a wide range of local environments, especially different kinds of forest (see Figure 4.1).

A complete description of the methods, with some guidance on background issues, analysis and data handling, has been published (Sheil et al, 2002, 2003a, 2003b, 2004) and is free and available online on the CIFOR website in Indonesian, English, Spanish and French (www.cifor.cgiar.org/mla/_ref/publication).

To develop a more complete understanding of many of the species in Malinau and the implications for management, we reviewed a considerable quantity of information available from other sources. This process has helped to define

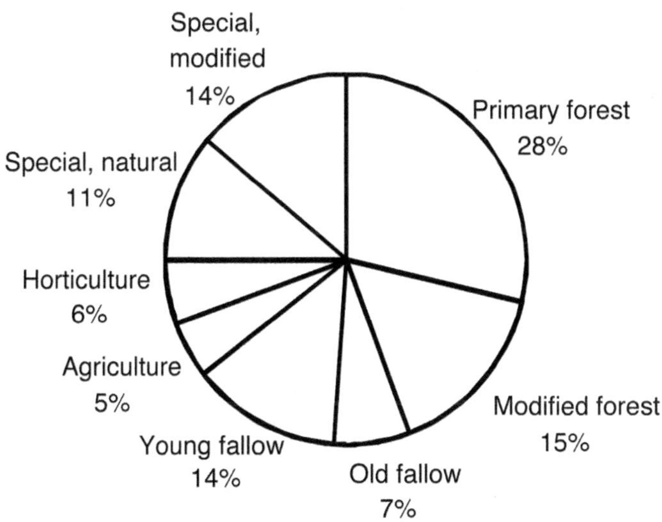

Source: Sheil et al 2002

Figure 4.1 *Distribution of samples by eight summary site-type classes*

principles for 'wildlife-friendly' land and forest management, and uncovered knowledge gaps where further research is required (Meijaard et al, 2005, 2006). Our engagement in Malinau also included various means to bring significant results to the attention of various stakeholders (see the section 'Local choices for local actions').

Local perceptions

The following sections focus on the local perceptions, village-based scoring exercises and some aspects of the fieldwork related to local views, and provide some synthesis and summary of the diverse results. Further details can be found online at CIFOR's website (www.cifor.cgiar.org/mla/_ref/result/index.htm) and in Sheil et al (2006), Sheil and Liswanti (2006) and Liswanti et al (2004).

Occupations and attitudes

Interviews with the head of each household and various key informants provided considerable information on the people in the seven communities and their views and aspirations; some of the basic results are summarized in Table 4.3. Most heads of households primarily considered themselves as 'farmers' (353 out of 576 households); 'eaglewood collectors'[2] (85 of 576) and 'labourers' (forestry/mining; 46 of 576) were also common. Despite many concerns, discussed below, most people considered their lives had improved during the last five to ten years (87 said it had improved, 22 said it had worsened, while 28 said their lives were neither better nor worse).

Community histories were often complex. All local communities have shifted locations several times over the past decades; in the past, inter-tribal wars were a frequent cause, but they have ceased for the last half century. Floods, disease and crop failure are still common reasons for communities to move territories. In recent decades, the government has successfully encouraged several communities (or a part of these communities) to leave their remote traditional territories and settle closer to Malinau Town.

Landscape units

How do people value their different land and forest types (hereafter referred to as 'landscape units')? Table 4.4 summarizes the scores that people gave to the different units and forest types, highlighting the highest mean score per exercise. For most uses, not only those requiring timber, people rated forest as the most important. Rivers were preferred for 'recreation' (mostly fishing) and secondary forests for firewood; but the unlogged primary forest (*hutan rimba*) was unquestionably the most important overall. Results were not necessarily intuitive; but discussions with community members provided explanations. People from Langap are sophisticated farmers, yet rated the forest more important than cultivation sites. They explained that the forest was their most important source of medicine and

Table 4.3 *Suggested actions to address environmental concerns*

	Response from local people/ total number of respondents in each community[a]				
	Lg	*LN*	*LJ*	*LM*	*GS I*
n =	*30*	*32*	*30*	*14*	*31*
Suggested action to be taken if forest resources are degraded or exhausted					
Replanting	1	4	3	–	1
Protecting trees – banning of cutting	4	3	7	4	2
Barring outsiders from entering community area	–	–	2	–	–
Keeping the forest as a protected area or customary forest	1	2	2	1	9
Limiting the logging area by the company	3	7	1	–	8
Growing tree crops	16	13	–	3	3
Didn't know	6	5	13	6	6
Factors important to maintaining forest value					
Birds because they spread forest seeds	4	–	17	1	4
Bats because they spread forest seeds	–	–	1	–	–
Wild animals as a heritage for grandchildren	1	2	3	–	1
Fruit trees as a heritage for grandchildren	4	11	5	–	13
Fig tree, or *Beringin*[b] (*Ficus* spp.) because it has mythical associations	–	–	2	–	–
Eaglewood or *gaharu*[b] trees (*Aquilaria* spp.), Sago (*Eugeissona utilis*), *meranti*[b] timber trees (*Shorea* sp.), resin trees or *damar*[b] (*Agathis borneensis*), etc, because of value for local people	2	1	3	–	1
Kempas[b] trees (*Koompassia excelsa*)	4	5	1	1	3
Trees in customary forest	–	1	–	–	2
Didn't know	17	13	9	12	11

Notes: Lg = Langap; LN = Laban Nyarit; LJ = Long Jalan; LM = Liu Mutai; GS = Gong Solok.

Unfortunately, due to a change in the questionnaire, there are no comparable data from Paya Seturan and Punan Rian.

a Each respondent could give as many responses as they wished.

b Indonesian name.

Table 4.4 *Mean scores of value categories per landscape unit and per forest type for all groups in all seven communities*

Value categories	*All*	*Food*	*Medicine*	*Light construction*	*Heavy construction*	*Boat construction*	*Tools*	*Firewood*	*Basketry/ cordage*	*Ornament-ation/ritual*	*Marketable items*	*Hunting function*	*Hunting place*	*Recreation*	*Future*
Landscape units:															
Village	12.7	10.2	15.5	1.4	2.3	0.3	1.8	1.6	2.7	13.2	9.2	7.0	0.1	17.8	13.0
Old village site	5.9	6.5	4.8	4.8	1.5	0.8	2.5	2.2	4.5	5.3	6.7	5.0	6.0	2.1	4.9
Garden	11.4	13.9	8.4	4.7	1.1	0.2	0.3	8.6	2.5	10.5	16.9	4.5	7.0	11.7	15.9
River	13.4	15.5	11.1	11.0	6.7	7.8	8.9	19.0	10.7	15.6	14.6	7.9	14.5	26.6	8.5
Marsh/swamp	7.4	6.8	5.7	9.2	9.2	11.5	10.6	3.9	7.9	3.8	4.4	5.6	7.3	1.5	7.2
Cultivation site	13.5	14.4	4.7	1.8	1.8	0.9	0.4	17.0	1.1	0.8	12.3	0.7	7.5	12.4	10.4
Young fallow	6.6	6.4	5.8	1.7	1.3	0.8	2.0	10.0	3.5	3.3	3.6	1.5	5.1	0.3	8.0
Old fallow	8.0	5.5	8.4	27.0	4.9	4.7	12.1	13.8	17.5	14.3	2.5	14.5	14.9	3.2	10.5
Forest	21.1	20.9	35.6	38.3	71.2	73.1	61.4	23.9	49.6	33.3	29.8	53.4	37.5	24.5	21.5
Total	**100**	**100**	**100**	**100**	**100**	**100**	**100**	**100**	**100**	**100**	**100**	**100**	**100**	**100**	**100**
Forest types:															
Unlogged forest	31.4	38.8	36.3	35.6	50.7	49.5	44.7	29.1	39.0	30.3	35.8	43.5	36.5	34.3	30.7
Logged forest	10.1	8.8	8.2	8.6	5.9	4.6	5.1	15.9	5.9	10.0	8.4	4.9	7.3	8.4	12.7
Secondary forest	15.8	11.2	15.1	23.0	4.0	2.0	4.8	35.6	15.6	26.8	7.1	9.1	11.8	15.3	23.6
Swamp forest	18.9	11.3	12.7	12.1	10.0	15.5	14.6	10.1	14.7	12.1	12.4	13.7	15.6	18.2	13.7
Mountain forest	23.8	30.0	27.8	20.6	29.4	28.5	30.9	9.3	24.8	20.8	36.4	28.7	29.0	23.8	19.3
Total	**100**	**100**	**100**	**100**	**100**	**100**	**100**	**100**	**100**	**100**	**100**	**100**	**100**	**100**	**100**

timber, and that much forestland could itself be converted to temporary cultivation.

If we look across communities, we generally get similar results. In terms of overall importance, there is little variation among communities in their rating of forest or cultivation sites. All agreed that forest merited the highest scores.

Importance over time

Another exercise elicited scores on the past, present and future importance of forest, and the relative importance of each category of use. A large variation was found among respondents, but the mean results (see Table 4.5) show an increasing relative role of forest for timber, saleable items and recreation, and a decreasing role for medicinal use, firewood and light construction. This might be viewed as an anticipated shift from subsistence and cultural needs to economic and material goals. Interestingly, the past value of the forest was scored lower than its perceived future value. Informants explained this as the result of their previously having 'taken the forest for granted' despite their dependence upon it: an acknowledgement of the subjective nature of these assessments.

Table 4.5 *Mean scores of past, present and future importance of the forest (all seven community responses)*

	30 years ago	*Now*	*In 20 years*	*Total*
All values	31.3	32.0	36.8	100
Per value category:				
Food	12.1	11.9	10.2	
Medicine	8.5	8.1	4.7	
Light construction	10.1	7.8	5.4	
Heavy construction	7.7	8.4	14.2	
Boat construction	5.5	7.0	7.5	
Tools	5.4	5.4	7.0	
Firewood	7.5	7.2	5.2	
Basketry/cordage	7.5	6.4	6.6	
Ornamentation/ritual	5.1	6.6	4.7	
Marketable items	6.7	8.1	8.7	
Hunting function	6.6	6.6	5.4	
Hunting place	8.4	7.7	6.4	
Recreation	2.8	3.1	5.1	
Future	6.1	5.7	9.0	
Total	**100**	**100**	**100**	

Table 4.6 *Mean scores of the importance of different sources of plants and animals in all seven communities (each result is the mean of four groups)*

Origin of plants and animals:	*Wild from forest*	*Wild from other places*[a]	*Cultivated*	*Bought*	*Wild from forest*	*Wild from other places*[a]	*Domesticated*	*Bought*
Long Jalan	19.75	7.75	16.25	14.25	21.75	12.25	8	0
Liu Mutai	4.75	10.25	27	8	18.5	12.25	11	8.25
Punan Rian	15.5	10.5	16	6	15.5	15.5	15	6
Paya Seturan	7.75	10	25	7.75	14.5	13	16.5	5.5
Laban Nyarit	9.75	6	18.75	8	15.5	12.75	22	7.25
Langap	8.5	12.5	28	7.5	10.25	9.5	18.75	5
Gong Solok	10.75	9.5	25.75	10.5	10.25	12.75	14.75	5.75
Mean value	11.0	9.5	22.4	8.9	15.2	12.6	15.1	5.1

Note: a It should be noted that many wild products that come from outside the forest are known to depend upon forests – for example, much hunting goes on in open areas and on field margins, and this hunting value is ascribed to these areas; but the respondents are well aware that the wildlife *originates* from the forest and it is simply easier to *hunt* in these non-forest sites.

Alternative sources of valued plants and animals

Another series of PDM exercises asked how people rated wild plant and animal resources compared to farmed or bought alternatives (see Table 4.6).

A striking result from this exercise is that all communities, even those of the most sophisticated rice farmers in Langap and Gong Solok, recognize a considerable dependency upon wild plant and animal resources. The mean score for the seven communities showed that approximately half (48 per cent) of all the importance of the plant and animal products that they valued were derived from wild sources; this never dropped below 40 per cent even in the most agricultural communities. Focusing on the results for Langap (Merap) and Long Jalan (Punan) helps to clarify the usefulness of this exercise: remote Long Jalan places more importance on wild forest products than Langap, and the reverse is true for cultivated plants and farmed animals. The Punan in Long Jalan are relatively dependent upon buying rice and crops from traders – using revenue based on the sale of eaglewood (*Aquilaria* spp.) and some birds such as *Pycnonotus zeylanicus* and *Buceros rhinoceros*. They find it relatively easy to gain animal protein by hunting and fishing. These patterns make good sense, which serves to strengthen our confidence that the patterns revealed are meaningful.

Reviewing each of our value categories provides a more complete picture: Table 4.8 provides a summary of the five highest scoring species per value category.

Table 4.7 (a) *Top ten list of plants and* (b) *animals when taking into consideration the combined scores of a species in all value categories*

Rank	*LUVI*	*Name in Indonesian (English)*	*Scientific name*	*Value categories*
1	98.5	Kayu ulin (Ironwood, timber)	*Eusideroxylon zwageri*	Heavy construction, hunting function, tools, marketable items, light construction, (ornamental/ritual, kayu bakar)
2	94.9	Kayu meranti (Meranti, timber)	*Shorea parvifolia*	Heavy construction, boat construction, tools, hunting place, marketable items, light construction (kayu bakar)
3	91.7	Kayu kapur (timber)	*Dryobalanops lanceolata*	Heavy construction, tools, hunting place, marketable items, light construction, boat construction
4	73.8	Kayu dao (timber)	*Elmerrillia tsiampacca*	Heavy construction, boat construction, tools, marketable items, (light construction)
5	62.1	Kayu puspa (timber)	*Sindora leiocarpa*	Heavy construction, hunting function, tools, boat construction, (light construction)
6	42.2	Kayu tengkawang (timber)	*Shorea pinanga*	Heavy construction, hunting place, boat construction, tools, marketable items
7	39.1	Kayu agathis (Agathis timber, resin)	*Agathis borneensis*	Heavy construction, tools, marketable items, (light construction)
8	36.6	Rotan sega (rattan)	*Calamus caesius*	Basketry/cordage, marketable items, ornamental/ritual
9	32.6	Aren (sago palm)	*Arenga undulatifolia*	Hunting function, food, hunting place
10	32.2	Palas biru/Daun wiru (palm)	*Licuala valida*	Basketry/cordage, light construction, ornamental/ritual

Table 4.7b *Top ten animals*

Rank	*LUVI*	*Name in Indonesian (English)*	*Scientific name*	*Value categories*
12	30.1	Rusa (Sambar deer)	*Cervus unicolor*	Food, marketable items, tools, ornamental/ritual (medicines)
16	25.2	Babi hutan (wild bearded pig)	*Sus barbatus*	Food, marketable items (medicines)
28	14.8	Beruang madu (Sun bear)	*Ursus malayanus*	Medicines, ornamental/ritual (food)
30	13.7	Landak (porcupine)	*Hystrix brachyura*	Medicines, food
34	12.6	Pelanduk / Kancil (Mouse deer)	*Tragulus javanicus/T. napu*	Food, medicines, pasar
45	11.7	Rangkong gading/Raja burung (Helmeted hornbill)	*Buceros vigil*	Ornamental/ritual, marketable items, medicines,
48	10.1	Rangkong badak (Rhinoceros hornbill)	*Buceros rhinoceros*	Ornamental/ritual, marketable items, (food)
49	9.8	Tiong emas (songbird)	*Gracula religiosa*	Marketable items, ornamental/ritual
50	9.8	Kijang (Borneon muntjac/ barking deer)	*Muntiacus muntjak*	Food, marketable items, ornamental/ritual, (medicines)
53	9.1	Lutung banggat (Grey leaf monkey)	*Presbytis hosei*	Food, tools, medicines, marketable items

Table 4.8 *The five highest scoring species per value category (scientific names and common English descriptions)*

Value category	*Score*	*Species 1*	*Species 2*	*Species 3*	*Species 4*	*Species 5*
Heavy construction	12.0	*Eusideroxylon zwageri* Timber (Iron wood)	*Dryobalanops lanceolata* Timber	*Shorea parvifolia* Timber (Meranti)	*Elmerrillia tsiampacca* Timber	*Sindora leiocarpa* Timber
Food	11.0	*Sus barbatus* Wild pig	*Arenga undulatifolia* Sago	*Cervus unicolor* Sambar deer	*Eugeissona utilis* Sago	*Durio* sp. *termasuk* *Durio zibethinus* Durian fruit
Marketable items	9.5	*Aquilaria beccariana/* *A. malaccensis* Gaharu	*Eusideroxylon zwageri* Timber (Iron wood)	*Dryobalanops lanceolata* Timber	Fish N.A.	*Shorea parvifolia* Timber (Meranti)
Boat construction	7.4	*Hopea dryobalanoides* Timber	*Palaquium gutta* Resin (for glue)	*Shorea parvifolia* Timber (Meranti)	*Elmerrillia tsiampacca* Timber	*Alseodaphne ceratoxylon* Timber
Future	7.3	N.A.	N.A.	N.A.	N.A.	N.A.
Hunting place	7.2	*Dryobalanops lanceolata* Tree fruit	*Lithocarpus sundaicus* Tree fruit	*Ficus* sp. Figs	*Shorea pinanga* Tree fruit	*Shorea parvifolia* Tree fruit
Medicines	6.8	*Ursus malayanus* Bear gall bladder	*Aristolochia* sp. Climber	*Hystrix brachyura* Porcupine pins	*Phyton reticulatus* Snake oil (Phyton)	*Formicidae* ant nests
Firewood	6.4	*Vitex pinnata* Small tree	*Nephelium ramboutan-ake/* *N. mutabile* Small (fruit) tree	*Cephalomappa lepidotula* Tree	*Tristaniopsis whiteana* Tree	*Lithocarpus cantleyanus* Tree
Light construction	6.2	*Vitex pinnata* Small tree	*Schizostachyum brachycladum?* Bamboo	*Phrynium parvum* Herb	*Licuala valida* Palm	*Macaranga gigantea* Pioneer tree
Tools	6.2	*Elmerrillia tsiampacca* Timber	*Dryobalanops lanceolata* Timber	*Eusideroxylon zwageri* Timber (Iron wood)	*Agathis borneensis* Timber (Agathis)	*Shorea parvifolia* Timber (Meranti)
Basketry/cordage	6.0	*Calamus caesius* Rattan	*Korthalsia echinometra* Rattan	*Licuala valida* Palm	*Pandanus sp.* Pandan	*Calamus javensis* Rattan
Hunting function	5.6	*Antiaris toxicaria* resin from tree bark	*Arenga undulatifolia* palm (arrows)	*Eusideroxylon zwageri* Timber (blowpipe)	*Eugeissona utilis* Sago palm (arrows)	*Sindora leiocarpa* Timber
Ornamental/ritual	4.7	*Cocos nucifera* Coconut palm leaves	*Licuala valida* Palm	*Cervus unicolor* Deer (antlers)	*Calamus caesius* Rattan	*Cyclemys dentata /* *Notochelys platynota* Turtles
Recreation	3.8	N.A.	N.A.	N.A.	N.A.	N.A.

Note: N.A. = not applicable.

The importance of plants

A large number of specific uses for plants were recorded in the course of our field surveys (18,058 records). This included 2141 separate uses/values for approximately 1457 species. Of these, 779 were trees and 620 were herbs and climbers. The number of specific species uses per value category is illustrated in Figure 4.3. The categories 'light construction' and 'hunting place' (species noted for providing food for hunted animals) elicited the longest species lists.

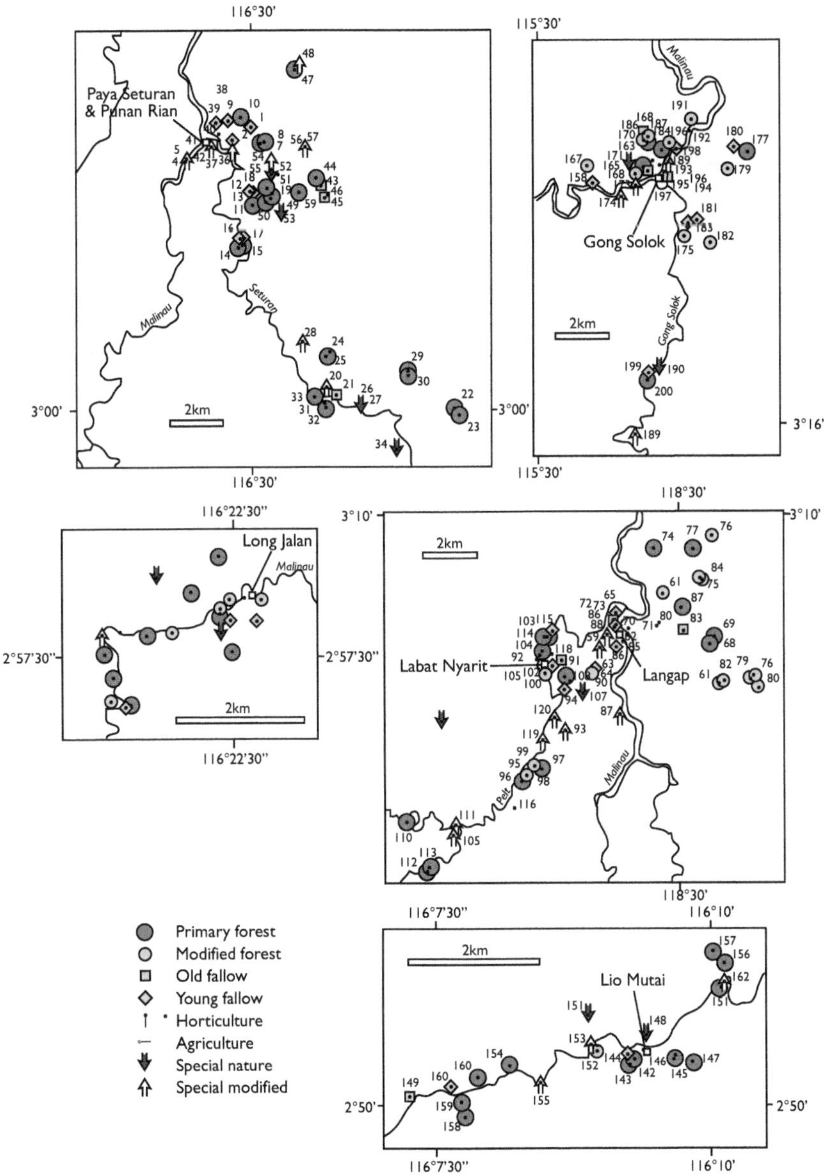

Figure 4.2 *Map of sample locations, by type and community*

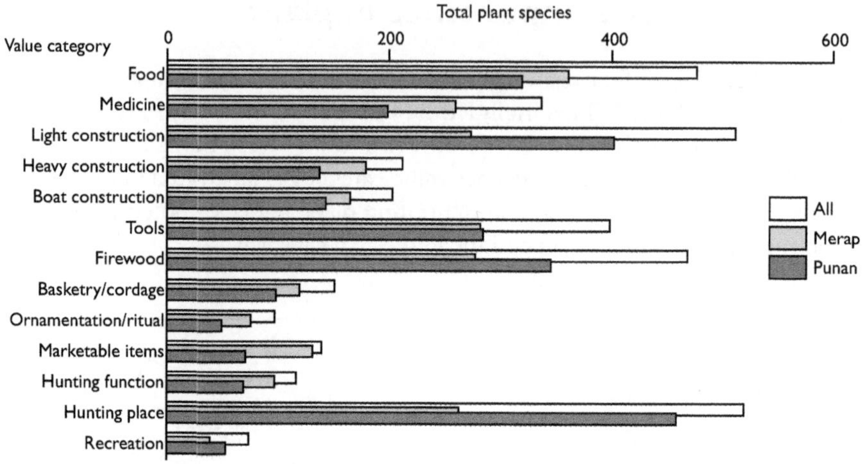

Figure 4.3 *Total number of specific species uses recorded, by value category*

Overall, 119 species were recorded as having uses that were in some way viewed by the informant as exclusive to that plant alone. According to the informants, these specific species uses cannot be substituted by other species – implying that their loss would cause a reduction in local well-being. These examples relate to diverse genera and encompass 85 families. Such records are mainly in the ritual/ornamental, medicinal and tools categories (37, 28, 25 species, respectively) and are much more commonly reported by Merap than by Punan informants. There were no exclusive uses/values recorded in construction or boat-building, although the number of 'best species' for building is restricted.

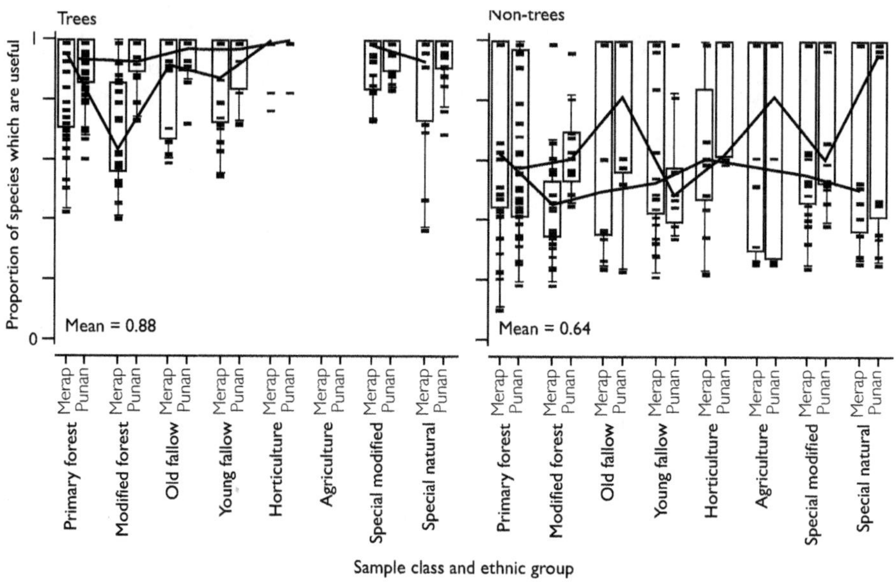

Figure 4.4 *Total proportion of useful species plotted by sample type and ethnicity of the informants*

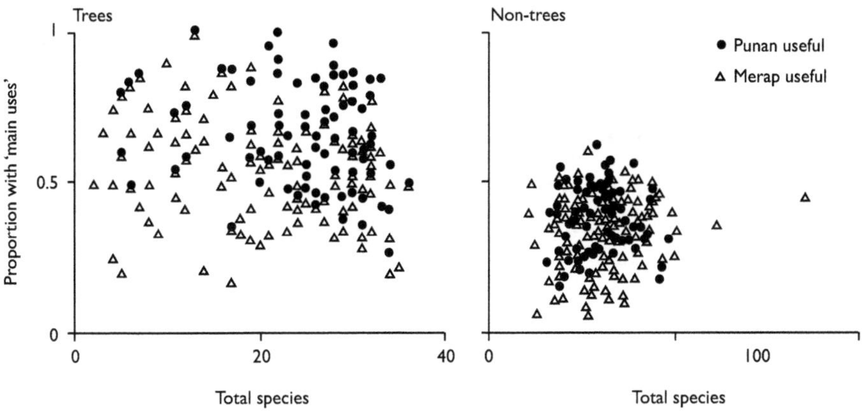

Figure 4.5 *The proportion of useful species plotted against overall species count*

A different perspective is gained by a 'per plot' basis. In Figure 4.4, the total proportion of useful species has been plotted by sample type and ethnicity of the informants. Most plots have such a strikingly high proportion of useful species.

Finally, the question may be asked if vegetation that is richer in species is also richer in uses. The answer is clearly yes (see Figure 4.5). The proportion of useful species increases in rough proportion to the overall species count (this monotonic relation is highly significant at $P < 0.001$, by tree and non-tree vegetation for both Merap and Punan informant groups).

The importance of the forest

Our results illustrate both the rich biodiversity of the Malinau forests and its considerable significance to their inhabitants. All seven of the communities live from the land; all trade for some commodities; and none appear totally self-sufficient in food. Yet, at the same time, all of these communities, including those that appear most agriculturally oriented, place a high importance on wild resources and forests in general.

Not all forest areas are valued equally. The values associated with different sites vary in subtle and complex ways. Within the landscape, the best soils and the most valuable resources are distributed unevenly. Some resources, such as birds' nest caves, are highly localized. Depletion of eaglewood trees (*Aquilaria* spp.) explains why it is now found principally in remote and less accessible regions. Complex histories of shifting settlements and community interactions are reflected in local perceptions of these landscapes. These histories have a very concrete influence on patterns of knowledge, claimed ownership and the personal and heritage values associated with the past. A more detailed study of the spatial aspects' importance was conducted in Liu Mutai and is available in Lynam et al (2007) and Cunliffe et al (2007). These studies do not contradict any of the conclusions offered here, but indicate that distance and accessibility – factors normally considered as 'costs' in extracting value from a location – have little influence on the values that the Punan ascribe to a location.

Logged versus unlogged forest

Unlogged forest is considered the most important land (see Table 4.4). Logged forest is given a low preference by local communities. There appear to be several reasons for this, including a diminished level of key resources, reduced physical accessibility and reduced access rights. It is certainly relevant that seven out of the top ten most important species overall are timber trees. Even if local people had the right to cut the timber, the best wood was often already taken and the damage to the forest made access difficult. Wild pigs are the most important source of animal protein and are known to be harder to find after logging. A key factor is understorey slashing (clearing all undergrowth following timber extraction).

The practice of understorey slashing in logged compartments has hurt local communities as well as forest biodiversity. Current regulations of the Ministry of Forestry (2007) on timber cutting (TPTI and TPTII) require timber companies to repeatedly slash all undergrowth and climbers – including many useful and prized species such as rattan – in an effort to encourage regeneration within the concessions. Observations show that understorey cutting has been widely applied in Malinau. Even if applied properly, its silvicultural benefits remain dubious and the biodiversity impacts are considerable. Also, it severely and unnecessarily depletes the value of the forest for local people. This policy of cutting undergrowth should be revoked (Sheil et al, 2006).

A closer look at the respective value categories

Let us consider the various value categories individually. These summary sections compile and synthesize information from the complete range of results from the seven communities.

Food

The forest is generally seen as the most important source of food. Unlogged forest and mountain forest scored highly; but logged forest was rated low for food. The communities did not generally expect their reliance on forest foods to decline by much in the next decades, but they did see some problems arising. The list of most important wild food species was headed by pigs (*Sus barbatus*), which were said to be less abundant in logged areas. Certain emergency forest foods such as sago were reduced unnecessarily by logging. *Eugeissona utilis*, an important sago, grows on ridge tops and is often damaged by skid trails. *Eugeissona* and *Arenga* palms were in the top five most important food species, along with Sambar deer (*Cervus unicolor*), which notably (in contrast to pigs) were said to be more common in disturbed forest – an observation also supported by ecological studies (Meijaard et al, 2005). Various fruiting trees also occurred among the top species (*Durio* spp., *Parkia* spp., *Nephelium* spp., *Musa* spp., *Diplazium* spp., *Artocarpus* spp., *Mangifera* spp. and *Baccaurea* spp.). Many of the animal species valued as food items were legally protected (e.g. the hornbills). A total of 476 specific uses were recorded in the plant surveys; but only 12 were exclusive to a specific species. These were primarily food items with secondary importance as medicine or for ritual purposes. The mean number of species per plot with food value was generally below 20 per cent in all landscape units except gardens.

Medicinal products

Although all landscape units provide some medicinal value, the forest was by far the most important, contributing over one third of all value. More than 15 per cent of medicinal products derive from gardens or the village itself. Logged forest was given a low rating. The communities generally expect the importance of forest-derived medicines to halve in the next two decades. Surprisingly, and in contrast to most value categories, several animal species were among the most important medicinal species, including Sun bear, porcupine, python and ants, the first two of which are protected by national law. Of the 336 medicinal plant uses recorded in the field survey, 28 were said to be unique to a given species, only 1 of which was reported by a Punan informant. Plants with medicinal use were typically uncommon and seldom encountered in field plots, contributing less than 10 per cent of non-trees and less than 5 per cent of tree species in most field samples. In addition, several were noted as vulnerable to understorey cutting.

Light construction

The need to improvise shelters and other temporary structures is an integral part of everyday life for Malinau communities. Forests rated highly as a source of such materials, but for the most part, the choice of species was broad, and any areas with woody cover appeared to yield some suitable lightwood and leaves for cover. Indeed, over 510 species had some use in this category, none of which were exclusive to the species.

Heavy construction

Forest, especially primary unlogged forest, was by far the most important source of heavy construction materials, contributing more than 70 per cent of the scored importance according to this category. The communities generally expected their reliance on forest-derived construction to increase in the next decades. They noted a conflict with the logging concessions that limited access to many of the key species. *Eusideroxylon zwageri* (known as 'Ulin' in Indonesian) is the most important species for the communities. At the time of our surveys, it was protected for local community use by national law, but nonetheless had a high market value. In addition, in Malinau, companies had gained a reputation for ignoring prohibitions on cutting it (today, it is still protected, but, officially, this prohibition also applies to local people). All of the other prized timber species wanted by the communities (i.e. *Dryobalanops lanceolata* and *Shorea parvifolia*) were also reserved for the timber concessions. When communities collect this timber, they do so surreptitiously because it is officially 'illegal logging'. These values have been depleted in logged areas. Notably, we find that *Eusideroxylon* and various other timber species are relatively plentiful in unlogged forest.

Boat construction

Almost three-quarters of the total importance attributed to boat-building products derives from the forest (73 per cent) and most of the rest (20 per cent) derives from swamps or riversides. The importance of unlogged forest was reported as more than ten times that of logged forest for these products. Clarifying discussions with local boat builders implied considerably increased searching time to collect suitable timber and sealants in nearby logging concessions. Certainly, there is a conflict over preferred and needed species since several are cut by concessions (e.g. *Hopea dryobalanoides* and eight species of *Shorea* were mentioned) and were depleted in logged areas. In the field survey, 203 relevant plant uses were recorded, of which none were unique to a given species.

Tools

Forest, especially unlogged and mountain, was the most important source of materials for making tools. All of the most important species noted in this category were timber-tree species (such as *Agathis borneensis*, *Dryobalanops lanceolata* and *Shorea parvifolia*) and have been depleted in logged areas. Only two animals were noted (*Cervus unicolor* and *Presbytis hosei*). In the field survey, 399 plant species with tool uses were recorded in the field survey, of which 25 were said to be unique to a given species.

Firewood

As with the light construction category, forests rated highly as a source of such materials; but for the most part, the choice of species was broad, and any areas with woody cover appeared to yield some suitable wood. In contrast with other values that people expressed for secondary forest, this was the one category most exclusively associated with secondary forests, probably a result of the lighter timber found inside and near old fallow areas. In this category, 469 recorded tree species were valued.

Basketry/cordage

The forest was most important in this category, containing half the mean scored importance. The most important were all palms and pandanus species: *Calamus caesius*, *Korthalsia echinometra* and *Calamus javensis* and the leaves of *Licuala valida* and large *Pandanus* spp. In the field survey, 151 plant uses were recorded, almost exclusively related to monocot species. Seven of these uses were said to be unique to a given species. Many of these species are negatively affected by understorey cutting and logged-over forest had little importance in this category.

Ornamentation/ritual

The forest was the most important source of products in this category, which includes both plants and animals. Some of the important plant species were very common: young coconut leaves (*Cocos nucifera*) are used to make decorations and leaves of *L. valida* and *C. caesius* are used for ornamentation. However, several of the valued animals, including clouded leopards, bears and hornbills, are protected by Indonesian law. In the field survey, 96 plant uses were recorded, of which 37 were said to be unique to a given species. Many of these species are lost or damaged by understorey cutting.

Marketable items

Forest, especially unlogged mountain forest, was the most important source for marketable items, and eaglewood (*Aquilaria* spp.) emerged as the most significant species in this category. Local people distinguish two eaglewood species whose distribution overlap: a lowland species (*Aquilaria beccariana*) and a small-leaved mountain species (probably *A. malacennsis*). Note that it is not every tree that has value. The only value is in the resin found in a small proportion of trees and appears to be produced as a response to fungal infection. It is the resin and the resin-infused wood that is the product traded as eaglewood and locally known as *gaharu*. Expert harvesters only take infected trees; but non-experts (including outsiders, such as forest concession workers looking to make extra money) often cut down healthy trees in their search for the resin. Notably, while larger trees are scarce, small *Aquilaria* trees are not uncommon and were often observed during the field survey.

Interestingly, rattan and birds' nests, often considered important trade items in the Borneo interior, although noted, did not score very highly. Rather, timber species and fish came out on top after eaglewood. Among the additional species identified in the 'top ten' exercise were various cage birds, hornbills (feathers and heads), monkeys (gall stones) and a diverse range of species for meat and fish. Various fruit species were also proposed. In the field survey, 138 plant uses were recorded, of which 7 were said to be unique to a given species. This was a difficult category for people to deal with in an abstract way – 'it depends upon what people wish to buy', they told us. They could not predict what the future demands might be. Several of these species are negatively affected by logging and understorey cutting.

Hunting function

Over half of the total importance of products identified in this class was contributed by forests. Poisonous *Antiaris toxicaria* comes out as the most important species, while the sago palms *Arenga undulatifolia* and *Eugeissona utilis* again emerged as important for their use in making darts, as did *Eusideroxylon zwageri*, the preferred wood for blowpipes. Most useful plants were trees or palms. Only one animal was recorded in this category. Although some Punan knew how to collect the poison of the king cobra (*Ophiophagus hannah*) for their blowpipe darts, their use is largely symbolic and associated with past warfare (Edmond Dounias, pers comm, 2005) and is no longer viewed as important. In the field survey, 115 relevant plant uses were recorded, of which 12 were known to be unique to a given species.

Hunting place

This category refers to the locations required for good hunting. Although hunting seems to occur in all landscape units when the opportunity arises, forests, especially unlogged and remoter mountain areas, were rated most important for hunting. An interesting aside is that some Punan informants implied that one of the values of cultivation sites was that rice and cassava drew animals into open areas where they could be hunted. The forest is still recognized as the source of these animals; the fields merely served as bait. The most important species that improved hunting were fruiting species (mainly dipterocarps, oaks, figs and palms) that draw animals to locations that allow hunting. In the field survey, 518 plant uses were recorded in the field survey, of which none were said to be unique. 'Salt-springs' and clays were also considered key areas for attracting animals, as were some old village areas where there was a concentration of fruit trees. Understorey cutting degrades a forest's suitability for hunting for food and diminishes cover for many animals.

Recreation, toys and fun

Rivers and forests rated higher than any other landscape units for recreation, each with around 25 per cent. Many people, especially older people, were initially unwilling to admit any recreation time at all. Yet the scoring exercises suggested that this was due to reticence. In addition, discussions suggest that some practical activities such as fishing, wood collection, hunting and, indeed, handicrafts are often pursued as much for the pleasure of the undertaking as for any final product. Communities expected their demands for recreation and leisure to increase in the next decades. Although no species were proposed in the community exercises as belonging to this class, 73 specific plant use combinations were recorded under 'fun or play' in the field survey (most relating to children's games), 1 of which was known to be unique to a given species.

The future

Communities saw the forest as the most important landscape type for the future. The forest was a major alternative source of food and life support. Interestingly, 'the future' was the value class in which logged forest scored highest (although the score is still low), which seems to acknowledge forest recovery. People in all seven communities reported episodes of crop loss due to floods, droughts or other problems that had forced them to seek food elsewhere. Most turned to the forest. There was some reticence to volunteer information about this: in our preliminary interviews, only 15 out of 576 respondents specifically noted the need to collect forest products for sale as a means of staving off the threat of hunger. Nonetheless, when a stronger rapport was built, most people acknowledged the considerable importance of being able to collect edible items from the forest, as well as items that could be traded to procure food. We worked with Punan Rian during a period of food shortage when their rice reserves were exhausted and observed this value of the forest firsthand. The extent to which the 'safety net' value permeated daily consideration of the forest was not clear, however, or the degree to which these considerations influenced the value given to forest and forest resources in our scoring. Discussion with local people suggested that the safety-net value is declining due to conversion of land to other uses and the impacts of logging and over-hunting. The future is a rich topic that is not easily summarized. The next sections look at some current trends and the choices that people are anticipating.

Options and changing circumstances

New developments – roads, concessions and mines – have accelerated. Many immigrants have come to the province seeking work. Accessible land is under pressure and conflicts are increasing. People are increasingly forced to consider future options, which reflects their insecurity over land ownership. The people in Gong Solok, Paya Seturan, Liu Mutai and even the most remote village, Long Jalan, all perceived a need to convert forest to agriculture as a means of claiming land and increasing the security of their tenure. The ownership of cultivated areas is clear to everyone, whereas forests are too easily claimed and taken by others.

In discussions with local people about the most important resources, it becomes clear that many perceive a decline, especially in the animals they hunt for food and in the plants that they rely on for daily needs and trade. Some other wild resources have declined in value, such as those used in making poison for hunting with poison darts due to blow pipes being replaced by the easier hunting made possible by guns.

Gong Solok is nearest to Malinau Town and has the highest local population density. Here, hunting is notably less successful than elsewhere – this was the only place where we found people regularly killing and eating monkeys, a less preferred and previously taboo source of meat.

A shortage of preferred construction materials (e.g. *Eusideroxylon zwageri*) and boat-building materials was already felt in Paya Seturan and neighbouring communities. An interesting response to this resource depletion was that some people in Paya Seturan promoted *de facto* protected areas as a source of building material where there was mutual agreement on the need to keep forest cover. Such areas are threatened by concessions, however, and need official status.

Not all trends were seen as negative for the future. In interviews, schooling was often highlighted as especially important for young people, providing them with opportunities of finding good work away from the village. Government work was seen as highly desirable in guaranteeing a reliable income. Improvements in healthcare were also noted.

The forest is not just a source of wild forest products, but also a reserve of land that can be converted to other land uses. Local people are interested in innovation and willing to experiment with new crops and modes of cultivation. Many of the questions asked us during our stay concerned opportunities to grow cash crops such as teak, cocoa and coffee.

As the data comparing past, current and future values show, people's sense of how their reliance on the forest is changing varies; but, overall, there is some consistency in the trend away from subsistence and cultural needs to economic and material goals. Despite the increasing importance of the cash economy, subsistence and cultural needs remain highly significant to most people.

Are forests becoming less important? Various trends and aspirations may be drawing people out of the forests. Certainly, the shift of many remote communities closer to schools and clinics has been a major change. Many informants stated that education for their children created opportunities to move to the city and have a 'better life'. Yet, aspirations are tempered by realities. One key informant told us that he thought the city would be a good place to live if you were wealthy enough to buy things, but it was expensive if you had to buy everything. 'As I am poor, it is better I live off the forest', he said.

Local people are pragmatic – they like the forest and can be nostalgic about their past, but they also make the best of new opportunities. History shows that people in Malinau have taken a flexible and adaptable approach to life that has allowed them to respond readily to the boom and bust of past forest product economic cycles (Sellato, 2001). Yet the adaptability of forest-dwelling people is based on their expertise in the forest and the options that the forest provides – once it is gone, this flexibility will be lost.

Modernity and political upheaval have brought many new challenges and opportunities to the people of Malinau. For the past five years, there has been considerable disorder relating to the laws and regulations governing rights over land and natural resources. These conditions have brought great insecurity but also new options for developing cash incomes. Both aspects of accessibility are in flux – the physical opportunities to reach a site and extract resources, and the legal and social controls that might influence choices to do so. The availability of boat engines, new roads, increasing trade and rising fuel costs all influence physical access, as do patterns of resource depletion associated with logged areas. At the same time, diverse and overlapping claims and controls by concessions, local government and communities have led to confusion. The threats to the forest are real since these conflicts and uncertainties seem to favour liquidation of forests over any long-term options. As previously discussed, this reflects both the impact of clearing land in order to claim recognized ownership, as well as a free-for-all, in which there is no incentive for personal restraint. The cost to the biodiversity that local communities value and to their livelihoods could be considerable.

On plantation developments

Local government in Malinau and the national government in Jakarta appear very interested in large-scale schemes to develop revenues. There is constant discussion of timber plantations, oil palm and other possibilities. For example, in 2005, the Indonesian minister of agriculture announced a plan for a huge oil palm plantation covering 1.8 million hectares of hills and mountains (areas regarded as inappropriate for oil palm cultivation) on the Kalimantan–Malaysia border (Wakker, 2006). This would have been the largest oil palm development in the world and would have included large areas of Malinau. After an outcry from non-governmental organizations (NGOs) and other stakeholders, the government appears to have retracted the plan (Wakker, 2006), although it may still be resurrected. Local views have been ignored on these large externally driven processes. We have identified some widely held misconceptions and various concerns, summarized below.

One of the most critical results of the biophysical component of our surveys was to underline just how limited opportunities for sustainable production of cash crops are in middle and upper Malinau. Without major and costly engineering to develop terraces, the vast areas of steep, erodible, nutrient-poor and acidic soils do not provide a viable basis for sustainable, large-scale, non-forest land uses. Only limited areas of floodplain next to the rivers are potentially suitable for a few crops such as rice and coconut (Basuki and Sheil, 2005). Unsurprisingly, these areas, a mixture of fields and fallows, are already cultivated by communities.

Large-scale plantation crops that have been considered include cocoa, pepper and, most recently, oil palm. Within Indonesia, oil palm has gained a reputation as a wonder crop that can be grown anywhere and makes high profits – a reputation partly deserved. Oil palm grows well on various 'problem sites' such as acid sulphate soils, deep peat and acidic high aluminium soils where few other crops are viable (Auxtero and Shamshuddin, 1991). Some plantations on these low-value soils have even produced higher yields of palm oil per hectare than others on good agricultural lands (Auxtero and Shamshuddin, 1991; Corley and Tinker, 2003). But this does not mean that oil palm can be profitably cultivated everywhere. It has specific requirements for water and humidity throughout the year and demands fertilizer and other management inputs. Among plantation companies, there is considerable technical knowledge on when and how palms can be cultivated. For example, Tailliez (1998) states that 'the choice of site is very important, combining a gentle topography, deep soils, rainfall well distributed throughout the year, environment-friendly land preparation techniques, crop management sequences, input use (intensive fertilization), integrated disease and pest control, oil mill quality and pollution risks'. Thus, considering access, erosion and low production, serious plantation developers avoid steep slopes and abide by laws prohibiting conversion of forests on slopes of over 40 per cent.[3]

Investors have promised easy wealth to the people in Malinau through oil palm plantations; alleged benefits have been promoted and potential risks have been played down. But why would investors promote schemes that are not economically viable? The answer is the one-off profits that can be derived from clearing land and selling the timber; many promised plantation schemes have actually never been planted after clearance. According to *Kompas*, an Indonesian newspaper, such fictitious plantations in East Kalimantan have already cost the state more than 3.5 trillion rupiah (US$372 million as at 2005) (Basuki and Sheil, 2005). The cost for local people is considerable since they rarely see any of these profits and have lost access to many forest resources and services.

Local opinions vary: our interviewees who have had firsthand experience as labourers in plantations in Malaysia or elsewhere in Indonesia are typically opposed to any such large-scale developments in Malinau; others, especially those with no firsthand experiences, welcome the promised benefits (Padmanaba and Sheil, 2007).

Almost everyone involved in our interviews believe that such developments should be carefully planned and integrated with other needs, including conservation. We have already taken some steps to inform people of our results and warn against hasty and ill-informed decisions (Basuki and Sheil, 2005). Given vested interests and widespread misinformation, plantation programmes remain a controversial topic for which additional public education is needed (see Padmanaba and Sheil, 2007).

Local choices for local actions

The methods we used to identify local values of biodiversity and landscapes can be applied to inform many processes, from the development of reduced-impact logging guidelines to better local land-use planning, international forestry, and local, national and global conservation policy. It is our hope that the documentation of local views may help people in Malinau to better communicate their needs to decision-makers and other outsiders. The studies indicate areas where small changes in forest management or policy can create win–win opportunities for all. These changes include the protection of sago and other forest values in reduced-impact logging; the prevention of understorey slashing; and the identification of protected areas that can be respected by different forest users or stakeholders, such as grave sites, birds' nest caves, springs and community timber reserves.

To help share the key results of the study with people in Malinau, we have used a variety of media and approaches. We prepared and distributed four colourful posters. These posters present survey results highlighting the views of the communities, while also giving a more technical 'outsider's view'. They are a record of the statements of importance that Malinau communities shared with us, for themselves as well as for outsiders. A follow-up survey shows that these posters have been well received and have resulted in significantly increased understanding of the issues among local stakeholders, such as villagers, town dwellers and civil servants (Padmanaba and Sheil, 2007).

Playing cards showing the most important species and outlining the issues involved in their conservation were also developed and widely distributed. A video presenting local people's views on the forest and the value of keeping forests has also been made and widely disseminated locally as well as in the region. It has been particularly valuable in expressing the local voice to external audiences. There have been showings to the Ministry of Forestry in Jakarta, a number of forest faculties in universities in the country, local NGOs and international environmental organizations based in Indonesia. The video is always used to stimulate debate and discussion. We find that while Indonesian audiences are not used to being presented with a range of differing views (some find the message confusing), the video is a powerful means of communicating the relevant issues more widely. In a young democracy, the value of this role is uncertain but may be considerable: it is only through a broad and informed discussion that a shared vision for Malinau's forests and people can be developed, articulated and implemented.

Our assessments have shown that all of these forms of information were appreciated by the communities we worked with and stimulated wider recognition of their concerns. We raised awareness of the potential significance of suitable conservation actions for the local people of Malinau. This is only a beginning. Awareness must now be turned into action.

Conclusions

Our methods provide a framework for documenting the complex relationships that local communities can have with their environment and the types of biodiversity and landscapes that they value most. In the past, these views were largely neglected by the government, forest management entities and conservation groups. Yet, for effective forest management, they *need to be respected, understood and taken into account in decision- and policy-making and implementation.* In many places, this message requires a paradigm shift for all the institutions and processes related to forest management.

Decentralization has opened many issues to more localized scrutiny than was previously possible. Numerous local institutions, both governmental organizations and NGOs, are seeking ways of integrating the needs and aspirations of local communities with national development strategies and conservation plans. The key point is to develop a dialogue by learning to understand each other, as needed between government agencies and their constituencies. We believe that we have at least identified some points of mutual interest and an approach that allows for improved communication and understanding. A crucial step is to ensure that communities are aware of the importance of their land, heritage and resources, and can express their priorities to decision-makers in local government.

Notes

1 Our biophysical studies have since borne out these predictions – indeed, our data suggest that the Malinau area contains the most species-rich forests in Indonesia (CIFOR, unpublished data).

2 Eaglewood is locally called '*gaharu*' and is the fragrant fungus-infected wood of Aquilaria spp, valued in trade for its use in perfume and incense. In Malinau, the main species are *A. malaccensis* and, to a much lesser extent, *A. beccariana.*

3 Land with slopes over 40 per cent cannot be converted to agriculture in Indonesia according to *Rencana Tata Ruang Wilayah Propinsi* (RTRWP, an Indonesian government regulation on land use) (Basuki and Sheil, 2005).

References

Auxtero, E. A. and Shamshuddin, J. (1991) 'Growth of oil palm (*Elaeis-Guineensis*) seedlings on acid sulfate soils as affected by water regime and aluminium', *Plant and Soil,* vol 137, pp243–257

Barr, C., Wollenberg, E., Limberg, G., Anau, N., Iwan, R., Sudana, I. M., Moeliono, M. and Djogo, T. (2001) *The Impacts of Decentralization on Forests and Forest-Dependent Communities in Malinau District, East Kalimantan: Case Studies on Decentralisation and Forests in Indonesia,* CIFOR, Bogor, Indonesia

Basuki, I. and Sheil, D. (2005) *Local Perspectives of Forest Landscapes: A Preliminary Evaluation of Land and Soils, and Their Importance in Malinau, East Kalimantan, Indonesia,* CIFOR, Bogor, Indonesia

Corley, R. H. V. and Tinker, P. B. (2003) *The Oil Palm,* 4th edition, Blackwell Science, Oxford, UK

Cunliffe, R., Lynam, T., Sheil, D., Wan, M., Salim, A., Basuki, I. and Priyadi, H. (2007) 'Developing a predictive understanding of landscape importance to the Punan-Pelancau of East Kalimantan, Borneo', *Ambio,* vol 36, no 7, pp593–599

Fimbel, R. A. and O'Brien, T. (1999) 'Faunal survey in unlogged forest of the INHUTANI II, Malinau timber concession', www.cifor.cgiar.org/

Gomez-Gonzalez, I. C. (1999) *Indigenous Management of Forest Resources in East Kalimantan, Indonesia: The Role of Secondary Forests,* MSc thesis, Tropical Forestry, Wageningen Agricultural University, The Netherlands

Kier, G., Mutke, J., Dinerstein, E., Ricketts, T. H., Kuper, W., Kreft, H. and Barthlott, W. (2005) 'Global patterns of plant diversity and floristic knowledge', *Journal of Biogeography,* vol 32, pp1107–1116

Liswanti, N., Indrawan, A., Sumardjo and Sheil, D. (2004) 'Persepsi masyarakat Dayak Merap dan Punan tentang pentingnya hutan di lansekap hutan tropis, Kabupaten Malinau, Kalimantan Timur', *Manajemen Hutan Tropika,* vol 10, pp1–13

Lynam, T., Cunliffe, R., Sheil, D., Wan, M., Salim, A., Priyadi H. and Basuki, I. (2007) *Livelihoods, Land Types and the Importance of Ecosystem Goods and Services: Developing a Predictive Understanding of Landscape Valuation by the Punan Pelancau People of East Kalimantan,* CIFOR, Bogor, Indonesia

Lynch, O. J. and Harwell, E. (2002) Whose Natural Resources? Whose Common Good? – Towards a Paradigm of Environmental Justice and the National Interest in Indonesia, Centre for International Environment Law (CIEL), Washington, DC, US, and Jakarta, Indonesia

MacKinnon, K., Hatta, G., Halim, H. and Mangalik, A. (1996) *The Ecology of Kalimantan, Periplus Editions*, Singapore

Meijaard, E., Sheil, D., Nasi, R., Augeri, D., Rosenbaum, B., Iskandar, D., Setyawati, T., Lammertink, A., Rachmatika, I., Wong, A., Soehartono, T., Stanley, S. and O'Brien, T. (2005) *Life After Logging: Reconciling Wildlife Conservation and Production Forestry in Indonesian Borneo*, CIFOR, ITTO and UNESCO, Bogor, Indonesia

Meijaard, E., Sheil, D., Nasi, R., Augeri, D., Rosenbaum, B., Iskandar, D., Setyawati, T., Lammertink, A., Rachmatika, I., Wong, A., Soehartono, T., Stanley, S., Gunawan, T. and O 'Brien, T. (2006) *Hutan Pasca Pemanenan: Melindungi Satwa Liar dalam Kegiatan Hutan Produksi di Kalimantan*, CIFOR, ITTO and UNESCO, Bogor, Indonesia

Padmanaba, M. and Sheil, D. (2007) 'Finding and promoting a local conservation consensus in a globally important tropical forest landscape', *Biodiversity and Conservation*, vol 16, no 1, pp1137–151

Puri, R. K. (1998) *Assessment of the Biodiversity of the Bulungan Research Forest: Ethnoecology of the Punan Tubu*, Consultant report, Wildlife Conservation Society (WCS)/CIFOR, Bogor, Indonesia

Puri, R. K. (2001) *The Bulungan Ethnobiology Handbook*, CIFOR, Bogor, Indonesia.

Sellato, B. (2001) *Forest, Resources and People in Bulungan: Elements for a History of Settlement, Trade and Social Dynamics in Borneo, 1880–2000*, CIFOR, Bogor, Indonesia

Sheil, D. and Liswanti, N. (2006) 'Scoring the importance of tropical forest landscapes with local people: patterns and insight', *Environmental Management*, vol 38, pp126–136

Sheil, D., Puri, R. K., Basuki, I., van Heist, M., Syaefuddin, Rukmiyati, Sardjono, M. A., Samsoedin, I., Sidiyasa, K., Chrisandini, Permana, E., Angi, E. M., Gatzweiler, F. and Wijaya, A. (2002) *Exploring Biological Diversity, Environment and Local People's Perspectives in Forest Landscapes*, CIFOR, Bogor, Indonesia

Sheil, D., Ducey, M. J., Sidiyasa, K. and Samsoedin, I. (2003) 'A new type of sample unit for the efficient assessment of diverse tree communities in complex forest landscapes', *Journal of Tropical Forest Science*, vol 15, no 1, pp117–135

Sheil, D., Puri, R., Basuki, I., van Heist, M., Wan, M., Liswanti, N., Rukmiyati, Sardjono, M. A., Samsoedin, I., Sidiyasa, K., Chrisandini, Permana, E., Angi, E. M., Gatzweiler, F., Johnson, B. and Wijaya, A. (2003) *Exploring Biological Diversity, Environment and Local People's Perspectives in Forest Landscapes*, 2nd edition, CIFOR, Ministry of Forestry and International Tropical Timber Organization, Bogor, Indonesia

Sheil, D., Puri, R. K., Wan, M., Basuki, I., van Heist, M., Liswanti, N., Rukmiyati, Rachmatika I. and Samsoedin, I. (2006) 'Local people's priorities for biodiversity: Examples from the forests of Indonesian Borneo', *Ambio*, vol 35, pp17–24

Sorensen, K. W. and Morris, B. (eds) (1997) *People and Plants of Kayan Mentarang*, WWF-Indonesia, Jakarta (printed in UK)

Tailliez, B. (1998) 'Oil palm: a new crop ... for what future?', *Ocl-Oleagineux Corps Gras Lipides*, vol 5, pp106–109

Wakker, E. (2006) *The Kalimantan Border Oil Palm Mega-Project*, Friends of the Earth Netherlands and the Swedish Society for Nature Conservation (SSNC), Amsterdam, The Netherlands

Whitmore, T. C. (1984) *Tropical Rain Forests of the Far East*, Clarendon Press, Oxford

Wilkie, P. and Amiril, S. (1999) 'The limitations of vernacular names in an inventory study, Central Kalimantan, Indonesia', *Biodiversity and Conservation*, vol 8, pp1457–1467

Wulffraat, S. and Samsu (2000) *An Overview of the Biodiversity of Kayan Mentarang National Park*, WWF-Indonesia Kayan Mentarang Project, Samarinda, Indonesia

Part II

Small-scale Timber Harvesting in Malinau

5

Impact of Small Timber Harvest Permits on Villages in the Malinau Watershed

Godwin Limberg

Small-scale timber harvest permits (*Izin Pemungutan dan Pemanfaatan Kayu*, or IPPK) provided the first formal opportunity for villages to receive significant financial benefits from commercial timber extraction. However, the issuance of IPPKs has led to much debate. Supporters of IPPKs, such as local government officials, local entrepreneurs and elites, mentioned the greater role of villages and the more equitable division of benefits of timber extraction as positive aspects. Opponents, non-governmental organizations (NGOs) and some community members pointed out that because IPPK permits are only valid for one year, companies have no incentive for long-term commitments. After explaining the local setting and the process leading to the IPPK operations, this chapter describes the social impacts of IPPKs upon villages, and impacts upon access to forest land and upon the forest itself. Impacts at the household level are described in Chapter 7. The implications for future community involvement in forest management are then analysed.

Development of IPPK in the Malinau watershed

Changes in forestry regulations in 1999 enabled district governments to issue permits for forest exploitation of 100-hectare plots to villagers. Regional autonomy gave districts the authority over forest management, but also the responsibility to increasingly generate their own income. In forest-rich areas, forest exploitation provided an easy means to increase locally generated revenues as well as income to local communities. At the same time, achieving sustainable forest management became the topic of debates in national and international fora, with increasing calls to reduce forest exploitation. Within these two debates, Malinau as a newly established autonomous region in a remote location had limited options. Its government therefore chose to focus on forest exploitation to increase both the district income and the villages' well-being. The local entrepreneurs and the local elite quickly realized the opportunity to (mis)use arguments of community development for their personal objectives of financial gain from forest exploitation.

IPPKs were originally intended for involving the local communities directly. Because villagers lacked the capital and skills needed, the actual implementation was usually done by local entrepreneurs after reaching an agreement with the local community. At some point in the negotiation process, a village meeting was held providing an impression of community empowerment and transparency typical after the political reform in Indonesia. Villages also recognized this window of opportunity to obtain at least some benefits from forest exploitation. As will be explained in more detail, communities seemed not so much concerned about obtaining legal recognition of their claims to forest and territory as with obtaining financial benefits from exploitation of forests in the areas they claimed.

From July 2000 to February 2001, seven entrepreneurs came to agreements with villages in the Malinau watershed (see Figure 5.1 and Table 5.1 for location of villages and IPPK). Unlike the previous large-scale logging concessions that were allocated by central government to big companies, the main actors in the IPPK deals were local entrepreneurs, village elite and the district government. In six out of seven villages, deals were negotiated by entrepreneurs already known to the villagers. These entrepreneurs or brokers had previously acted as traders in non-timber forest products (NTFPs) or buyers of illegal timber. They were based in Malinau or Tarakan and had wide networks in the area (see also Chapter 6). Through their previous business, they established patron–client relations throughout the area.

The first logging operations in the Malinau watershed started in the territory of Bila Bekayuk in July 2000. All companies were active for approximately two years, except for the company operating in Bila Bekayuk, which closed its operations during early 2001.

In general, the process before logging operations started was as follows:

- The entrepreneur approaches the district government to come to a principle agreement.
- The entrepreneur comes to the village, or the village leader(s) meet(s) entrepreneur in Malinau Town or Tarakan.
- An offer is made by the entrepreneur or a demand is made by the village leader(s).
- The offer is discussed, sometimes only among village leaders and sometimes in a community meeting.

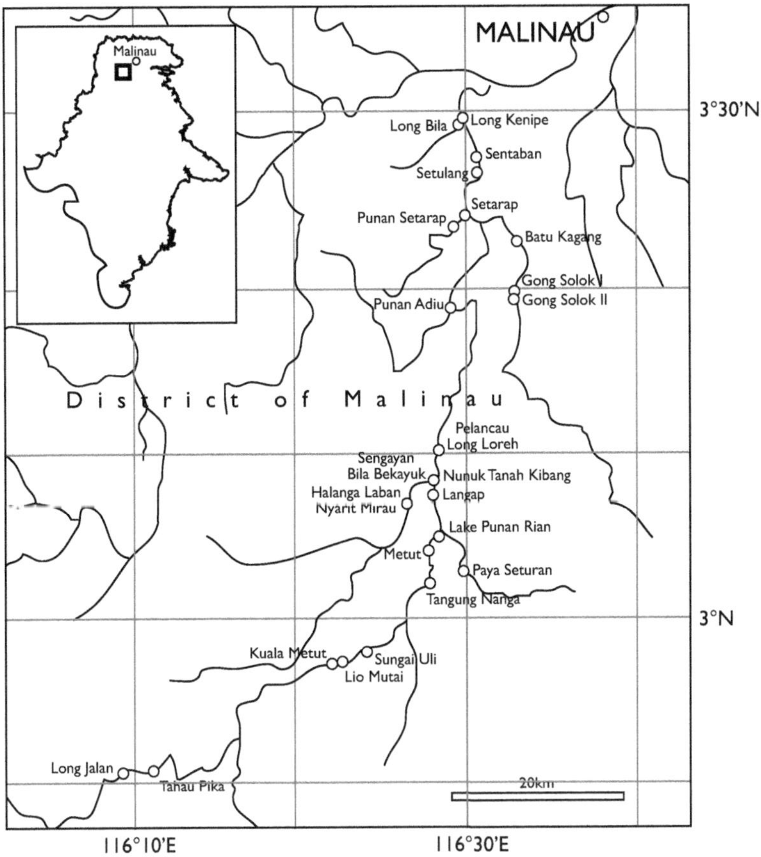

Source: CIFOR GIS Unit, 2007

Figure 5.1 *Location of the 27 villages of the upper Malinau River*

- If accepted, a letter of agreement is made between the village and the entrepreneur. If the offer is not accepted, there is a second round of negotiations between the village and the entrepreneur.
- The benefits for the village are formalized in a notarized contract.
- The entrepreneur takes the letter of agreement from the villagers to the district government.
- The District Forestry Service provides technical advice related to the status of the area requested and harvestable timber.
- The district head officially approves the permit.

Negotiations between villagers and entrepreneurs focused on payment of a fee per cubic metre of timber extracted, in-kind benefits and employment for community members. In addition to these benefits, in the early negotiations, entrepreneurs also promised to establish cash crop plantation as in the cases of

Table 5.1 *List of villages in the Malinau watershed with small-scale timber harvesting (IPPK) operations*

	Village	*Company*
1	Setarap-Punan Setarap	CV Gading Indah
2	Adiu-Punan Adiu	CV Wana Bakti
3	Bila Bekayuk	CV Sebuku Lestari
4	Sengayan	PT Trisetia Abadi
5	Langap	CV Hanura
6	Nunuk Tanah Kibang	CV Putra Surip Wijaya
7	Tanjung Nanga	CV Meranti Wana Lestari

Source: field data

Bila Bekayuk, Langap, Setarap–Punan Setarap and Adiu–Punan Adiu. Later, companies no longer made such promises (Sengayan, Nunuk Tanah Kibang, Tanjung Nanga). Table 5.2 outlines the benefits that were negotiated by individual villages (see Chapter 6 for more details on negotiations).

During negotiations, agreements on the exact location to be logged were vague. Several villages included verbal agreements on local employment and community control on operations, especially monitoring of timber production and logging operations. For example, Sengayan, Nunuk Tanah Kibang and Tanjung Nanga specifically requested that the companies applied selective logging of trees with diameters greater than 50cm.[1] Tanjung Nanga also requested that lesser known species be cut to increase the volume of timber extracted.

Over time, the fee per cubic metre demanded by the villages increased. In Bila Bekayuk, the first village where a company started logging, there was no clear agreement on the fee per cubic metre. In Tanjung Nanga, on the other hand, the last village in the Malinau watershed to strike a deal, the amount was set at 50,000 rupiah per cubic metre, the highest fee per cubic metre. Villagers learned from each other's experience and tried to get better deals. The outcome of negotiations on the fee was influenced primarily by access to information and the negotiation skills of village leaders. In one case, villagers were surprised that the entrepreneur actually offered a higher fee per cubic metre than they had intended to ask. They also realized that they might have gotten more had they known the amount paid in other villages in the Malinau watershed and in other parts of East Kalimantan.

Entrepreneurs often tried to co-opt or intimidate influential people in the village to reduce amounts paid. Prominent villagers were offered personal payments per cubic metre; even a small fee would create reasonable personal wealth. Through these unofficial arrangements, companies ensured that influential people had a personal interest to keep the flow of logs going. Entrepreneurs also made use of instructions by the East Kalimantan governor for logging concessionaires – *Hak Pengusahaan Hutan* (HPHs) – to pay compensation to villages in whose village territories the companies had logged (*Surat Keputusan Gubernur Kalimantan Timur nomor 20 Tahun 2000 tentang Penetapan Dana Kompensasi kepada Masyarakat di dalam dan sekitar Hutan di Propinsi Kalimantan*

Table 5.2 *Benefits promised by IPPKs to villages in the Malinau watershed (in cash and in kind)*

	Fee per cubic metre (rupiah)	*Benefits in kind*	*Starting date (in permit)*	*Date operations started*
Setarap–Punan Setarap	40,000	Access road, levelling of area for village expansion, village office, church, rattan plantation, irrigated rice field	10 August 2000	October 2000
Adiu–Punan Adiu	20,000	Access road, levelling of area for village expansion, village office, village meeting hall, health clinic, rice huller, rattan plantation	23 June 2000	September 2000
Bila Bekayuk	Fee per cubic metre not defined	House for each household, village office, village meeting hall, health clinic, school, TV, coffee plantation	29 May 2000	July 2000
Sengayan	30,000	Village meeting hall, rice huller, 5000-sheet iron roofing	17 November 2000	January 2001
Langap	7500[a]	Levelling of area for village expansion, truck, bush cutter, pepper plantation, rice	7 July 2000	October 2000
Nunuk Tanah Kibang	30,000	30 houses, village office, village meeting hall, water pump, water tank (two), bush cutter, scholarships	17 November 2000	January 2001
Tanjung Nanga	50,000	–	25 January 2001	June 2001

Note: a This amount was renegotiated to 15,000 rupiah per cubic metre in 2001.
Source: field data based on interviews and regular monitoring visits to villages between 2000 and 2003

Timur: Governor's Decree defining compensation payments to communities in and near forests in East Kalimantan). This decree required logging concessionaires (HPHs) to pay compensation retroactively starting from 1995 to 2000 at a rate of 3000 rupiah per cubic metre. Citing this figure, the entrepreneurs claimed that they showed extraordinary generosity by paying more than was required by the governor, even though it was at a rate set for compensation to be paid by HPH companies that had nothing to do with the IPPK system.

What did villages get out of it?

As shown in Table 5.2, villages expected to receive substantial amounts of money, in addition to benefits in kind and employment from the IPPKs. This section analyses how much villages actually benefited from the presence of the IPPKs in terms of financial payments and social developments.

Money

Villagers complained that HPH companies never contributed to the development of the villages where they operated. The total amounts received by the villages from the IPPKs is very large compared to previous community development aid by HPH companies or government assistance, which was annually, per village, 10 million rupiah in cash plus minor community development projects. In comparison, from 2000 to 2002, the following governmental village development projects were allocated to villages where IPPKs were operating: the construction of cement paths in Setarap; piped water in Adiu; the construction of cement paths in Tanjung Nanga; and a new elementary school in Langap. Bila Bekayuk, Sengayan and Nunuk Tanah Kibang received no projects during this period. In this light, it is not surprising that villages were very enthusiastic about the financial payments and social development they were promised.

The initial negotiations resulted in villagers expecting high financial gains from the IPPKs. After two years, were they right? Table 5.3 shows that all villages but one (Bila Bekayuk) received large sums of money.

Payments were made once every two to four months based on the volume of timber logged. The company disbursed the money to the village leader who distributed it according to a variety of systems different for each community. Some divided the total amount equally among the number of households in the village. Other villages set different rates for different groups such as widows, students and bachelors, and sometimes even included village members working outside the village. Tanjung Nanga initially divided the money among head of households, widows and community members working outside the village. Over time, this system changed (see Table 5.4). Households received a total of 500,000 rupiah to 2 million rupiah per payment.

Villages faced several problems related to the fee payments. Having no accurate data on log production, they had to rely on company statements to calculate the total fee to which they were entitled. Companies sometimes used vague stories to delay or reduce payments. A second problem was that companies tended to pay the full amount to one person, usually the village leader or elite. Both of these problems created suspicion among villagers.

In two locations, Setarap–Punan Setarap and Adiu–Punan Adiu, an additional problem was that the community did not reach a consensus on the distribution

Table 5.3 *Sum paid by IPPK to villages in the Malinau watershed from 2000 to June 2003*

Village	*Number of households*[a]	*Total amount paid (rupiah)*[b]	*Average per household (rupiah)*
Setarap–Punan Setarap	94	640 million	6.8 million
Adiu–Punan Adiu	42	800 million[c]	19.0 million
Bila Bekayuk	52	27 million	0.52 million
Sengayan	65	1 billion	15.4 million
Langap	104	320 million	3.1 million
Nunuk Tanah Kibang	36	500 million[d]	13.9 million
Tanjung Nanga	130	1 billion[d]	7.7 million
Total	523	4287 billion	8.2 million

Notes: a Population data from December 2002; information obtained per village from village head or secretary.
b Amounts obtained through regular interviews with village leaders and informal leader during 2000 to 2003.
c Estimate based on information that four payments were made; each payment was made after 10,000 cubic metres had been produced at 20,000 rupiah per cubic metre.
d The data in Chapter 6 of this volume are even higher, at 900 million and 2 billion, respectively (see Chapter 6).

Source: field data based on interviews and regular monitoring visits to villages between 2000 and 2003

Table 5.4 *Fee distribution in Tanjung Nanga*

	First payment	*Second payment*	*Third payment*
Head of household	900,000 rupiah per household	800,000 rupiah per household	1 million rupiah per household
Widow or widower	450,000 rupiah per household	400,000 rupiah per household	500,000 rupiah per household
Youngster[a]	Did not receive any payment	Did not receive any payment	200,000 rupiah per youngster
Elementary school pupil	Did not receive any payment	Did not receive any payment	100,000 rupiah per pupil
Villager working outside community (as far as Malaysia)	Received unknown amount per person[b]	Did not receive any payment	Did not receive any payment

Notes: a Youngster is defined as unmarried children who do not attend school but still live with their parents.
b Paid to relatives in the village.

Source: field data based on interviews and regular monitoring visits to villages between 2000 and 2003

mechanism between the two villages in the location. The village with a low number of households wanted to divide the fee equally between the two villages. The village with a high number of households wanted to divide the fee per household. Because there was no prior consensus, once the payments started several debates took place to discuss the appropriate mechanism. In one case, the fee was equally divided between the two villages, and in the other case, the village with more households got a slightly higher portion of the fee.

The amounts in Table 5.3 only show the amount distributed at the village level. Different sources of information indicate that various village leaders received individual payments not publicly accounted for. Villagers could not influence this system as these promises were made secretly.[2] However, villagers did not protest against this arrangement as long as the money divided among the villagers was equal. People only protested if payments were late or if the discrepancy between the amounts per household compared to the suspected sum received by village leader(s) was too large, as occurred in three villages. Protests were directed at the company as villagers realized that they had little means to pressure their leaders. Villagers sent delegations to the company office to protest and also seized heavy equipment. In one case, the village secured a higher fee per cubic metre (twice the amount of the original agreement). It is impossible to know whether this increase went at the expense of the amount paid to the village leaders or reduced the profit of the company.

In 2000, both village leaders and villagers were very eager to explain to outsiders about their negotiations and agreements with the individual companies. Over time, however, people directly involved with the IPPKs controlled information more, which, in turn, became more distorted. Villagers only knew about the payments made to the individual household and sometimes the amount that was to be paid. Village leaders became more evasive when asked about payments by and arrangements with the companies. This practice resulted in increasing difficulty obtaining accurate information and indicates the lack of transparency at the village level.

Box 5.1 Using IPPK money for village development: An example from Punan Adiu

When the community of Punan Adiu received its first payment from the IPPK company, village leaders set aside 50 per cent and put it in a bank account. Shortly thereafter, villagers used part of it to buy 3 cubic metres of sawn timber for each family to enable them to improve their houses or start building new houses.

This arrangement benefited the community as a whole and contributed to long-term development. However, the village leader admitted that he was under constant pressure from fellow villagers to divide the remaining money. During consequent fee payments, the total amount was equally distributed among all households and no additional money was put into the village treasury.

The lack of transparency also affected the use of fees for general village development purposes. Villages set little aside for public needs, as shown in Box 5.1. Some village leaders explained that even if they suggested that part of the money be saved or allocated for community development, most villagers objected due to lack of trust. In many cases, villagers have ample experience with leaders using village funds without discussion with, or agreement from, villagers. This lack of transparency and accountability resulted in the villagers choosing the 'safe option': divide it all!

Most households spent their money on immediate needs such as food, medical care, education or luxury goods (e.g. electronic devices). Especially in the early days, people were easily lured by itinerant traders to buy a variety of goods; some even provided them on credit. The traders kept track of when the companies made payments and arrived with their goods at that time. Over time, the number of itinerant traders increased. A teacher in one village told how a family had just spent their fee on luxury items when, a few days later, they had to borrow money to pay for medical care for one of their children. He stated that this experience increased people's awareness of the need to be more careful in spending their money, especially given that they have relatively few alternative opportunities for cash generation.

The perception from inside the villages and from neighbouring villages is that little of the money from the IPPKs has had a long-lasting impact. Some of the neighbouring villages might make comments on the lack of impact partly out of jealousy. However, influential people within the villages where IPPKs were operating also acknowledged that the IPPK money contributed little to improving the living conditions of individual households.

In-kind benefits from IPPK

In six out of seven villages, the agreements included promises by the companies to provide a variety of infrastructural development and, in four villages, to establish plantations of cash crops. Table 5.5 shows the in-kind benefits resulting from these agreements.

Table 5.5 *Benefits in kind for villages in the Malinau watershed from IPPKs from 2000 to June 2003*

	In-kind benefits received	*Unfulfilled promises*
Setarap–Punan Setarap	Access road, levelling of area for village expansion, village office	Church, rattan plantation, irrigated rice field
Adiu–Punan Adiu	Access road, levelling of area for village expansion, village office, village meeting hall, health clinic, generator, rice huller	Village office, village meeting hall, health clinic, rice huller,[a] rattan plantation
Bila Bekayuk	–	House for each household, village office, village meeting hall, health clinic, school, TV, coffee/beetle nut/pulp plantation
Sengayan	Village meeting hall, rice huller, 5000 sheets iron roofing	–
Langap	Levelling of area for village expansion, bush cutter, rice	Truck,[b] pepper plantation
Nunuk Tanah Kibang	30 houses, water pump, two water tanks, bush cutter, scholarships	Village meeting hall and office; scholarships only provided for short period
Tanjung Nanga	–	–

Notes: a The company had promised to provide each village with its own village office, meeting hall, generator and health clinic. However, only one village received all facilities; the other village received none.
b The truck was returned because its high price was deducted from the total amount of the fee.

Source: field data based on interviews and regular monitoring visits to villages between 2000 and 2003

Companies provided assistance for any heavy and relatively expensive equipment needed (i.e. road construction and levelling of village expansion area). The roads constructed were necessary for timber extraction, so they involved no additional company costs. Landscaping around the settlement did incur additional costs and villages had more trouble in ensuring that this assistance was provided. The construction of other village infrastructure such as houses, the village office or meeting halls was only partly accomplished.

As with the negotiations on the fee, villagers also learned from experience concerning demands for in-kind benefits. Initially, many demands for in-kind benefits were made but few materialized. Over time, the number of demands decreased, but the percentage of demands fulfilled increased. Villages involved early, such as Bila Bekayuk, had a long list of demands, but did not receive any facilities, and Langap received little in-kind benefits. The villages of Setarap-Punan Setarap, Adiu–Punan Adiu and Nunuk Tanah Kibang received approximately half of promised facilities. In Sengayan, the few items in addition to the fee payment were all fulfilled and Tanjung Nanga, with the highest fee, had no additional benefits promised. However, no company made any effort to establish plantations! Surprisingly villagers did not seem to be too concerned. Initially they appeared to have high hopes that they would not only receive money from the IPPK, but also a cash crop plantation providing a stable source of income for the future. Most likely, once the villagers experienced difficulties in obtaining the fee payments, they realized that enforcing the establishment of cash crop plantations would be even more difficult. Several villagers also mentioned that because the plantations were to be established in remote parts of their village territory, the chances of successful establishment would be limited in any case.

Employment

Initially, villages envisaged that the IPPKs would provide many opportunities for local employment as many villagers had previous work experiences with logging companies in Malaysia. During the negotiations, the entrepreneurs promised that the logging companies would provide employment for villagers. However, in three cases, villagers were quickly disillusioned when the companies brought in complete crews and withheld employment opportunities for villagers. In two of these cases, the villagers demonstrated against the company, which then provided at least some employment for local people. About 40 local people worked in logging operations as chainsaw and heavy equipment operators in 4 villages.

Villagers were also quick to learn that employment with the companies was less lucrative than anticipated. Payment rates were comparable to other parts of East Kalimantan (e.g. 2750 rupiah per cubic metre for a chainsaw operator) despite the fact that costs of living are higher in Malinau. Second, several companies did not pay their employees for several consecutive months. Jobs on the sideline, such as cutting and selling timber for local consumption, actually provided a better source of income.

In one village, villagers supplied sawn ironwood to the entrepreneur earning 400,000 rupiah per cubic metre. The logging company's tractors pulled the ironwood logs out of the forest for free. The local chainsaw operators claimed that they could produce approximately 1 cubic metre of sawn timber per day. Since the entrepreneur bought several tens of cubic metres, this provided substantial additional income to these chainsaw operators. The difference between the fee payment per cubic metre in this village (20,000 rupiah) and the income from sawn timber (400,000 rupiah per cubic metre) is large. However, this source of additional income is not accessible to all households due to a lack of skills or equipment. It should also be noted that this source of timber is available with the assistance of the heavy equipment of the IPPK; otherwise, a large portion of the ironwood would be difficult to extract. The extraction of ironwood also provided more benefit to the entrepreneur, who could sell the ironwood at over 1 million rupiah per cubic metre in Malinau, compared to the operator (doing the hard work) obtaining 400,000 rupiah per cubic metre.

The presence of IPPKs provided some additional benefits, such as transport. In all villages, people regularly used company vehicles to travel to their fields, go to Malinau or to transport timber for personal use. Individuals in most villages benefited from the opportunity to saw timber for local sale. The presence of company labourers provided a temporary market for vegetable, meat and fish.

Impact upon the forest

Villagers observed that the logging had a significant impact upon the forest. However, as Table 5.6 shows, villagers had little notion of how large an area was

Table 5.6 *Comparison between actual and estimated acreage of forest affected by IPPK operations*

	Acreage according to informants[a] (ha)	Acreage according to permit (ha)	Acreage based on satellite image[b] (ha)
Setarap–Punan Setarap	3000–10,000	1200	207
Adiu–Punan Adiu	1500	1500	532
Bila Bekayuk	N.A.	1050	299
Sengayan	1300	2000	18
Langap	5000	3000	554
Nunuk Tanah Kibang	2000	2000	220
Tanjung Nanga	1600	2000	121

Notes: a In June 2003, one to four village leaders were asked to estimate forest area logged by IPPK.
b Interpretation of Landsat image, January 2003.
N.A. = not available.
Source: field data based on interviews and regular monitoring visits to villages between 2000 and 2003 and interpretation of Landsat image, January 2003

actually affected. Most estimates referred to the official acreage as assigned in the government permit. Some claimed that the impact was limited due to the application of selective logging. Yet, in other villages, community members acknowledged that very few trees with diameters greater than 30cm were left and that the logging operations caused much damage to the soil. In four villages, logging operations damaged people's gardens and agricultural fields without prior consultation.

Table 5.6 shows that the area actually affected by the logging operations is significantly smaller that the concession area allocated to the IPPKs. This might seem surprising given that there was little control by government agencies of the small-scale timber harvesting. In the case of Malinau, three factors minimized the impact of IPPK operations on the forest. First, in the Malinau watershed, marketable timber is still readily available at a short distance from the logging roads previously established by logging concessionaires. Second, many IPPKs used second-hand heavy equipment in a poor condition, preventing major logging operations due to the high frequency of equipment breakdown. Third, during this period the timber price was very low, further forcing the IPPKs to minimize costs in order to maintain their profit margin. An additional problem that IPPKs faced was the rugged terrain in Malinau, which has slowed down forest exploitation over the last 20 years.

Further ground truthing of the areas affected by logging proved that some caution is needed, relying on the satellite image to estimate the total forest area affected. The survey revealed that some of the areas that were logged early (in 2000 to 2001) did not show up on the satellite image since re-growth was fairly rapid.

Although the immediate impact upon the forest over the last two years is relatively small, IPPKs have further extended the road network in a substantial part of the Malinau watershed. On the satellite image (Landsat, January 2003) main and logging roads accounted for an average of 69 per cent of the still visible

Table 5.7 *Share of road construction on total forest area affected by IPPK operations*

Village	*Area affected by road construction of IPPK (ha)*	*Total forest area affected by IPPK (ha)*	*Roads as a percentage of total forest area affected*
Setarap–Punan Setarap	81	207	39
Adiu–Punan Adiu	387	532	73
Bila Bekayuk	299	299	100
Sengayan	12	18	67
Langap	384	554	69
Nunuk Tanah Kibang	125	220	57
Tanjung Nanga	96	121	79
Average			70

Source: interpretation of Landsat image, January 2003

impact upon the forest (see Table 5.7). This expansion of the road network creates opportunities for future forest exploitation.

In June 2003, the village elite and community members involved in IPPK tended to downplay the negative impacts of the IPPK operations on the forest. This occurred after the IPPKs stopped logging and many villages were negotiating new deals with companies hoping that logging and the flow of fee payments would continue. Obviously, the benefits from payments outweigh the above-mentioned problems. One cause is that the villages have few cash-earning alternatives and an increasing demand for cash and consumer goods. In November 2002, the district head of Malinau informed all IPPKs that their permit would not be extended. During this transitional period, entrepreneurs tried to obtain new permits to convert IPPKs into small-scale timber concession of a maximum of 50,000ha – *Izin Usaha Pemungutan Hasil Hutan Kayu* (IUPHHK) – which involve larger areas and longer-term permits regulated following the model of the large concession of the previous regime. The entrepreneurs approached the villages to gain their support for continued logging in their village territory. We know at least three cases where the villagers thought that downplaying the impact of logging operations upon the forest might increase chances that new permits will be granted. At the same time, in villages that received limited benefits or villages without IPPK, people tended to be more outspoken about the negative impact of logging operations on forest condition.

Impact upon territory

In an early stage, villages recognized that IPPKs only wanted to operate in village territories that were undisputed. Thus, in order to attract an IPPK, villages had to settle any outstanding conflict over village boundaries. In some cases, this condition was positive and stimulated neighbouring villages to come to boundary agreements. In other cases, it actually aggravated disputes over territory and resource control. In three cases, the conflict was between two villages located in one settlement. Before the arrival of the IPPKs, access to the forests in village territory was regulated with relatively loose customary (*adat*) rules. Upon the arrival of the company, villages debated who had the most rights to certain forested areas and thus was entitled to a greater portion of the benefits. In one case, this resulted in the total exclusion of one village with the argument that they had only settled in the area approximately 30 years ago. In two other cases, initially the benefits were more or less equally shared. However, over time, the discussion concerning distribution of benefits and control over certain parts of the village territory became increasingly heated. In one case, one of the reasons given for early termination of logging operations was because no agreement could be reached between the two villages.

In two other areas issues of control over resources and territory expressed itself in a different manner. Certain influential families (i.e. chief customary leaders or descendants from sultans' own birds' nest caves; see Chapter 4) and their rights have been recognized since the Dutch colonial period. Using the recognition of rights over these caves, they claimed large tracks of forest and were able to obtain IPPK permits. In one case, the community in whose village territory the caves are located received approximately half of the monetary benefits from the logging operations.[3] However, community members suspected and complained that they received only a small part of the benefits compared to the cave owners. In the second case, the community's right to this territory was completely ignored and only the cave owners received benefits from the IPPK.

It is still too early to draw conclusions about how far the issuing of IPPK permits has influenced access to, and rights over, certain areas. The IPPKs were issued for logging in customary (*adat*) territories, so it could be interpreted as an indirect recognition by the district government of claims by the communities. The local government so far has been hesitant to deal with the issue of the recognition of traditional rights because there are many different ethnic groups and different interpretations of *adat* (see Chapter 11). On the community side, there are no examples to date of where they used the issuance of the IPPK permit as an argument to strengthen their claim over a certain territory. The birds' nest cave owners, however, are more inclined to use the IPPK as a tool to expand and strengthen their claim on forests adjacent to the caves.

What lessons have been learned?

Short-term gain versus long-term interest

The period during which IPPK permits were granted was one of rapid change and great uncertainty. The district of Malinau had just been established, regional autonomy was to be implemented, authority over forestry policy was debated and the district government's view on issues such as tenure, *adat* and community involvement in forest management was not clear. This situation of uncertainty provided little incentive for villages, especially village leaders, to be too concerned about long-term interests. For example, it was not clear whether the district government would recognize village tenure based on *adat* claims. On the other hand, the district government might follow a centralized system of resource control and allocation similar to the New Order era. An additional factor was that the companies put pressure on villages to come quickly to a deal. In many cases, short-term gain prevailed over long-term interest (see Chapter 9 for an exceptional case).

How to make agreements

Villagers were aware that written agreements with the companies were essential. However, due to a lack of experience and, possibly, optimism about the companies' trustworthiness (due to previous relationships with brokers and strong network effects), the agreements drawn were ambiguous. For example, they did not specify the size of buildings or sites to be levelled, the schedule or conditions for payments. Many companies suggested finalizing the agreement with a notary deed, allegedly to provide a stronger legal basis. However, the villagers never questioned whether or not they knew how to use the notary deed to take legal action if necessary. Eventually, the notary deed was never used to take action against a company despite the fact that certain promises had not been fulfilled.

The ambiguity in the written agreements provided ample opportunities for the companies to interpret promises to their advantage. For example, a community received a payment before the company started logging. At the time the money was handed over, it was presented as an additional gift; but the amount was later deducted from the first fee payment. No clear terms were defined for the timing of payments, resulting in many delays of fee payments.

Lack of transparency

Throughout the IPPK operation villagers faced difficulties in obtaining information about what was going on. Brokers and companies were in a hurry to use the opportunity created by decentralization, whereas at the community level, people and institutions were not prepared for the sudden changes. The role of the village leader became crucial as companies preferred to deal with one representative only. They were not interested in accountability mechanisms at the village level. It can be imagined how big the temptation for corruption/collusion was for village leaders. During the initial negotiation, opportunities for personal gain appeared with little chance of being exposed due to a lack of control mechanisms.

Transparency was also badly needed during the logging. In some cases, companies agreed to allow for community control on timber production. In practice, however, the companies tried as much as possible to prevent any trips to the logging area. This resulted in a lack of clear production data. This lack of reliable production figures increased suspicion towards the village leader who received the fee payment. It was impossible for villagers to check whether or not the village head deducted anything from the amount. The independent inspection trips that some villagers made to the logged areas informed them about the actual condition. Since this was a personal initiative, however, little collective action was taken if violations were reported.

Support for communities

During this phase of rapid change, villages had few opportunities to seek information or support for considering alternative economic options or drawing contracts with the logging companies. There are hardly any active NGOs in Malinau District. One NGO was actually accused of being a broker for a logging company. During this initial stage, the district government had limited capacity and often made only one brief visit to the community accompanying the broker or company staff. Villagers therefore had little option to seek assistance when needed. In one case, when a village did not receive any benefits from the IPPK, a neighbouring village supported them and the IPPK was eventually paid. Companies normally put time pressure on villages to make deals, further reducing the chances for villages to consult other parties concerning important issues, such as payment distribution, size and location of area.

Community empowerment

Initially villages had the impression that their role in IPPK was crucial. It was the first time businessmen would contact villagers and that they could bargain directly. The letter of agreement between a community and the company seemed crucial in the permit procedure. The possibility of having community members control logging operations further added to their feeling of empowerment. After the initial euphoria, however, villagers quickly realized that their role was actually marginal and difficult to enforce on their own.

Several local people worked to control the operations, such as control over size of trees cut and measuring of total volume cut. In some cases, however, the company paid the salaries of the community members assigned to check on logging operations and they discouraged them from visiting the logged area. Even in Tanjung Nanga, where the community paid the salary of the community members, the company tried to discourage or obstruct them from performing their tasks. On the other hand, villagers went to the logged areas to hunt or collect forest products. During these trips they observed the actual logging operations. However, there are no cases where reports from this informal control led to action against the logging company if violations were observed.

Once villagers realized that they had little control, they concentrated on obtaining as many tangible benefits as possible. Their main concern was to receive cash payments. Although villages had little means of controlling logging operations, they had one powerful tool to force companies: stopping logging operations through road blockades or confiscating heavy equipment. Some villages had tried to enforce agreements through negotiation or letters, but to little avail. Based on previous experience, they then relied on demonstrations.

The learning curve

The experience in the Malinau watershed shows that villages quickly learn from the know-how of neighbouring villages. As mentioned above, the fee demanded per cubic metre payment gradually went up. The negotiation process also became more sophisticated, including a village meeting to discuss the offer made by the company in the negotiation procedure and keeping lists of villagers attending meetings to show wide support for agreement and with better defined arrangements. Villages quickly adjusted their negotiation techniques with the companies. Initially, they tried to enforce agreements through dialogue by village leaders with company personnel or sending letters to the company with copies to the district government. Because these methods yielded little result, villages used road blockades or threats to put up road blockades to ensure that the company would respond quickly, which usually happened.

What does this mean for the future?

Villagers want to have a say in managing the forest in their village territory. The experience with IPPKs has increased their desire to be involved in the negotiation process about forest management or use. Small-scale entrepreneurial logging does not, however, enhance community management. The role of the villages is restricted to granting permission of access to their territory; but control remains with the district government. The negotiation process should not only focus on benefits that villages obtain from forest exploitation in their territory, but also needs to address forest use and management issues. In general, villages in the Malinau watershed want to maintain some forest near their settlement for different uses and they envisage that more remote forest can be exploited for timber harvesting.

Villagers have shown to be quick learners in dealing with the IPPKs. Additional information on current regulations could assist in reducing conflicts and improved implementation. For example, villages could be provided with information on standard benefits or silviculture practices that logging companies have to apply. The villages then might be more confident in checking logging operations in the forest and reporting on violations of regulations. More information at the village level might avoid unrealistic expectation from the villages and thus reduce potential sources of conflict.

Many problems at the village level related to IPPK can be traced back to lack of reliable information and transparency. Greater transparency might seem to be against the interest of the village elite. Experience shows (both in Malinau and Kutai Barat) that villagers have no problem if people involved in the management of forest exploitation receive additional benefits as long as the distribution of benefits is acceptable.

When asked about the positive aspects of IPPKs, the first thing mentioned is the economic gain. The opportunity for villages to gain directly from forest exploitation has changed their perception of the value of forest resources. For future forest management systems it will be important to balance the short-term economic gain that villagers have become used to with maintaining long-term options.

If no mechanisms are created to resolve conflicts quickly and adequately, villagers might continue to rely on demonstrations. Villages have become disappointed with current mechanisms to deal with conflict, and have started to rely on a more radical way of solving problems. To avoid conflict escalation, effective alternative mechanisms have to be put in place.

Our study shows that during the two years of IPPK operations in the Malinau watershed, the direct impact upon the forest was relatively small. It was more a lucky combination of factors that contributed to this than a well-designed and implemented forest management policy. Changing circumstance, such as depletion of timber sources elsewhere or a dramatic rise of timber prices, can quickly result in a rapid, uncontrolled exploitation of the area if no system of checks and balances is developed. The expansion of roads has opened up new areas with high volumes of timber that will continue to attract interest for forest exploitation and increased pressure on forest resources.

Two years of IPPK operations in Malinau District provides important learning experience both at the village and the district level. The current suspension in forest exploitation can provide a good opportunity for both parties to rethink and discuss issues such as village involvement in forest management, access and tenure, transparency and information dissemination from the village to the district level, and appropriate mechanisms for conflict management.

Notes

1 The IPPK permits were not consistent in defining the diameter of trees to be logged. In four cases, the permit defined diameter as greater than 8cm; in two cases as greater than 40cm; and in one case as greater than 50cm.

2 In a case known from the district of West Kutai, the amount that the *pengurus* (organizer or broker) was entitled to was stated in the official documents, which was high: 10,000 rupiah per cubic metre. The *pengurus* was not a village leader, and in many cases was selected by the community to manage the distribution of logging fees.

3 This is based on information from regular visits to this village. However, it is suspected that additional payments were made directly to the cave owners since one of the lower-ranking company staff members originating from that village acknowledged that he had, indeed, received a personal fee per cubic metre.

6

Who, How and for What? Negotiations and Outcomes from IPPK Agreements between Communities and Brokers in Malinau

Charles Palmer[1]

The legislation of small-scale timber harvesting – *Izin Pemungutan dan Pemanfaatan Kayu* (IPPK) – enacted in 2000 by the Malinau District government, followed similar legislation enacted in all districts of East Kalimantan and many forest areas of Indonesia to enable local people to access benefits from small-scale commercial logging. Thus, for the first time ever, forest-dependent communities in Malinau were granted the opportunity to capitalize on their natural assets. This chapter looks at the factors that led some communities to obtain greater benefit in negotiations under the post-decentralization, small-scale timber regime in Malinau.

Communities are likely to reap more benefits under certain conditions. Well-organized, more powerful communities who have access to information and are more experienced in dealing with outsiders and directly with the logging companies are expected to negotiate higher benefits than those who are fragmented, marginalized and disorganized, and who have low expectations. By being well-organized, communities may set up strong networks that exchange information on opportunities and prices. Good organization may be reflected in high rates of participation in village organizations and meetings over access to, and use of, community forests.[2] Moreover, it is expected that communities with higher incomes and higher dependence upon forest products for incomes will negotiate better deals than poorer communities and those with little interaction with the forest for income generation. Poorer communities are expected to have higher discount rates (Pearce and Warford, 1993).

In order to explore the issue, a number of themes are covered, including subcontracting, the marginality of certain ethnic groups in Malinau and the role of people claiming the rights to birds' nest caves. Discussions will concern who negotiated with whom, on behalf of whom, how, and for what kinds of trade-offs. The negotiated outcomes are then presented, together with the actual pay-offs from these agreements, followed by an analysis of what was driving these outcomes. Some comparisons are drawn with similar processes surveyed from another study area, Kutai Barat. The chapter then assesses some of the factors that explain important differences between the negotiation processes and their subsequent outcomes, both the negotiated and actual ones. The chapter then concludes with a final section.

Research methods

Between September 2003 and January 2004, we undertook fieldwork in Malinau in East Kalimantan, conducting interviews at both the community and household levels.[3] In every community, one community survey (for community leaders who had been involved in the original negotiations, such as the village head and *adat* head) and at least five household surveys (for family heads) were undertaken. In Malinau, 22 communities and 189 households were surveyed, covering a total of 15 IPPK agreements, including 7 joint agreements. For analytical comparability, we surveyed only those cases where the company commenced operations or entered community territory after negotiations were concluded.[4] Moreover, only the first agreement (and any changes to this) was surveyed in cases where there were multiple agreements. A complete list of the communities surveyed for this study can be seen in Table 6.1, together with some basic data on community location, land claim, ethnicity and size.

The organization of negotiation in Malinau

As explained in Chapter 5, before a company operationalized the harvesting of a community's forest claim, it followed a number of steps. This chapter focuses on the role of the broker in this process as the key liaison between the government, logging contractors and communities. Brokers – all of whom lived in Malinau Kota or further away – were involved in all first negotiations surveyed in Malinau. The broker was typically known as an 'investor', although the capital for logging deals tended to originate from timber buyers. The latter, due to the proximity of the enormous timber industry in Sabah, were mostly Malaysian and had links with logging contractors, who also tended to be Malaysian.[5] This grouping of buyer, contractor and broker could be loosely defined as the 'company', although a more accurate definition might be the locally used *Commanditaire Vennootschap*

Table 6.1 *Communities surveyed in Malinau District, September 2003–January 2004*

Village	*River*	*Distance from Malinau Town (km)*	*Community land claim (ha)*	*Year community was founded in current location*	*Approximate location of land claim (river)*	*Number of family heads*	*Dominant ethnic groups (30% or more of the population)*
Sub-district of west Malinau							
Batu Lidung	Bengalun	9	1700	< 1963	Bengalun	156	Tagel, Lundaye
Sesuà	Bengalun	15	1500–3000	1948	Bengalun	250	Berusu
Punan Bengalun	Bengalun	30	4600	1941	Bengalun	43	Punan
Sub-district of south Malinau							
Setarap	Malinau	38	9000[a]	1920s	Malinau	65	Kenyah, Lundaye
Punan Setarap	Malinau	38	Don't know[a]	1940s	Malinau	38	Punan
Long Adiu	Malinau	50	70,000	1880	Malinau	26	Merap
Punan Adiu	Malinau	50	Don't know	1900s	Malinau	17	Punan
Langap	Malinau	80	35,000	1914	Malinau	114	Merap
Nunuk Tanah Kibang	Ran	80	50,000	< 1971	Ran/Malinau	34	Punan, Merap
Tanjung Nanga	Malinau	85	50,000	1974	Malinau	145	Kenyah
Sub-district of Mentarang							
Long Sulit	Sesayap	5	3000	1982[d]	Mentarang	32[e]	Lundaye
Long Simau	Mentarang	20	Don't know	1980[d]	Mentarang	41[e]	Lundaye
Paking Lama	Sesayap	40	122,000[b]	1940s	Sesayap/Mentarang	87	Abai, Putuk
Sub-district of north Malinau							
Malinau Seberang	Sesayap	1	5000	1920s	Sesayap	320	Tidung
Tanjung Lima	Sembuak	3	70,000[c]	1900s	Sembuak	60	Abai
Tajan	Sembuak	4	70,000[c]	1960s	Sembuak	80	Abai
Sebatiung	Sembuak	5	70,000[c]	1900s	Sembuak	30	Abai
Semolon	Sesayap	5	Don't know	1972[d]	Semolon	16	Punan
Kaliamok	Kaliamok, Semeladung	5	20,000	1968[d]	Mentarang	111	Lundaye
Kelapis	Keliamok	9	1000	< 1977	Semendurut	104	Lundaye
Salap	Semendurut	13	20,000	1927	Semendurut	70	Tagel
Selidung	Semendurut	16	40,000	1930s	Semendurut	85	Abai

Notes: All data in the table are derived from community interviews. '< denotes that this community was only recognized that year, but had lived in the area before then (respondents were not sure exactly since when). a Setarap and Punan Setarap share a forest claim. Punan Setarap was unaware of the size of its forest claim. b This includes forest claimed separately by Semolon. c 70,000ha are shared by these three villages. d Long Sulit, Long Simau, Semolon and Kaliamok have all relocated far from the original land claim, along the Sesayap and Mentarang rivers and nearer to Malinau Kota. e Subsets of the populations of Long Sulit and Long Simau continue to live in their original villages with 10 and 25 family heads in each village, respectively. These figures are included in the totals.

Source: Interviews with community leaders and households in Malinau, September 2003–January 2004

(CV) in Dutch, otherwise known as 'limited partnership'. This chapter will focus only on the negotiations between the broker and the community, the former acting as representative for the 'company' or CV, as a whole; although in some cases the contractor accompanied the broker in negotiations with the community. Here, the terms 'CV' and 'broker' will be used interchangeably.

The broker decided in advance with district government officials on certain elements of the community–company agreement, such as the size of the IPPK concession area,[6] any logging borders and logging rules, and plantation requirements once harvesting has been completed.[7] Nevertheless, some communities still discussed and even negotiated for some elements, particularly since there appeared to be much ambiguity over the precise nature of these in the regulations (e.g. the types of plantation to be established).

The broker initiated negotiations with the community through the village leader(s), either in the village or elsewhere.[8] Sometimes, as in the six communities that we surveyed (27 per cent of the sample), negotiations began during this meeting and concluded with an agreement. In other cases, offers and/or demands were exchanged by the two parties, which then led to further discussions, meetings and negotiations. Some of these meetings also included other community members, as mentioned below.

Usually, at some point before any agreement was concluded, a public forum would be organized within the community, although this rarely involved the broker directly. Soon after, a formal letter of agreement was written up, typically leading to a notarized contract between the two parties, but not necessarily. All communities surveyed in Malinau made notarized agreements with the broker. How negotiations led to this notarized contract varied from community to community, although some patterns can be discerned from this survey and are described in more detail below.

After making an agreement, containing the names of the families and their signatures, it would be brought by the broker from the village to the Forestry Office in Malinau Kota to obtain the IPPK permit. The broker was also responsible for the work letters and other administrative duties. The villagers themselves did not apply for the permits directly; but their names were used so that the broker could claim ownership of the permit. For example, in Adiu, CV Wana Bakti in effect transferred the right of 15 families (10 from Long Adiu, 5 from Punan Adiu, each owning 100ha each) to the name of the broker, transferring a total of 1500ha for the IPPK. One of the reasons that this might have occurred was because the cost of permit application was probably too expensive for many villagers; in Malinau, it was around 200,000 rupiah per hectare.

Themes in the organization of negotiations

This section describes a number of key themes underlying negotiations in Malinau. It begins with the role of brokers and subcontracting. Next, negotiations are described from the standpoint of communities, some of whom were dominated by powerful local families who claimed rights to the harvesting of birds' nests. Moreover, in addition to local vested interests, large communities and joint arrangements among communities often resulted in more indirect negotiations between communities and brokers. This is followed by a focus on Punan representation during negotiations and on communities with multiple land claims. The latter tended to be more of an issue for resettlement villages in Malinau.

Brokers and subcontracting

Brokers in Malinau were *the* key players in IPPK negotiations. While not employees of vertically integrated logging companies (i.e. ones owning capital), they had the necessary links to contractors (who undertook the actual logging), timber buyers (who owned capital), government officials and, of course, local communities. The IPPK broker could be found in the centre of a large web or network of people, all of whom (communities included) greatly depended upon him to develop and arrange all of the deals at the production end of the timber chain: the permits, the buyers, the equipment and the concession areas.

A small number of brokers dominated IPPK deal-making in Malinau, as elsewhere in Kalimantan. Many of the key brokers in Malinau were well known, at least to community leaders. They tended to be local men who had previously acted as traders for non-timber forest products (NTFPs) or had worked in illegal logging networks. The former saw IPPKs as a lucrative new opportunity, using well-established trading networks in the area, while the latter saw IPPKs as a means by which they could legitimize their businesses. Community leaders, on the other hand, saw these brokers in a position to help extract timber rents for district-level groups. This was in contrast to the pre-decentralization period when rents were almost exclusively captured and retained by large centrally controlled forest concessionaires, known as *Hak Pengusahaan Hutan* (HPHs).

In our sample, the key brokers were Spiner (CV Putra Mentarang), Sudjarwadi (CV Putra Surip Wijaya) and Tengkawang (CV Wana Bakti) (see Table 6.2). The first two were mainly active in the north of Malinau (Selidung, Salap, Kelapis, Kaliamok, Malinau Seberang, Tanjung Lima, Tajan and Sebatiung), while Tengkawang was more active in the east (Batu Lidung, Sesuà and Punan Bengalun) and south of the district (Long Adiu and Punan Adiu). The backgrounds of these men are reflected in their spheres of influence. For example, Spiner was a Dayak Lundaye man with family relations in Kaliamok, Kelapis and Salap, and Tengkawang, a Chinese Dayak Berusu, had many family relations in Sesuà. Overall, at least seven communities (32 per cent of the sample) had historical ties or a previous relationship with their brokers.

These brokers dominated their respective areas of influence in Malinau and there was never any serious competition with other brokers for making agreements with any of the villages, as seen in Table 6.2.[9] Alternative brokers did enter nine villages (41 per cent of the sample) while discussions and negotiations were taking place for their first agreements. Actual alternative negotiations did not, however, take place at the same time. Brokers were scouting around for communi-

Table 6.2 *Negotiations for IPPK agreements in Malinau*

Community	Community members who knew about IPPK before negotiations	How they knew about IPPK	Location of first meeting between company and community	Company representative(s)	Position	Community representative(s)	Previous relationship between representatives	Others from outside community involved in first meeting	Was agreement made in the first meeting?	Further discussions among the community?	Any other companies approaching the community?	Any other outsiders involved in negotiations after first meeting?
Sub-district Malinau West												
Batu Lidung	Don't know	–	In village	Tengkawang	Owner (CV Wana Bakti)	Village head, *adat* head, village leaders	No	No one	No	Yes	Yes	Yabin, village head of Sesuà
Sesuà	*Camat*, village head, *adat* head, most of the village	Company, district government	In village	Tengkawang, Aliang	Owner (CV Wana Bakti)	Village head, *adat* head, some Sesuà community members	Yes, family	No one	No	Yes	No	District head, village leaders: Punan Bengalun
Punan Bengalun	Most of the village	Other village	In village	Yabin, Sadirin	Ex-village head Sesuà, village head Sesuà	Village head, *adat* head, village leaders	No	No one	No	No	Yes	The whole community of Sesuà, NGO Yayasan *adat* Punan
Sub-district Malinau South												
Setarap	Village head, *adat* head, most of the village	District government	In village	A. Liong, Titi	Owner (CV Gading Indah), friend	Village head	Yes, work	No one	No	Yes	Yes	The whole community of Punan Setarap
Punan Setarap	Village head, *adat* head	Other village	Setarap	A. Liong, Titi	Owner (CV Gading Indah), friend	Village head, *adat* head, all family heads	Yes, work	Village head, *adat* head, the whole community Setarap	No	Yes	Yes	None
Long Adiu	Most of the village	Company	Malinau Kota	Tengkawang	Owner (CV Wana Bakti)	Village head, *adat* head	No	No one	No	Yes	No	The whole community of Punan Adiu

Punan Adiu	Village head, *adat* head	District government	Long Adiu	Tengkawang	Owner (CV Wana Bakti)	Village head, *adat* head, community	No	Village head, *adat* head, the whole community of Long Adiu	No	Yes	No	
Langap	Village head, *adat* head, most of the village	Other village	In village	Haji Hamid	Owner (CV Hanura)	Village head, *adat* head, village leaders	Yes, recognition	No one	No	Yes	No	Some members of the community of Punan Rian
Nunuk Tanah Kibang	Village head	Company	Tarakan	H. Sudjarwadi	Owner (CV Putra Surip Wijaya)	Village head, village leaders	No	No one	No	No	Yes	No one
Tanjung Nanga	Village head, most of village	Other village, district government	Malinau Kota	H. Rosalie AC	Camp manager (CV Meranti Wana Lestari)	Village head, *adat* head	No	No one	No	Yes	No	No one
Sub-district Mentarang												
Long Sulit	Pewaris	Company	Malinau Kota	Lintung	Owner (CV Wana Prima Mandiri)	Pewaris	No	No one	No	No	No	Don't know
Long Simau	Village head, *adat* head, pewaris	Company, district government	Malinau Kota	Ase Atak	Owner (CV Sinar Busakan)	Village head, *adat* head, village leaders, *pewaris*	Yes, family, work	No one	No	Yes	No	No one
Paking Lama	Village head	Company	In village	Leti, Suma	Directors (PT Lestari Timur Indonusa)	Village head, *adat* head, the whole community of Paking Lama	No	Pewaris Semolon	Yes	No	No	The whole community of Paking Baru

Table 6.2 *continued*

Community	*Community members who knew about IPPK before negotiations*	*How they knew about IPPK*	*Location of first meeting between company and community*	*Company representative(s)*	*Position*	*Community representative(s)*	*Previous relationship between representatives*	*Others from outside community involved in first meeting*	*Was agreement made in the first meeting?*	*Further discussions among the community?*	*Any other companies approaching the community?*	*Any other outsiders involved in negotiations after first meeting?*
Sub-district Malinau North												
Malinau Seberang	Some of the village	District government	In village	Alvend, Spiner	Owner (CV Tripela, CV Putra Mentarang)	Village head, *adat* head, village leaders	Yes, family	DPRD, district government, the Forestry Office	Yes	No	No	Camat, village leaders: Kaliamok
Tanjung Lima	Village head, *adat* head, village leaders	District government	Tanjung Lima	H. Sudjarwadi	Owner (CV Putra Surip Wijaya)	Village head, *adat* head	No	Lembaga *adat* Dayak Abai Sembuak (LADAS)	No	No	No	Village heads, *adat* heads: Tajan, Sebatiung
Tajan	Camat, village head, *adat* head	Company	Tanjung Lima	H. Sudjarwadi	Owner (CV Putra Surip Wijaya)	Village head, *adat* head	No	Lembaga *adat* Dayak Abai Sembuak, (LADAS)	No	No	No	Village heads, *adat* heads: T. Lima, Sebatiung
Sebatiung	Village head, *adat* head	Company, district government	Tanjung Lima	H. Sudjarwadi	Owner (CV Putra Surip Wijaya)	Village head, *adat* head	No	Lembaga *adat* Dayak Abai Sembuak (LADAS)	No	No	No	Village heads, *adat* heads: T. Lima, Tajan
Semolon	Village head	Other village	Paking	Don't know	–	Pewaris	No	Farmers' group Paking	Don't know	No	No	The whole community of Paking Lama

Kaliamok	Some of the village	Other village	Malinau Seberang	Alvend	Owner (CV Tripela)	Village head, *adat* head, village leaders, village staff	No	Sub-distrct head	No	Yes	No	Forestry Office, the district government, village leaders of Malinau and Seberang
Kelapis	Village head, *adat* head, some of the village	District government	In village	Asiong, Agus	Director, Manager (PT. Bakti Bumi Perdana)	*adat* head, the whole community of Kelapis	No	No one	No	Yes	No	No one
Salap	Village head, *adat* head	Company, district government	In village	Spiner	Owner (CV Putra Mentarang)	Village head, *adat* head, all family heads	Yes, family	No one	No	Yes	No	No one
Selidung	*Adat* head, pewaris	Company	In village	Asiong, Agus	Director, manager (PT. Bakti Bumi Perdana)	Village head, *adat* head, the whole community of Selidung	No	No one	No	Yes	No	No one

Source: interviews with community leaders and households in Malinau, September 2003–January 2004

ties with whom to make deals, but, on the whole, did not enter negotiations with those who were already in the process of making a deal with someone else. They were careful not to get in each other's way, particularly if they were local to the watersheds or villages concerned. Nevertheless, there was first-mover advantage in securing as many deals as quickly as possible in Malinau, which would explain the rush to secure agreements and permits from 2000 until 2001.

In late 1999 or early 2000, for example, the *adat* head of Selidung personally claimed all of the forest near the source of the Semendurut River, an estimated 180,000ha. He contacted CV Putra Mentarang to obtain permits for this area, which were also claimed by the villages of Luba, Seruyung, Putat and Salap. Community members from these villages protested to the district government and formed a united working group, Persatuan Dayak Sewilayah Semendurut (PADUS), to counterclaim and then divide the forest into equal parcels (20,000ha) for their villages. The district government working with PADUS also allocated the forest to the *adat* head of Selidung as a conciliatory gesture and partial recognition of his original claim. A further 1000ha were allocated to Kelapis, a small village that split from Selidung in 1977.

The rush to secure permits and agreements meant that brokers were holding numerous options on potential concession areas, but did not always have the capacity to take them up. But these options still held value. For example, CV Putra Mentarang still had the IPPK option (if not the permits) for Selidung and Kelapis, which were then sold or subcontracted to PT Bakti Bumi Perdana (BBP). In addition to all the other permits that it held for forest areas on the Semendurut River, BBP may have been forming a strategy to hold contiguous concession areas, thus giving a much larger concession overall.[10] BBP approached Selidung and Kelapis independently to negotiate contracts and then arranged the IPPK permits.

Another web of subcontracting can be seen in 2000, when CV Putra Mentarang entered into negotiations with the village of Malinau Seberang. CV Putra Mentarang used another broker, CV Tripela, who then arranged for a subcontractor who could undertake logging operations (PT Nelly Permata Wood Industry).

Pewaris: The heirs and their land claims

From the community side, there were usually individuals or even a single individual who met the broker first and negotiated on behalf of the community (see Table 6.2 and Chapter 12). Who these people were and who they actually represented depended upon the composition of the community and the nature of the land claim (see Table 6.2). In this survey, it varied from a single family (usually, the *pewaris*) to three separate communities sharing a single forest claim.

Pewaris, which means inheritors, were powerful local families who claimed to have inherited rights to birds' nest caves and, sometimes, the timber around them.[11] These rights were inherited either from the former Dutch colonial authorities or the former sultanate, and were widely recognized by many people in the local area, including government officials. *Pewaris* were involved in at least 5 of the 15 IPPK negotiations reviewed in Malinau, all of which were heavily influenced by their strong land claims.

In Long Sulit, the *pewaris*, who lived in a different village, Pulau Sapi, claimed entitlement to forest also claimed by a number of communities in the area. While Long Sulit's main forest claim was in a different area altogether, the community still had a vague claim to forest in its locality. Community members consented to giving their signatures for the IPPK permit in exchange for an oral promise of a share in the IPPK fee. The community was hardly involved in the negotiation process, with all negotiations undertaken directly between the *pewaris* and the broker, Lintung (CV Wana Prima Mandiri). There was never a village meeting and the broker never visited the village.[12]

The *pewaris* of Semolon were among the members of Semolon who lived in the village of Paking Lama. The *pewaris* of Semolon, together with some other Paking-based Semolon community members, were apparently involved in the original negotiations held in the village of Paking Lama, although they did not formally represent the community of Semolon in any way (see Table 6.2). Paking Lama made an agreement alone with PT Lestari Timur Indonusa, and the company operated for two months before it entered Semolon's territory. Apparently, the company did not originally have any plans to work in Semolon's territory and therefore did not formally negotiate with Semolon community members or visit the village beforehand. At this time, the village head of Semolon (who did not live in Paking Lama) was unaware of the company's operations on community territory, and the suspicion was that the *pewaris* of Semolon was involved to some degree. When the village head found out, he asked to make a deal with the company, and Paking Lama's agreement was formally shared between the two communities. Semolon did not attempt to make a deal with another company and was content to share Paking Lama's agreement.

In Long Simau, there was a *pewaris* who was aware of IPPKs before anyone else in the community. This *pewaris* was also a powerful family with a private historical claim on this community's land. Even though this family did not even live in Long Simau (neither the original nor the newer settlement), it informed village leaders about IPPK. It then undertook negotiations with the village head on behalf of the whole community. In this case, the *pewaris* of Long Simau met the broker Ase Atak (CV Sinar Busakan) in Malinau Kota and held regular discussions on IPPK with all community members in the village. By contrast, community members in Semolon and Long Sulit decided to take up an IPPK agreement (see Table 6.3), despite never having been directly involved in the negotiations with the broker.

Indirect negotiations

Typically, however, in most cases and in the absence of a *pewaris*, the village head and the *adat* head, and sometimes other prominent community members or elders, would represent the community in negotiations. Thus, most if not all of the

Table 6.3 *Outcomes from IPPK agreements in Malinau*

Community	Final decision-making on contract	FIRST AGREEMENT CONTRACTUAL OUTCOME							FIRST AGREEMENT ACTUAL OUTCOME						
		Agreement shared with another community?	Proposed logging area (ha)	Fee (rupiah per cubic metre)	Social developments	Plantation development/ replanting	Date of agreement	Length of agreement (months)	Area logged (ha)	Total amount of fee received (million rupiah)	Length of operations (months)	Problems as a result of fee distribution	Problems with other communities	Problems with company	Changes to original agreement
Sub-district Malinau West															
Batu Lidung	Vote in community	No	1500	20,000	Yes	Yes	September 1999	36	1400	550	36	None	None	Yes	None
Sesuà	Consultation among community	Punan Bengalun	500	20,000	Yes	Yes	June 2000	Open	5000	> 2000	36	Yes	Yes	Yes	None
Punan Bengalun	By village leaders in Sesuà	Sesuà	Don't know	20,000	Yes	Yes	June 2000	Open	> 500	300	36	None	Yes	Yes	None
Sub-district Malinau South															
Setarap	Vote in community	Punan Setarap	9000	40,000	Yes	Yes	July 2000	60	5000	910[c]	24	Yes	Yes	Yes	None
Punan Setarap	Vote in community	Setarap	9000	40,000	Yes	Yes	July 2000	Open	2000	650[c]	24	None	Yes	Yes	None
Long Adiu	Vote in community	Adiu	1500	20,000	Yes	Yes	June 2000	Open	1200[a]	800[d]	36	Yes	Yes	Yes	None
Punan Adiu	Vote in community	Punan Adiu	1500	20,000	Yes	Yes	June 2000	Open	1200[a]	235[e]	36	Yes	None	Yes	None
Langap	Vote in community	No	15,000	7500	Yes	Yes	2000	24	Don't know	Don't know	24	Yes	Yes	Yes	Fee to 25,000[g]
Nunuk Tanah Kibang	By village leaders	No	20,000	30,000	Yes	No	2001	24	< 20,000	900	28	Yes	Yes	Yes	None
Tanjung Nanga	By village leaders	No	20,000	50,000	No	No	2000	24	< 20,000	2000	24	Yes	Yes	Yes	Fee to 35,000[g]
Sub-district Mentarang															
Long Sulit	By *pewaris*	No	10,000	20,000	Yes	Don't know	2000	Don't know	7000	> 900	24	Yes	Yes	Yes	None
Long Simau	Vote in community	No	10,000	20,000	Yes	No	2000	Open	5000	450	12	Yes	Yes	Yes	Fee to 50,000[h]
Paking Lama	Vote in community	Semolon	3000	25,000	Yes	Yes	July 2000	Open	2000	> 500	28	None	Yes	Yes	None

Sub-district Malinau North														
Malinau Seberang	By village leaders	Kaliamok	1500	40,000	Yes	No	2001	12	1500b	> 1000	24	None	Yes	None	None
Tanjung Lima	By LADAS	Tajan, Sebatiung	3000	20,000	Yes	Yes	October 2000	12	< 3000	> 870[e]	12	None	None	Yes	Fee to 50,000[h]
Tajan	By village leaders and LADAS	Lima, T. Sebatiung	3000	20,000	Yes	Yes	October 2000	12	3000	> 1000[f]	12	None	None	Yes	Fee to 50,000[h]
Sebatiung	Vote in community	T. Lima, Tajan	3000	20,000	Yes	Yes	October 2000	12	2500	> 1000[f]	12	None	None	Yes	Fee to 50,000[h]
Semolon	By Paking Lama and *Pewaris Semolon*	Paking Lama	3000	25,000	Yes	Yes	July 2000	Open	> 1300	100	28	Yes	None	Yes	None
Kaliamok	Vote in community	Malinau Seberang	1500	40,000	Yes	No	September 2001	12	1500[b]	200	12	None	Yes	Yes	None
Kelapis	Vote in community	No	1000	30,000	Yes	Yes	August 2000	Open	> 200	200	4	None	None	Yes	None
Salap	Vote in community	No	15,000	30,000	Yes	No	2001	Open	< 1000	120	12	None	Yes	No	None
Selidung	Vote in community	No	20,000	30,000	Yes	Yes	August 2000	Open	500	> 300	30	None	Yes	No	None

Notes: All first agreements were notarized contracts. 'Open' indicates that there were no time periods agreed in the contract. a All forest claimed by Long Adiu. b 900ha claimed by Malinau Seberang, 600ha by Kaliamok. c Each figure gives fee for both villages. d For both villages. e For Punan Adiu only. f For all three villages. g Both agreements renegotiated when the *pewaris* of Langap claimed that forest in the IPPK area was in the vicinity of birds' nests owned by this group. For Langap's agreement, the total fee increased to 25,000 rupiah per cubic metre with 15,000 rupiah for the community and 10,000 rupiah for the *pewaris*. For Tanjung Nanga's agreement, the fee remained the same, although the fee for the community decreased to 35,000 rupiah per cubic metre while the *pewaris* claimed 15,000 rupiah per cubic metre. h For all of these villages, the agreements were renegotiated with the fee, reaching 50,000 rupiah per cubic metre in exchange for all social developments being dropped out of the contract.

Source: interviews with community leaders and households in Malinau, September 2003–January 2004

negotiations surveyed in Malinau involved a few well-placed people dealing with the broker, at least during the initial meetings. In 16 communities (73 per cent of the sample), most community members were only involved in negotiations indirectly and their impact on decision-making was not always easy to discern. In negotiations where community members were involved directly, the communities tended to be relatively small (e.g. Paking Lama and Punan Setarap).

Malinau Seberang was the largest community in the sample, with approximately 1500 members. It is unique in that it is located next to Malinau Kota and has increasing immigration from the town itself. Decision-making within such a large community was devolved down to the heads of sub-village units – known as *Ketua RT* or *Rukun Tetangga*, of which there were eight in Malinau Seberang – including that for the proposed IPPK. These men were responsible for consulting with the rest of the community with regard to all aspects of IPPK before reporting back to the village and *adat* heads. Thus, in larger communities, the gap between the ultimate decision-makers and the rest of the community was larger than in smaller ones, which would explain why community members in places such as Malinau Seberang felt less involved in decision-making concerning a collective resource. Certainly, many people in Malinau Seberang did not feel that they had been consulted on IPPK during negotiations.

Similar to Malinau Seberang and Kaliamok and many other cases, the negotiations among the communities of Tajan, Sebatiung and Tanjung Lima on the Sembuak River were not held directly between the people in the village and the broker, CV Putra Surip Wijaya. In 2000, three well-connected and powerful individuals from these three villages formed a not-for-profit institute (*lembaga*) known as Lembaga Adat Dayak Abai Sembuak (LADAS). It acted as lead negotiator for the joint forest claim of the three villages of around 70,000ha located at the source of the Sembuak River. The entire negotiation process took a month and followed these steps:

1 LADAS arranged a meeting between the broker and representatives from the communities. In Tanjung Lima, CV Surip Wijaya met the village heads and *adat* heads from Tanjung Lima and its neighbouring villages, as well as LADAS for preliminary discussions.
2 The village heads and LADAS organized an open one-day meeting for all three communities at the Catholic church in Malinau Kota for a discussion of the company's intentions and community demands. Over 80 per cent of surveyed households participated in this meeting. The broker was not present for this meeting.
3 The village heads, *adat* heads and LADAS met with the broker again, discussed the communities' demands and initiated negotiations.
4 The village heads, *adat* heads and LADAS consulted with the communities separately for further discussions on what the company might offer with respect to community demands.
5 The village heads, *adat* heads and LADAS returned to the broker and made an agreement.

6 CV Putra Surip Wijaya entered Tajan and met all the communities for the first and last time. LADAS organized a party in which the advance fee was distributed among all members of the three communities.

There was disagreement among the three villages about how the final decision on the IPPK was made, although it was clear that there was some consultation among community members and their leaders about what they wanted. There was also some disagreement over the precise role of LADAS, accusing them of meeting the broker alone and away from village leaders. In any case, coordination among all the groups within these ethnically homogenous communities (all are Dayak Abai) caused some problems during the negotiation process.

Joint arrangements were relatively common in Malinau, accounting for over one third of agreements surveyed and nearly always involving Punan communities. The involvement of, and consultations with, different groups varied from case to case. For example, village leaders from Kaliamok held joint negotiations in Malinau Seberang with the leaders from the host village. While Kaliamok claimed that all community members were consulted and a vote was taken on the decision over whether or not to get involved with the proposed IPPK, the final decision in Malinau Seberang was taken directly by its village leaders.

In the village of Nunuk Tanah Kibang, all negotiations were led by the village head, both in the town of Tarakan and in the village. Negotiations in the village also involved other village leaders, with regular consultations with the rest of the community. Again, and as was common in many cases in Malinau, community members, although consulted, were not directly involved in negotiations. They only actually met the company for the first time once the deal had already been made, during 'socialization' (*socialisasi*). This occurred when the broker visited the community and got involved in public discussions about IPPK and sometimes paid out an advance payment of the fee agreed. As can be seen from Table 6.3, the final decision on this agreement was made by the village head.

For Selidung and Kelapis, negotiations were held in the villages and all members of the communities were involved, with votes on the final decision on whether to have an IPPK or not. A similar process also took place in Salap. All of these communities decided to make agreements with those holding the IPPK permits. Overall, 12 out of 22 communities (55 per cent) surveyed claimed that a vote was taken in the community to determine the final decision on making an IPPK agreement. In the remainder, the final decision tended to be taken by members of the village elite (see Table 6.3).

Punan representation

With the exception of Nunuk Tanah Kibang, all of the Punan groups surveyed were involved in joint arrangements with non-Punan communities. Many semi-nomadic Punan groups were located near non-Punan, 'sponsor' communities, who tended to take the lead in IPPK negotiations. Sesuà and Punan Bengalun, for example, did not share forest claims, although they shared an agreement brokered by CV Wana Bakti. Also, as with many Punan communities surveyed, they were uncertain as to the exact size of their land claim (see Table 6.1). Moreover, as was common to other arrangements that involved Punan groups, the broker entered Sesuà first and held meetings with village leaders before involving any Punan representatives. However, unlike the case of Long Adiu on the Malinau River, village leaders from Sesuà entered Punan Bengalun and discussed potential IPPK arrangements without the broker present. Punan Bengalun did not negotiate directly with CV Wana Bakti, but indirectly through Sesuà village leaders. While the Punan were involved in discussions before any agreement was made, the final decision with respect to the IPPK was made by the village leaders of Sesuà alone.

Similar to Punan Bengalun, the Punan people of Semolon were not involved in negotiations at all and only shared an agreement with Paking Lama after the company had already entered community territory.

People in Long Adiu were among the first to arrange an IPPK deal on the Malinau River in April 2000, after the village head and *adat* head went to Malinau Kota to try to find a broker. They became familiar with Tengkawang (CV Wana Bakti) and may have already known him through historical ties. He was invited to Long Adiu to meet community members along with those from Punan Adiu. A joint arrangement was discussed. Unlike the Sesuà–Punan Bengalun and Paking Lama–Semolon arrangements, members of Punan Adiu, along with Long Adiu community members, were directly involved in negotiations. While this occurred early in the process and before any agreement was made, it was clear that the village and *adat* heads had the first meetings alone with the broker and excluded the rest of the two communities, including all the village leaders of Punan Adiu. The latter only became involved once the broker visited the village of Punan Adiu. The final decision was apparently taken by a vote of all community members from both villages.

In addition, on the Malinau River, negotiations between Setarap and Punan Setarap followed a similar pattern as those in Adiu. However, here the initial meetings with the broker (CV Gading Indah) all took place in Setarap village and not in Malinau Kota first, with the village and *adat* heads (of Setarap), before public discussions took place involving members from both communities. Moreover, similar to the Long Adiu–Punan Adiu deal, the final decision on the IPPK agreement was taken by a vote of all community members from both villages.

Villages with multiple lands

A number of communities in the sample had forest claims in areas where they no longer resided, while others made new claims in areas also claimed by other communities (see Table 6.1). For example, two of the three villages surveyed in the sub-district of Mentarang, Long Simau and Long Sulit (both predominantly Lundaye) were resettlement villages that still maintained populations in the original villages near the source of the Mentarang River. These old settlements were hundreds of kilometres from where most of the members moved in the 1980s.

Kaliamok also had a historical land claim near the source of the Mentarang River and near the location of the original settlement (see Table 6.1). However, in this case, Kaliamok no longer had any community members living in the original settlements and claimed forest for its IPPK in the vicinity of the current settlement near Malinau Kota. This forest was also claimed by Malinau Seberang and a joint agreement was made with most of the benefits flowing to the larger village.

Some Punan groups also claimed multiple lands. Since 1972, the Punan settlement of Semolon has been in North Malinau and very close to Malinau Kota. Semolon also had a land claim on the Mentarang River and, as mentioned above, still had community members living in this area, mainly in Paking Lama close to Semolon's original land claim.

Negotiated outcomes: What were the communities negotiating for?

In all cases surveyed in Malinau, there appeared to be no real negotiations over logging rules, the size and location of the concession area, or plantations. Typically, the broker would appear at meetings with a mapped-out area that had already been decided, in principle, by the district government. Communities did have a say in the composition of the proposed concession area and logging borders. Long Adiu and Punan Adiu negotiated the size of the IPPK to be logged within their respective forest claims and the time frame for doing so. Where plantations were discussed and negotiated, considerable variation was also observed on IPPK contracts from village to village, in terms of the type of plantation crop, the size of the plantation and the nature of promised assistance for the development of the plantation. However, in a number of cases (see Table 6.3), there were very few logging rules or replanting requirements in the actual agreement between the community and broker even if they were initially discussed before any agreement was made. Some community respondents (e.g. in Tanjung Nanga) explained that plantations and logging rules were the responsibility of the district government and, hence, a 'legal requirement'. Therefore, these elements were not considered to be necessary for inclusion in their contract with the broker.

Half of the agreements surveyed in Malinau contained no time limits on when the company was to complete logging operations (see Table 6.3) but did require that they commence six months from the conclusion of negotiations. None of the official IPPK permits issued by the district government mentioned time limits either. Moreover, in general, no agreements contained any proposed timber production schedules. Overall, the agreements examined gave IPPK companies a high degree of freedom over the conduct of their operations.

In negotiations, the communities were usually asked what they wanted in terms of financial benefits (advance payments, fee per cubic metre), jobs and social developments, such as scholarships for children or village administrative offices. In some cases, particularly the first communities to be involved in IPPK negotiations, community expectations were either non-existent or very low. Long Adiu and Punan Adiu, for example, only asked for a fee of 5000 rupiah per cubic metre because their only reference point was instructions from the governor of East Kalimantan (SK 20/2000) that firms pay compensation of 3000 rupiah per cubic metre to *adat* communities. All communities covered by CV Wana Bakti were paid 20,000 rupiah per cubic metre for reasons that are not clear (see Table 6.3 and Figure 6.1) given that, in theory, the company could have offered anything between 5000 rupiah to 20,000 rupiah per cubic metre.[13]

In other cases, such as Malinau Seberang, communities asked for much more. Here, the village originally asked for 70,000 rupiah to 80,000 rupiah per cubic metre, and eventually were offered 40,000 rupiah per cubic metre, the second highest first agreement rate in our sample of Malinau communities. Only Seterap and Punan Setarap also negotiated a fee to this level, while Tanjung Nanga negotiated the highest fee: 50,000 rupiah per cubic metre (see Figure 6.1). Although Tanjung Nanga negotiated this fee, they received no promises of social development. Every single other community within this sample negotiated some social development, although there was considerable variation from one case to the next. In the case of Long Sulit, the promised fees were unknown although, unlike all other agreements, they were not arranged on a production basis (i.e. per cubic metre). In joint arrangements, social benefits were usually shared.

Some communities managed to increase the fee level to 50,000 rupiah per cubic metre during subsequent rounds of renegotiation after the original agreement had been made. For example, in Long Simau, community members asked the village leaders and the *pewaris* to change the agreement to drop all the facilities and increase the fee to 50,000 rupiah per cubic metre after the agreement had already been made but before the village received the first payment. This was due to the concern that the facilities built by the company would only be experienced by those living in one of two locations that constitute Long Simau. Also, community members described how they had learned from the experiences of Tajan, Tanjung Lima and Sebatiung, which at that time had just managed to increase their fee to 50,000 rupiah per cubic metre, again in exchange for dropping all social facilities from their agreement.[14]

In other cases featuring renegotiations, the cases of Langap and Tanjung Nanga further illustrate the power of the *pewaris*. In Langap, there was a family consisting of around ten family heads that had a longstanding claim to the harvesting of birds' nests, a high income-earning NTFP. The most powerful man in this family also happened to be *Kepala Adat Besar* (*adat* or, according to customary law, head of the watershed) at the time of the negotiations. After Langap had already concluded negotiations and made an agreement with CV Hanura, and the company had begun operating, Langap and the *pewaris* renegotiated for more money in order that the latter could claim for forest (*tanah warisan*) surrounding the location of the birds nests. The village doubled its fee to 15,000 rupiah per cubic metre, still the lowest of all villages surveyed in Malinau, while the *pewaris* claimed 10,000 rupiah per cubic metre, giving a total fee of 25,000 rupiah per cubic metre.

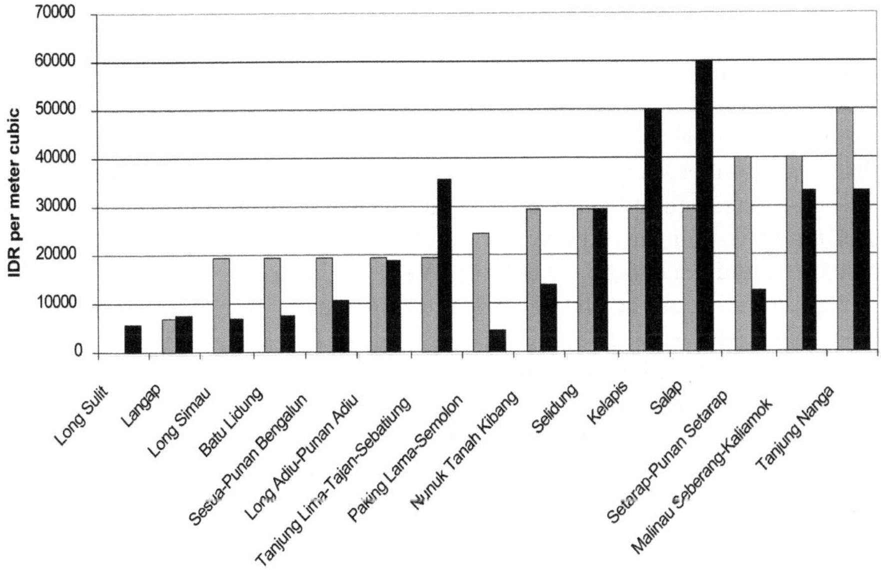

Note: Renegotiated fees are not included.
Source: Charles Palmer, 2003–2004

Figure 6.1 *Promised and actual fees paid for the first IPPK agreements reviewed in Malinau*

The same *pewaris* also complained when the IPPK operating on behalf of Tanjung Nanga, CV Meranti Wana Lestari, logged in the vicinity of its birds' nest cave. Renegotiations of Tanjung Nanga's agreement with the company led to the community sharing its fee with the *pewaris* of Langap: the total fee was increased to 64,000 rupiah per cubic metre, of which 40,000 rupiah were paid to Tanjung Nanga and the remaining 24,000 rupiah to Langap. The 24,000 rupiah per cubic metre for Langap was to be split 50:50 between the *pewaris* and the community of Langap. However, the money was paid to the *pewaris* and it is not clear if the Langap villagers received any of it.

From case to case, there appears to be some variation in negotiated benefits in Malinau, especially with regard to social developments. However, there is less variation where a single broker was involved in making multiple deals.[15] For example, all the deals brokered by a combination of CV Putra Mentarang and BBP resulted in a fee of 30,000 rupiah per cubic metre (Kelapis, Selidung and Salap), and on the Bengalun River, both the Batu Lidung and Sesuà–Punan Bengalun agreements resulted in a fee of 20,000 rupiah per cubic metre. Certainly, the negotiation process appears to have been rigged in parts of Malinau, with little or no competition among brokers, resulting in cartel-like behaviour. This price-fixing appeared to have resulted in relatively little variation in negotiated benefits in a number of agreements.

While there is large variation in negotiated fees, the inclusion of promised social benefits may result in some parity from agreement to agreement, particularly when considering the renegotiated fees of Tanjung Lima–Sebatiung–Tajan, Long Simua and Tanjung Nanga (all 50,000 rupiah per cubic metre). None of these agreements contained promises of social developments, unlike those of, for instance, Selidung or Long Adiu–Punan Adiu, which contained many promises. Certainly, as people in the district learned from their own experiences and that of other communities, they renegotiated their agreements or made new ones that reflected their newfound confidence in dealing with brokers and other outsiders. After renegotiations, the lowest fee level in the district was 15,000 rupiah per cubic metre and the highest, 50,000 rupiah per cubic metre.[16]

Concentrating on the first negotiated fees, the lowest levels were negotiated in communities who were less likely to come together and discuss access to and use of forests, and where there tended to be less participation in village organizations. Furthermore, in communities where there were higher proportions of households with experience working outside the community and greater mobility (ownership of a long boat or other form of transport) before negotiations, negotiated fees tended to be higher.

Actual outcomes from IPPK agreements

In April 2003, all IPPK logging operations in Malinau were stopped by order of the district head after the Indonesian government had issued a decree the previous year for the cancellation of all small-scale logging permits in Indonesia. In most cases, according to survey respondents, IPPK companies logged less than the area agreed in the contract with the exception of Sesuà–Punan Bengalun. Here, it was estimated that around 5000ha were logged in almost three years since June 2000 (see Table 6.3). Unlike the other villages, many of which had rolling open contracts with large areas proposed for multiple permit extensions, Sesuà and Punan Bengalun's contract was originally only for 500ha. The broker dealing with Sesuà, Tengkawang, also operated PT Malinau Jaya Sakti. He negotiated a second agreement of approximately 3000ha with Sesuà and Punan Bengalun, which would explain the logging of 5000ha of these communities' forest.

While most agreements in Malinau were not completed simply because the IPPK companies were forced to stop in early 2003, some operations had barely started even for agreements that were made nearly three years previously. For example, Selidung agreed to a 20,000ha concession with BBP in late 2000, and by 2003, only around 500ha around the village had been logged. In general, BBP was allegedly too busy logging on behalf of the private contracts arranged with the *adat* head of Selidung and not on behalf of the communities in the area.[17]

All communities reported receiving fee money, with a large variation in the quantity of money received and how it was used and distributed within communities. As can be seen in Figure 6.1, there appears to be considerable differences between what was promised and what was actually received by communities in Malinau. Long Sulit received the lowest fee per cubic metre, followed by Langap, and the highest rates were received by Kelapis and Salap. Langap and Tanjung

Lima–Sebatiung–Tajan all received higher rates than were actually negotiated, mainly because the original fees were renegotiated upwards. In the cases of Kelapis and Salap, actual rates exceeded promised rates mainly due to both communities receiving large advance fees to be followed by relatively little production by the companies involved. In all other cases, except Selidung, monies received were less than what was promised due to company or broker non-compliance. Alternatively, monies received were under-reported due to rent-seeking within the recipient communities. In all cases where social development was promised, there was some compliance, although in every case, companies promised more than was actually delivered in the duration of the contract. However, as demonstrated in Engel and Palmer (2006), the actual fees received by communities depend upon their ability to self-enforce contracts in a situation of weak property rights, as in East Kalimantan.

In most cases surveyed in Malinau, there were problems with neighbouring communities over territorial borders and conflicting land claims. Table 6.3 shows that at least 16 out of the 22 communities reported problems with other (usually neighbouring) communities due to IPPK deals. The most serious problems involved the Punan communities who, given their semi-nomadic status, have had the most problematic and contestable land claims. Other problems with land claims can be seen with negotiations involving the *pewaris*. The power of these families varies in Malinau, although their claims to land have almost certainly increased since the beginnings of the decentralization process in Malinau. This is possibly mainly because of the influence they exert over local government officials in addition to having family members who were actual government officials or members of the local legislative assembly, the *Dewan Perwakilan Rakyat Daerah* (DPRD).

After agreements were made in Malinau, an IPPK company usually paid out an advance payment. In almost all cases, this money was 'borrowed' from the fees generated from future wood production (*uang potong*, *uang pinjam* and *uang muka*). The only exception to this was the Setarap–Punan Setarap agreement. Furthermore, in many cases, the building of facilities has also been taken out of fee payments (resulting in underpayments) or agreements to build were simply not fulfilled (e.g. Setarap). In some cases, these 'under-payments' or non-compliance with promises have occurred without community members being informed.[18] In many cases, there have been problems and conflicts between the communities and the brokers or contractors.

Of the 22 villages surveyed in Malinau, 19 reported having had problems with the IPPK companies. In addition to late or non-payment of fees, and no or little development of social facilities, the most common problems were related to breaking logging rules, logging in different places than agreed upon or the non-cultivation of plantations. No plantations have been developed by any of the IPPK companies in Malinau. Of the 19 communities reporting problems with the IPPK companies in Malinau, 11 communities undertook some kind of activities against IPPK companies within the duration of the contract, such as a demonstration or the confiscation of company equipment.

From the findings of 22 communities surveyed in Malinau, it is clear that they have benefited from IPPK, at least financially and in the short term. The long-term benefits are not so obvious, particularly since many community members perceived widespread environmental damage from the IPPK operations and since there has been no plantation development whatsoever following these operations. Furthermore, a number of inter-village conflicts relating to IPPK and forest claims are still to be resolved.

Discussion

In Malinau, communities who were better organized and who had higher participation rates in public meetings over access to, and use of, forest land claimed by the community were more likely to negotiate better deals than more disorganized communities. Participation and regular meetings allowed community members opportunities to compare experiences and exchange ideas and information. Moreover, communities who also had greater mobility and higher proportions of households working in other locations outside the village were among those who negotiated the better agreements in Malinau. Relative levels of household incomes, the proportions of these derived from the trade and sale of forest products, and levels of education varied little from community to community in Malinau and, hence, appeared to have little impact upon negotiated outcomes.

Many of the communities surveyed suggested that the joint agreement made by Tajan, Sebatiung and Tanjung Lima was the one that led to the greatest net benefits. Certainly, collective negotiating action by three villages and the renegotiation of the agreement for a higher fee led to one of the highest levels of financial benefits for members of all three communities.[19] While problems remained regarding accusations of corruption (with respect to LADAS and the role of the Catholic church in relative material terms), these communities probably gained the most and had no problems with other villages, such as land or border conflicts. Other communities with relatively positive experiences, such as Long Simau, also had a *pewaris* involved in the negotiations. These families complicated the issue of land claims (e.g. the *pewaris* of Langap) and played a very powerful role overall in many negotiations surveyed in Malinau. In addition to the *pewaris*' claims, many negotiations were complicated further by multiple landholdings and by indirect negotiations between many communities and the brokers.

The 'biggest losers' were probably the Punan communities, who tended to be involved in joint agreements with neighbouring non-Punan communities. Punan communities tended to be the least involved in negotiations and were often not treated as 'equal partners' by their neighbours. Furthermore, they probably had less awareness of concepts of area (as defined by the measurements used in negotiations) and contractual obligations than other groups. In most cases, the Punan only seemed to have a vague awareness of the exact terms of their agreements with the brokers and of how much forest in their territory has been logged. The Punan had the weakest property rights of any of the groups and a greater vulnerability with respect to dealings with brokers and other outsiders, which is reflected in the survey data for these groups seen in Tables 6.1 to 6.3.

In Malinau, most of the contractors were Malaysian and were probably less likely to be able to make deals alone due to a lack of local knowledge and the dubious legality of many of their operations. As a result, they became more dependent upon working partnerships and brokers to arrange the deals with those who control access to commercially valuable stands of timber – the communities and local government officials. The brokers may have had the greatest influence on negotiated outcomes in Malinau. By contrast, in a different study area, only one quarter of the communities surveyed in Kutai Barat had experience with a Malinau-style middleman or broker, and even then they acted more as facilitators and tended to receive direct fees from the contractors, separate from the community fee. The main difference is that the brokers in Malinau were nearly all from outside the communities with whom they made deals, which meant that a sizeable proportion of timber rents remained with the brokers outside the communities. In Kutai Barat, a greater proportion of rents flowed to the communities as a whole.

Brokers in Malinau typically received a fee per cubic metre that was to cover all arrangements, including the agreement with the community and dealings (legal and illegal) with officials working for the district government. In other words, the typical broker in Malinau had a direct incentive to maximize his own benefits by minimizing any payouts to the community. In Kutai Barat, timber fees rarely passed through a middleman first, which had consequences for their level.

One of the key roles of the brokers in Malinau was to arrange for the IPPK permits. When the system was originally established in 2000, the rules stated that each family who applied for a permit was entitled to utilize 100ha. The process of application required the formation of a farmers' group (*kelompok tani*) to manage the activities of the IPPK area, and to obtain the signatures and copies of identity cards from each applicant, among other things. All districts in East Kalimantan established similar systems in 1999 and 2000. The high cost of permit application probably necessitated the use of brokers in Malinau, although similar material poverty did not stop community members from applying and owning permits in Kutai Barat.[20]

Many brokers also bought and sold permits once they were acquired, such as CV Putra Mentarang to BBP. During 2000 and 2001, it was clear that brokers were active in arranging as many deals as possible so that they could, at the very minimum, hold the option for the future. It seems, however, that in later arrangements, the brokers did not even bother to collect names and instead just made an agreement with the community as quickly as possible in order to have the 'option'. For example, in North Malinau, BBP and CV Putra Mentarang made deals with a number of other villages not analysed here for IPPKs that never became operational. Thus, a system that was originally established to involve local communities evolved to a point where they actually had relatively little involvement and instead appeared to benefit a small number of brokers in Malinau. As a result, none of the communities surveyed in Malinau applied for permits directly (with the possible exception of the *pewaris* of Long Sulit) and none of the survey respondents even knew the cost of IPPK permit application and had relatively little awareness of how the system was intended to work.

In Kutai Barat, the system worked differently: permits (known as *Hak Pemungutan Hasil Hutan* – HPHHs) were usually obtained before the company would make an agreement with the community. Furthermore, in many cases, it was usually community members and village leaders who approached the district government and applied for the permit (Palmer, 2006). Again, similar to Malinau, a permit for 100ha was allowed per family, and signatures and names were required, in addition to a series of fees usually totalling around 20 million rupiah per permit. By this stage, village leaders would have already made direct contact with a logging contractor, sometimes already known to the community, sometimes not. Those communities who used a broker were mainly responsible for arranging work letters and other administrative aspects, and again tended to deal with the company rather than the communities. Many meetings took place outside the villages, in the district capital, Melak, and even further away, in Samarinda, the provincial capital. In making the agreement, communities in Kutai Barat would ensure that permit application costs were covered in the fee structure. While advance payments were not common in Kutai Barat, the contractor made direct fee payments to the community and not via a middleman or broker. Ironically, the IPPK experience in Malinau, which many villagers considered the best, was also the one in which the contractor made a direct agreement with the Tanjung Lima, Tajan and Sebatiung communities, with no involvement of a broker.[21] The only other case not involving a broker, Tanjung Nanga, also negotiated the highest fee surveyed in Malinau: 50,000 rupiah per cubic metre.

The implication of permit ownership was that in applying for and owning the 100ha permits, communities in Kutai Barat had a much clearer understanding and awareness of what they were negotiating away (i.e. size of concession area), as well as what they were negotiating for. Thus, communities had a much greater sense of control and ownership over the negotiation process than those in Malinau, even while there was a similar level of uncertainty over land claims. Groups or co-operatives were formed in many communities to pool 100ha permits and create larger concession areas that might be more commercially attractive to prospective logging contractors. Pooling may also have allowed for a sharing of permit costs, although in some cases it was obvious that only those community members with access to cash or credit could feasibly apply for HPHH permits. Credit was sometimes provided in the form of *dana pinjaman*, where permit costs were borrowed from the company against future log production and community payments by timber investors so that members could participate in the permit application process.

With brokers assuming the responsibility for permits in Malinau, the communities had less understanding with regard to the deals they were making: the size of the proposed concession areas, the value of the timber, where the timber was to be shipped or even the validity of the land claim, in some cases (location of proposed concessions). This is one factor that may have led to a higher incidence of community–community conflict over forest claims and borders in Malinau than with Kutai Barat. In Malinau, all IPPK maps and proposed concession areas were supplied to the broker by the Forestry Office, which indicates how little control the communities had over the location and size of IPPKs. By comparison, there were some cases in Kutai Barat in which the company negotiated for a smaller area than indicated by the total size of the permits obtained previously by the community. While all agreements surveyed in Malinau were notarized contracts, only around 30 per cent of those surveyed in Kutai Barat were notarized, with a further 60 per cent in a written form and the remaining 10 per cent orally made. This is perhaps surprising given the relative weakness of Malinau agreements compared with those made in Kutai Barat.[22]

While there appeared to be greater transparency in Kutai Barat due to the inclusion of the fee structure in negotiated agreements, there were still problems of rent-seeking within communities, as in Malinau. One problem with our survey approach is that we could never observe everything occurring in the villages and in the deals being made with brokers. This may have affected the levels of actual benefits that were declared to the survey team, which is one reason why this chapter focuses on the determinants of negotiated benefits. Another problem with a systematic survey approach is that we were not always able to discuss the negotiations with people who were directly involved in the process. Moreover, there were always factors and pieces of information that did not always fit comfortably into our survey format, which sometimes makes a direct comparison very difficult. A great deal of 'extra survey' information collected by the team in Malinau was included in this chapter.

Conclusions

Research undertaken in Malinau on negotiations between communities and brokers has shown variation in negotiated and actual outcomes. It is clear that there was less variation in negotiated outcomes, at least compared with other study sites, such as the district of Kutai Barat. The main reason for this was the dominance, and cartel-like price-fixing behaviour, of the brokers in deal-making. They formed an essential component of the powerful local timber networks, one which also involved local government officials, the *pewaris*, and now, many community leaders as well. In the absence of formal law enforcement, these networks are self-reinforcing and necessary for the conduct of business, including timber deals. As noted, all agreements surveyed in Malinau were notarized contracts; but without community self-enforcement (requiring strong community networks for effective collective action), they were simply worthless pieces of paper.

These networks were also conduits for illegal payments; hence, in a survey such as this, it was impossible to determine their importance in decision-making regarding IPPK. Therefore, it was very difficult to judge how much control and influence most ordinary community members had over negotiations in their own communities.

Yet, with the exception of brokers, there were some community characteristics that seemed to shape the eventual agreements made in Malinau. First, community participation in village organizations and discussions at the community level on access to timber in forests claimed by communities were higher in communities with better deals. Moreover, these communities also had greater experience in working with, and hence dealing with, outsiders, and greater mobility. Incomes and forest-product dependence did not appear to have important impacts on negotiated outcomes.

A comparative analysis with the processes in Kutai Barat and their outcomes show that there are considerable differences between community experiences in both districts. The next question to be asked is why? Even when factoring in the greater use of brokers in Malinau,[23] the question still remains: why did communities in Kutai Barat appear to benefit more from small concessions than those in Malinau? Ignoring the obvious short-term material benefits and the much less understood environmental costs from logging, this question may be better framed in terms of control and community experiences.

Most of the communities surveyed in Kutai Barat – from those situated near the district capital, Melak, to those hundreds of kilometres away in Long Bagun – have all had more intensive exposure to logging networks in the area. From 1969 onwards, many communities on the Mahakam River have had some experience of manual logging (*banjir kap*) and many of the first HPH operations in East Kalimantan were situated on the Mahakam. By comparison, people have had less intensive exposure to logging in Malinau, at least until relatively recently. On average, each surveyed community in Malinau has been exposed to commercial logging activities for almost 15 years, while those in Kutai Barat have experienced logging for around approximately ten years more. Moreover, slightly fewer people in Malinau (58 per cent of households) have had experience of working in the logging industry than in Kutai Barat (64 per cent of households). More importantly, perhaps, people in Malinau tend to work for logging firms in Malaysia and have fewer contacts with the Indonesian logging scene. By contrast, people in Kutai Barat have only worked in Indonesian firms and therefore have an established network of contacts in the business, which may give them some leverage in negotiations.

On the whole, Malinau is much more isolated than the communities on the Mahakam River. For many years, there has been a great deal of trade up and down the Mahakam, much more exposure to coastal markets and much more pronounced influence from the regional capital, Samarinda. Samarinda, the 'capital of logging' in Kalimantan and the provincial capital of East Kalimantan, contains a huge timber industry. Despite being hundreds of kilometres away from some of the communities participating in HPHH agreements, the Mahakam River provides a direct channel to contacts in the industry. Only during the last few years has Malinau opened up to other parts of East Kalimantan and beyond, including Malaysia. Malinau has no comparable timber processing industry, with most contacts coming via brokers and the smuggling routes into Sabah.

People in the larger communities surveyed in Kutai Barat tend to have a slightly higher level of education and greater expectations and awareness than their counterparts in Malinau. More people in the former have university degrees, and while they may not return to their villages, they are well positioned in the towns to network with the logging industry and obtain information (e.g. on timber prices), which may be channelled to family members residing back in the villages.

While there were relatively few roads in either district, there were more markets and infrastructure in sub-district capitals in Kutai Barat. In addition, while poverty is a fact of life for people living in both districts, many of the communities in Kutai Barat appeared to be slightly better off than those in Malinau. This may be a contributing factor to the relative levels of expectations, along with a higher level of education and direct access and exposure to the city of Samarinda and beyond. All of this, together with the more prominent role of communities in leading negotiations, seems to have resulted in a high degree of control over the negotiation process compared with the experiences of communities surveyed in Malinau.

Notes

1 The research for this chapter was funded by the Robert-Bosch Foundation in affiliation with CIFOR, Bogor, Indonesia. The author thanks Rini Kusumawati and Dodi Hernawan for field assistance in Malinau.

2 Baland and Platteau (1996) and Agrawal (2001) provide detailed reviews of the determinants of successful collective action in the context of community management of natural resources.

3 Research was also undertaken in Bulungan and Kutai Barat.

4 There were also a number of communities in the north of Malinau District who had made agreements in 2000 and had also received advance payments, but where the company did not actually undertake any logging. These were not included in the survey. Most of the communities who had participated in active IPPK deals in Malinau were included in this sample.

5 See Obidzinski and Palmer (2002), and Palmer and Obidzinski (2002).

6 This was almost always larger than the area on the IPPK permits, which could only be utilized for one year, but could be easily extended. It seems that some communities wanted a long-term arrangement in order to ensure a constant future stream of benefits, as has also been observed in Bulungan (Palmer, 2004).

7 IPPKs are intended to be 'forest clearance' permits, so according to IPPK regulations, companies are required to establish plantations once the forest has been cleared. A single IPPK permit had a size of 100ha and typically a single community (or at least its broker) would apply for a number of these permits, giving a much larger logging area in sum total. Many of the actual agreements between communities and companies gave areas substantially larger than that listed on the permit, and permits would simply be extended every year to accommodate further operations.

8 Usually, a partner village on the shared agreement (should this be the case), or the district capital, Malinau Kota (see Table 6.3 for details).

9 Other brokers did, however, enter villages in north Malinau once agreements had been concluded. For example, PT Glory Sejahtera Mandiri obtained IPPK permits

for 2000ha and made agreements with the Sembuak River communities in late 2001, a year after the agreement was made with CV Putra Surip Wijaya.

10 This may have been a cost-minimization strategy. In total, BBP held permits for at least 6000ha and made separate agreements with the villages of Luba, Selidung, Putat, Kelapis and with the *adat* head of Selidung, all on the Semendurut River.

11 Land claimed by *pewaris* was called *tanah warisan*, although it was part of a different community's *hutan adat* (customary forest) claim. In many cases, the *pewaris* was more powerful than the village head in a single community. The *pewaris* sometimes represent particular ethnic groupings across different villages and/or have family members resident in these villages, thus having land/resource claims that transcend village boundaries. There has been a recent evolution from a cave heir (*pewaris gua*) to an heir to the land in which the cave is situated (*pewaris tanah*).

12 In this case, it seems that the *pewaris* was seeking signatures from members of the community with the weakest claim on the local forest just in order to obtain the IPPK permit from the district government. No one in Long Sulit ever received any fee money from the *pewaris*.

13 Brokers may have agreed to pay more in communities where expectations were very low due to the desire to develop a profile of being concerned with community welfare.

14 This trade-off between direct financial benefits and social developments has also been observed in the district of Bulungan (Palmer, 2004).

15 This pattern is even more pronounced in the neighbouring district of Bulungan, also surveyed during the course of this fieldwork. For example, in other Dayak Berusu communities located very near to these Bengalun River communities in Malinau, a fee of 20,000 rupiah per cubic metre was recorded in every single case.

16 It is very likely that additional payments were made in secret to village leaders. In Bulungan District, for example, it was common for village leaders to receive separate payments of 5000 rupiah per cubic metre or more (Palmer, 2004, 2006).

17 For forest land that had been previously allocated on the Semendurut River by the district government and PADUS to individuals such as the *adat* head of Selidung who had been co-opted into the local patron–client logging networks. It seems likely that these deals were more cost effective than those made with the communities and, hence, were prioritized by BBP.

18 On the other hand, it is possible that while the village leaders are aware of this, they neglected to inform their people because they receive a 'special fee' in exchange for dropping social developments from the contract. Of course, like all rent-seeking behaviour, this cannot be directly observed in the field.

19 Note that this discussion focuses on per-contract net benefits and is not concerned with distributional issues per se.

20 The one for Kutai Barat was called HPHH (*Hak Pemungutan Hasil Hutan*). On paper, the only real difference between the IPPK and HPHH systems was that permits for the former could be given out directly by the district head. Also, IPPK rules, unlike those for HPHH, explicitly require that plantations be developed in concession areas. For HPHH permits, approval from the provincial government was required first; and the rules stated that the concessionaire pay a reforestation tax per cubic metre of timber production to the community.

21 This situation only arose when the broker for these villages (CV Putra Surip Wijaya) refused to pay the communities, leading to a confiscation of the contractor's equipment. In exchange for having its equipment returned, the contractor was forced to make a deal with the communities, including back-payment for all non-paid fees and an increase of the fee to 50,000 rupiah per cubic metre.

22 From 21 first agreements surveyed in Kutai Barat, the average fee per cubic metre was just over 80,000 rupiah per cubic metre, ranging from 30,000 rupiah to 150,000 rupiah per cubic metre. The lower range fees tended to be for communities more dependent upon middlemen or brokers, which compares favourably with the average for first agreements in Malinau: 27,000 rupiah per cubic metre. To make a fairer comparison, however, one would need to factor in social developments, which were only a feature of around half of the agreements in Kutai Barat (and those with fees at the lower end of the scale as well), compared with almost all first agreements in Malinau. The only exception was Tanjung Nanga, which negotiated for the highest fee (50,000 rupiah per cubic metre) and for no social developments by the IPPK company, a type of contract with more in common with agreements made in Kutai Barat than in Malinau. Fees received in Kutai Barat, however, tended to include other costs that were mainly borne by the broker in Malinau, such as the costs of the permit application and even personal fees for members of village elites. While Malinau does have the latter, they tended to be outside of the formal fee received by communities. However, even if all other costs are factored in, including social developments, average fees in Kutai Barat are still higher and more transparent than those negotiated in Malinau (Palmer, 2006).

23 Most timber from Malinau was exported to Sabah in Malaysia, where a higher price on average can be obtained compared with the destination for most timber from Kutai Barat, the provincial capital Samarinda, or other parts of Indonesia.

References

Agrawal, A. (2001) 'Common property institutions and sustainable governance of resources', *World Development*, vol 29, no 10, pp1649–1672

Baland, J. M. and Platteau, J.-P. (1996) *Halting Degradation of Natural Resources: Is There a Role for Rural Communities?*, Clarendon Press, Oxford

Engel, S. and Palmer, C. (2006) 'Who owns the right? The determinants of community benefits from logging in Indonesia', *Forest Policy and Economics*, vol 8, no 4, pp434–446

Obidzinski, K. and Palmer, C. (2002) *How Much Do You Wanna Buy? A Methodology for Estimating the Level of Illegal Logging in Indonesia*, Research Report, CIFOR, Bogor, Indonesia, Presented at the 13th Annual Meeting of the European Association of Environmental and Resource Economists (EAERE) in Budapest, 25–28 June 2004

Palmer, C. (2004) *The Role of Collective Action in Determining the Benefits from IPPK Logging Concessions: A Case Study from Sekatak, East Kalimantan*, CIFOR working paper 28, CIFOR, Bogor, Indonesia

Palmer, C. (2006) *The Outcomes and Their Determinants from Community–Company Contracting over Forest Use in Post-Decentralization Indonesia*, Peter Lang, Frankfurt, Germany

Palmer, C. and Obidzinski, K. (2002) *Case Study Prepared for the Report: 'Higher International Standards or Rent-Seeking Race to the Bottom? The Impacts of Forest Trade Liberalization of Forest Governance'*, Food and Agriculture Organization (FAO) and the International Institute for Environment and Development (IIED) Project GCP/INT/775/JPN, Mimeo, Centre for Development Research, Bonn, Germany

Pearce, D. W. and Warford, J. (1993) *World without End: Economics, Environment and Sustainable Development*, Oxford University Press, New York, NY

7

The Effect of IPPKs on Household Livelihoods in Malinau District, East Kalimantan

Kewin Kamelarczyk and Uffe Strandby

Introduction

What is the role of forests for rural livelihoods? Do they have the potential to enhance livelihoods and mitigate poverty? Some researchers are doubtful (see Wollenberg et al, 2001; Wunder, 2001; Sunderlin et al, 2003; Levang et al, 2005). However, a large number of people rely, to some extent, on the forest for subsistence or cash income (Byron and Arnold, 1999; Angelsen and Wunder, 2003; Levang et al, 2005). Forest dependency varies widely among and within communities, and the role of forest products is usually aimed at filling gaps and complementing other incomes than being the primary income source (Byron and Arnold, 1999; Angelsen and Wunder, 2003). Several researchers state that the only way to use the forest to cushion poverty and to improve livelihoods is to convert forest capital into more profitable alternatives (Wunder, 2001; Angelsen and Wunder, 2003; Levang et al, 2005). The transferring of forest capital into more industrialized alternatives will therefore inevitably help poor forest-dwelling people. Possibly, in order to benefit rural livelihoods, a more commercialized and intensive use of forest resources needs to be accompanied by secure forest access and ownership, the removal of regulatory political barriers, the opening of market access and the strengthening of facilitating and supporting organizations, etc. (Lindsay, 1998; Balland and Platteau, 1999; Byron and Arnold, 1999; Steins and Edwards, 1999; Colfer and Wadley, 2001; Campbell et al, 2002; Angelsen and Wunder, 2003; Scherr et al, 2003). However, even if this is achieved, the poorest of the poor with low opportunity costs of labour and capital might not benefit from such development schemes and might continue relying on the possibilities of generating some income from the forest resource as well as harvesting subsistence products (Lindsay, 1998; Byron and Arnold, 1999; Wollenberg et al, 2001; Angelsen and Wunder, 2003).

As described in previous chapters, the decentralization of administrative authority gave districts the opportunity and incentive to issue small-scale forest licences known as *Izin Pemungutan dan Pemanfaatan Kayu* (IPPKs). Although the Ministry of Forestry made IPPKs illegal at the end of 2000, districts have continued issuing hundreds of IPPKs (Barr et al, 2001; FWI/GFW, 2002) and indirectly supporting illegal logging.

The size of each IPPK ranged from 100 to 3000ha (Barr et al, 2001). The aggregated area influenced by IPPK is not large; but considering the short period of operation, the process has been aggressive in terms of the extent and rate of forests affected (Barr et al, 2001).

The permits not only provide royalties to the district government, but also endow the communities with a range of benefits, such as volume-based fees, in-kind contributions and employment (Barr et al, 2001; FWI/GFW, 2002). Many communities have found IPPKs attractive because they expected to enjoy significant direct benefits from forests for the first time. However, in their eagerness to capture benefits, their lack of information and inability to control the logging companies, they have often been vulnerable to companies' abuse of contractual agreements (Warner, 2000; Barr et al, 2001; Colfer and Wadley, 2001; Sellato, 2001; Resosudarmo and Dermawan, 2002; see also Chapter 6 in this volume). In addition, the lack of transparency, unequal distribution of benefits, limited and badly paid employment opportunities, conflict, and unsustainable forest activities associated with IPPK operations suggest that they have not been a successful model of forest management (Warner, 2000; Colfer and Wadley, 2001; Sellato, 2001; Anau et al, 2002; Casson, 2002; Resosudarmo and Dermawan, 2002; see also Chapters 8 and 10 in this volume). Furthermore, the elite of each village have become clients of the IPPK holders and administer their timber resources with more focus on economics than on environment and conservation (Edmunds and Wollenberg, 2003; Levang et al, 2005).

This chapter examines the extent to which local people in three villages in Malinau District have been influenced by the local government's tendency to issue IPPKs. The previous chapters reported on the benefits at the village level; this chapter reports on the distribution of benefits to households in villages.

This chapter is structured as follows. First, three research sites are described with respect to history, ethnicity and aspects of IPPK (i.e. negotiation, implementation, development and extent). This is followed by a summary of our research methods. We then present our findings and discuss their implications. Since the study was built on a modified version of the sustainable rural livelihood framework (the SRL framework) of the UK Department for International Development (DFID), the presentation of the findings and the discussion are structured around four assets: financial capital, social capital, human capital and natural capital. The implications of IPPK on physical capital are not covered separately, but will be discussed together with each of the four mentioned assets.

Study locations

The three communities in which the field study was carried out are situated in the Malinau River Basin and were selected to represent contrasting conditions regarding ethnic affiliations, number of households, geographical remoteness and time of IPPK implementation. These factors were assumed to influence the impacts of IPPK on livelihood conditions. The villages Tanjung Nanga, Adiu and Sengayan were selected.

Tanjung Nanga

With more than 450 residents and 135 households, Tanjung Nanga is by far the largest of the three villages. Except for a few Punan families who have moved to the village from upstream settlements during the last decade, the village population mainly belongs to the ethnic group Kenyah. The village was established in 1974 when the community moved from the village Long Pua' at the Pua River (Sellato, 2001; Made Sudana, pers comm, June 2003) to its current location along the Malinau River, which is around a ten-hour boat trip upstream from Malinau (or three hours by car).

The social characteristic of the Kenyah people, with their highly influential aristocrats and relatively weak commoners, are pronounced in Tanjung Nanga. *Kepala Desa* (village head) *Kepala Adat* (customary leader) and their closest relatives tend to hold the most powerful social positions and also many of the favourable positions for gaining the most lucrative ventures. In contrast to the commoners, the aristocrats are well educated and informed. Even though the strong social hierarchy might appear discriminatory, Tanjung Nanga appears to be functioning very well institutionally.

Although the Kenyah have a reputation for being the foremost agriculturalists in the area, this is not reflected in the income-generating activities in Tanjung Nanga. The dominant activities in this sense are the selling of forest products and off-farm labour work, each activity involving around two-fifths of the respondents. Trading of agricultural products involved around one quarter of the respondents, as did activities such as handicrafts sales, local government employment, entrepreneurship and employment in Malaysia.

Adiu

Among the three villages, Adiu is located furthest downstream on the Malinau River, four to five hours upstream by boat from Malinau. The river is still the main way of transportation, although the village – as part of the IPPK operations – has been connected to the main logging road heading south from Malinau. With a total of only 223 residents, Adiu is the smallest of the three villages. It embraces two physically separated sub-villages: Long Adiu with 28 households, inhabited by the Merap ethnic group, and Punan Adiu, with only 18 households, inhabited by Punan. The Merap arrived at Long Adiu in 1920 after moving from further upstream areas at the secondary River Ran. In 1962, the Punan arrived and were granted a part of the forest area claimed by the Merap (Kaskija, 2002; Made Sudana, pers comm, June 2003). The relatively close distance of Adiu to Malinau Kota seems to influence the income-generating activities chosen by the villagers: four-fifths are engaged in trading agricultural products and around half sell forest products, one third are in timber harvesting and one fifth are in local government employment.

Sengayan

The Merap village Sengayan comprises 272 residents (around 60 households) and was established in the early 1980s after having moved from the upper Sengayan River (Kaskija, 2002). It makes up part of the larger settlement Long Loreh, which contains three other villages – Pelancau (Punan), Bila Bekayuk (Punan) and Loreh (Kenyah) – and has a total population of approximately 1200. Long Loreh is located along the Malinau River and at the end of the main logging road, a two-hour drive south of Malinau. It has become the main village in the area due to heavy coal mine operations taking place in the vicinity, which has created more traffic and trade. Apart from those in Malinau, the area's only secondary school and permanent health clinic are situated in Long Loreh. The coal mine plays a significant role in the area since it maintains the main road, contributing regular wages for a large proportion of the households and opportunities for additional income. Many employees originating from Sulawesi and Java have now settled down and married local villagers in Sengayan and Long Loreh, in general, which seems to separate Sengayan socially. Furthermore, the intrusion of outsiders and the strong connection to the 'outside world' have influenced the villagers towards a more modern lifestyle and behaviour. Roughly half of the villagers are connected to selling agricultural products and around two-fifths are engaged in trading forest products.

Historical experiences with logging operations

The introduction of IPPK was generally the first direct experience with concessionaires' logging and business practices. In the case of Tanjung Nanga, the negotiations concerning IPPK were initiated late in the history of IPPK, which allowed the village to learn from other villages' experiences. Another matter placing the village in a favourable negotiation position was that it had already learned from earlier experiences gained during logging concession – *Hak Pengusahaan Hutan* (HPH) – operations.[1]

The HPH operations, which took place at Adiu, neither benefited the village directly nor provided it with any useful experiences with regard to the IPPK process. In addition, when IPPK was introduced in the district, Adiu was among the first communities to be involved. This lack of experience in contract negotia-

Table 7.1 *Data at village level concerning IPPK activities in the forest area claimed by Tanjung Nanga*

Duration of IPPK activities[a,b]	Permit issued in January 2001. Logging operations began in August 2001 and were still active in August 2003.
Permit holder[b,c]	CV Jaya Nanga Bersama is the name of the IPPK co-operative. There are two IPPK permit holders in Tanjung Nanga, but they share the same permit and are owned by the same investor. The companies that hold the permit are Meranti Wana Lestari and Koperasi Serba Usaha Meranti, both owned by Ibu Rosalie. The actual logging operations are presumably carried out by a larger logging entrepreneur (Godwin Limberg, pers comm, June 2003).
Size of area in contract[a]	20 blocks (2000ha) in total: 11 blocks in Tanjung Nanga area and 9 in the area of the neighbouring village, Langap.
Area logged[a]	11 blocks (1000ha) in Tanjung Nanga area and 9 blocks (900 ha) in Langap area.
Location l	Western side of Temalang River.
Fee per cubic metre[a,d]	US$6.25 during the first three payments. Due to land conflicts, the neighbouring village of Langap claimed a share of the fee. Subsequently, the fee of Tanjung Nanga has been US$4 (for timber extracted from the blocks of Tanjung Nanga) and US$3.13 (from timber extracted from the blocks of Langap).
Production[a]	According to the contract, 100,000 cubic metres, but only 80,000 cubic metres actually extracted: 60,000 cubic metres extracted in Tanjung Nanga area and 20,000 cubic metres in the Langap area.
Total fee and number of payments[a]	US$225,000 in six payments, but more will follow.
Other benefits received[a]	Gas for generator (200 litres per month), transportation to the field, maintenance of road and assistance for ceremonies (e.g. for Christmas).
Common investments[a]	Student accommodation in Malinau (US$5000 in total), church (US$1250 of each payment), supporting village organizations (art, sport, security, school equipment, women's groups; total of US$1.25 per cubic metre from the two first payments).
Extraction per ha[e]	40 cubic metres

Notes:

a Based on focus group interview in Tanjung Nanga.

b Based on data obtained from Pemerintah Daerah Kabupaten Malinau (Malinau District Government Office).

c CIFOR, Long Loreh staff, pers comm, June 2003.

d Langap received US$3 per cubic metre extracted in Tanjung Nanga area and US$3.38 per cubic metre extracted in their own area.

e Calculation based on the estimated total production and size of area logged.

tion placed the village in an underprivileged position, which was partly reflected in the relatively small amount of fee received per cubic metre compared to other villages. The two sub-villages in Adiu were initially cooperating in the negotiation of the IPPK contract and the IPPK permit covered both of their respective claimed forest areas. However, it turned out that logging operations never reached the more remote forest area claimed by the Punan, which later caused several conflicts.

In the case of Sengayan, no experiences had been gained from former logging operations. Nevertheless, the coalmine and the dynamic structure and ethnic diversity of the village caused by the development of Long Loreh have caused households to become accustomed to external influences and monetary benefits.

Table 7.2 *Data at the village level on IPPK activities in the forest areas claimed by Long Adiu and Punan Adiu*

	Long Adiu	*Punan Adiu*
Duration of IPPK activities[a,c]	Permit issued in June 2000. The logging operations began in December 2000 and ended in October 2002.	Permit issued in June 2000. Logging never begun.
Permit holder[a,b]	Two IPPK permits issued to CV Wana Bakti and to CV Kelawing Jaya. Only the former was active. The logging operations were carried out by Pro Desa and, later, PT Trian Jaya Pratama.	Two IPPK permits issued to CV Wana Bakti and CV Kelawing Jaya.
Size of area in contract[a]	7 blocks (700ha).	8 blocks (800ha).
Area logged[a]	7 blocks (700ha).	No area logged.
Location[a]	Along the new road, around the village and on the western side of Malinau River south of Adiu.	Do not know.
Fee per cubic metre[a]	US$2.5.	US$2.5.
Production[a,d]	43,678 cubic metres.	No production.
Total fee and no. of payments[a,d]	US$109,250 in four payments. Punan Adiu received a share of US$29,375 of this amount.	US$29,375 in three payments.
Other benefits received[a]	Village office, health clinic, area levelling and 5km of roads.	Area levelling and 5km of roads.
Common investments[a]	Village saving (US$0.125 per cubic metre or 5% of received fee), generator, electricity cables, rice husk, student accommodation in Malinau, village hall, church, transportation of timber and consumption during community activities.	5 cubic metres of sawn timber and 5kg of nails for each household, with total costs of US$3750; US$2500 were saved.
Extraction per ha[e]	61 cubic metres.	No production.

Notes: a Based on focus group interviews in Long Adiu and Punan Adiu, and informal interviews with Kepala Desa in Long Adiu.
b Based on data obtained from Pemerintah Daerah Kabupaten Malinau (Malinau District Government Office).
c According to Barr et al (2001), the logging operations began in October 2000.
d The calculations of production and total fee are based on the production receipts received by *Kepala Desa* in Long Adiu from the logging operator.
e Calculation based on the estimated total production and size of area logged.

Elaborate information on IPPK activities at village level is presented in Tables 7.1 to 7.3.

Research methods

The data collection was carried out during July and August 2003, and included both qualitative and quantitative methods. For each of the three villages, semi-open questionnaires, including preference ranking, were applied at the household level. Additionally, group and informal interviews were conducted with key

Table 7.3 *Data at village level concerning IPPK activities in the forest area claimed by Sengayan*

Duration of IPPK activities[a,b]	The permit was issued in November 2000. Logging ended in June 2003, but extraction was still taking place in August 2003.
Permit holder[b]	CV Putra Surip Wijaya.
Size of area in contract[a]	20 blocks (2000ha).
Area logged[a]	13 blocks (1300ha).
Location[c]	Tebu River up to Batu lu'ung River.
Fee per cubic metre[a]	US$3.75.
Production[a,d]	Between 26,666 cubic metres and 45,500 cubic metres.
Total fee and number of payments[a,e]	Between US$100,000 and US$170,625 in 12 payments.
Other benefits received[a]	Village hall, rice husk, timber, 5000 sheets of roofing; timber and sheets were distributed to the households.
Common investments[a]	None.
Extraction per ha[f]	Between 20.5 cubic metres and 35 cubic metres.

Notes: a Based on focused group interviews in Sengayan.
b Based on data obtained from Pemerintah Daerah Kabupaten Malinau (Malinau District Government Office).
c CIFOR, Long Loreh staff, pers comm, June 2003.
d The lowest volume estimate is based on a focused group interview where it was informed that the total fee received was US$100,000 and the fee per cubic metre was US$3.75. The highest estimated is based on another statement from the same focused group interview that approximately 35 cubic metres were extracted from each of the 1300ha, which is the maximum extraction level allowed according to the IPPK permit.
e The total fee of approximately US$100,000 is an estimate based on focused group interview. Another estimate on approximately US$125,000 is made by CIFOR staff. A third estimate can be made by assuming that 45,500 cubic metres have been logged and the fee per cubic metre is US$3.75, which gives a total fee of US$170,625.
f Calculation based on estimates for total production and size of area logged.

informants. Wealth ranking was utilized prior to household selection, and households were selected using a stratified sampling method. Wealth was used as the sampling criterion since it was identified as having causal links to forest use, involvement in community decisions, and use of and access to the IPPK fee. A sampling intensity of approximately 25 per cent was the objective; in the end, it resulted in 23 per cent, 33 per cent and 25 per cent in Tanjung Nanga, Adiu and Sengayan, respectively.

Importance of the IPPK fee to the household economy

What was the importance of income from IPPK compared to other income-generating activities? Respondents were asked to list all of their income sources and compare these pair-wise with regard to the size of income generated during a whole year. Not surprisingly, the analysis revealed that the annual income received from IPPK fee in Tanjung Nanga and Adiu was considered superior to other sources of income. However, in Sengayan, household income from farming activities was held to be superior (see Figure 7.1). When all three villages are considered as one population, the IPPK fee was preferred in approximately one third of the pair-wise rankings.

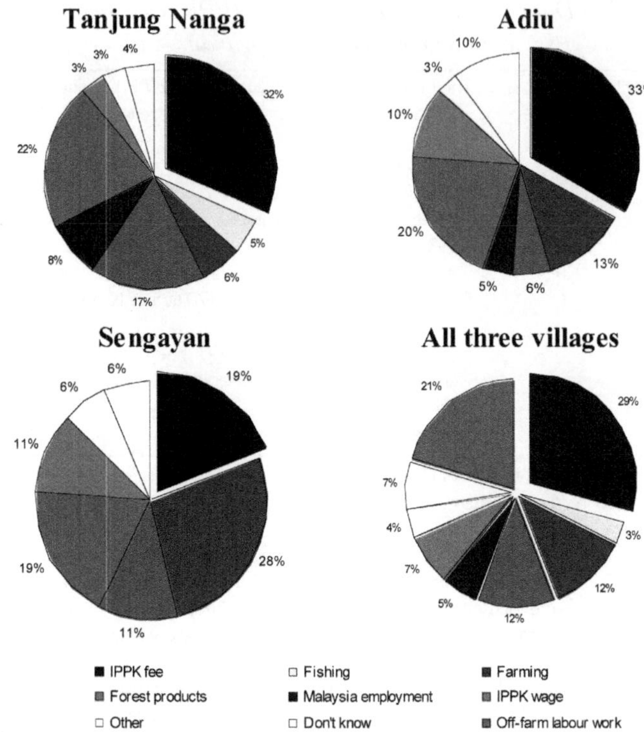

Figure 7.1 *Percentage allocation of pair-wise ranking of income-generating activities: The percentage indicates how often an income activity was considered highest relative to the total number of comparisons/pair-wise rankings*

In order to get a truthful picture of the significance of the IPPK fee, availability and size of households' income activities were investigated (see Figure 7.2). Looking at the total study population, IPPK was part of the 52 per cent of the pair-wise rankings and can therefore be considered the most available income source during the time of IPPK operations. In 56 per cent of these instances, it was perceived as superior to other sources of household income. Only income from employment in Malaysia was considered larger; but contrary to the IPPK fee, this was only an option for a few households. Income from off-farm labour work, sale of forest products, IPPK wage and sale of farm products were all considered less significant in terms of annual income and were also less available income options.

In the same context, the respondents were asked if they would have had access to another income source that could replace the IPPK fee. In Tanjung Nanga and Adiu, a significant majority of households stated that they would *not* have had access to such a substitute, while in Sengayan, most of the respondents claimed the opposite. The income sources mentioned by respondents in Adiu and Sengayan as potential substitutions for IPPK fees are similar in being stable and continuous. In Adiu, employment as teachers, with the IPPK contractors or in

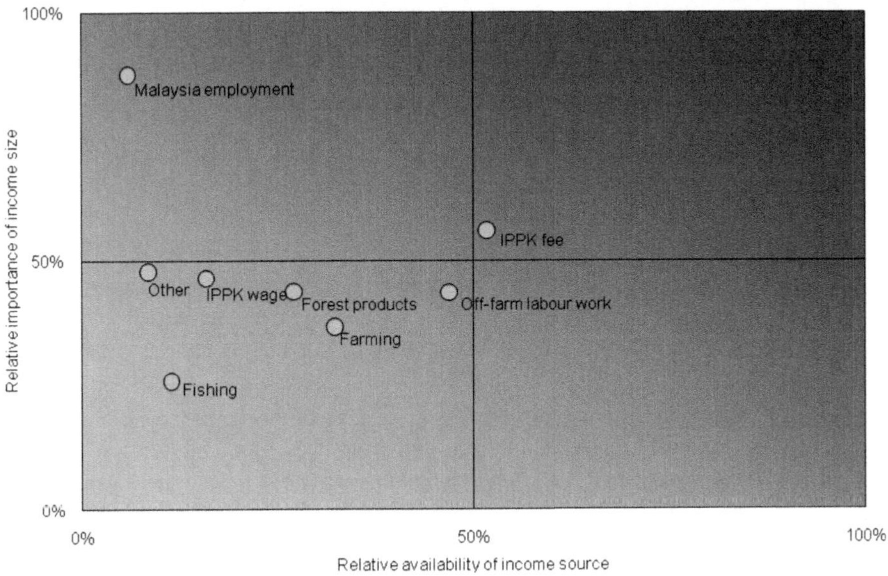

Figure 7.2 *Frequency of income activities represented in pair-wise rankings (relative availability) and the relative importance of income size: All three villages are considered as one population*

Malaysia was mentioned, while in Sengayan, the most important source of income was employment at the IPPK and coal mine concessionaires. In Tanjung Nanga, where the access to concessionaires is poor, government salary, income from handicrafts and *gaharu* (eaglewood) collection were considered potential substitutes for the IPPK fee. Incomes from farming and collection of forest products were only considered potential substitutions in a few instances.

Households in all study sites generally perceived their financial situation to be better to some extent during the period with the IPPK fee than before its introduction, although households in Sengayan felt a little less influenced (see Box 7.1). The fact that the IPPK fee was not a stable income was perceived as the main

Box 7.1 Respondents expressing their perception of the IPPK fee

- 'Before IPPK came to this village, I only earned little money. I have received a lot of money from the IPPK fee. Prior to IPPK, I only earned money from selling peanuts, which only supported my household with sugar and food. Now I can buy a TV and chainsaw'(respondent in Adiu).
- 'During the hunting season I earn more money selling pigs than from receiving a fee' (respondent in Adiu).
- 'There is nothing above the fee!' (respondent in Tanjung Nanga).
- 'There is only one cash income: IPPK fee!' (respondent in Tanjung Nanga).

Box 7.2 Examples of six respondents in Sengayan expressing the significance of the IPPK fee on the household's financial situation and the importance of income stability

- 'The IPPK fee is not so important because it is irregular, whereas income from my shop is regular.'
- 'The IPPK fee is not so attractive because it is not a stable income.'
- 'Farming income is more important than the IPPK fee because it is more regular/stable, although the IPPK fee has been higher than farming income. I have earned much more due to IPPK.'
- 'The salary from the IPPK company is more important than the IPPK fee because it is regular.'
- 'The IPPK fee is only temporary and therefore not as important as regular income.'
- 'Before IPPK, I earned money by selling farm products, but with IPPK I could get millions. Before IPPK I had no engine, now I have bought a long tail engine which might give me the chance to earn more money.'

disadvantage, especially in Sengayan (see Box 7.2). In Adiu and Tanjung Nanga, most households held the fee responsible for the enhancement of their financial situation, while some added the additional benefits received from being employed by the IPPK operator and the sale of agricultural and forest products to IPPK employees.

Households' allocation of fee disbursements

But how did the households use the IPPK fee received? Did they think in terms of long-term investment or in immediate and short-term needs? The answers were structured into 15 different groups of commodities and services, with no attention paid to the purchased quantities (for example, if a household had purchased more than one electronic device, it only counted as one disbursement within that group). In order to analyse the data in a broader perspective, the commodities and services were subsequently grouped into five main categories or investment strategies – 'maintenance of human capital', 'luxury goods', 'providing opportunity for additional income', 'savings' and 'other' (see Figure 7.3). It is important to emphasize that the interest has only been on the types of disbursement. Due to time limitations and respondents' lack of recall, the expenses attached to each disbursement were not investigated.

Variations in disbursements of commodities and services

Even though significant differences between the villages were only found in a few cases, there was a tendency for households in Tanjung Nanga to make the less expensive disbursements (see Figure 7.3). For instance, all households spent money on convenience goods such as food and clothes, and contrary to Sengayan and Adiu, some households paid off debts to local shops for buying subsistence

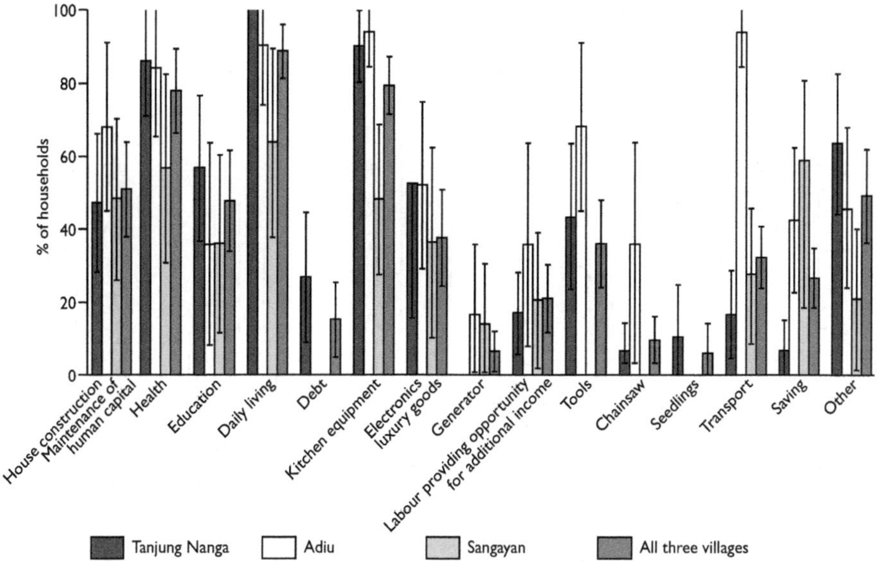

Figure 7.3 *Percentage of households purchasing different commodities and services (percentage is illustrated)*

products. In both Tanjung Nanga and Adiu, basic kitchen equipment was a significantly more common purchase item than in Sengayan. Even though saving does seem attractive in many of the villages, significantly fewer households in Tanjang Nanga used part of their fee on this opportunity (see Box 7.3).

The trend in Adiu seems to be that households have invested their fees in relatively more expensive items such as equipment, labour for house construction (or improvements), electronics (mainly TV, satellite dish, VCR and stereo), chainsaws and tools (mainly axes, wheel barrows and machetes). However, unlike in Tanjung Nanga and Sengayan, almost all households in Adiu had consistently purchased a long-tail boat engine.

Box 7.3 Examples of households' responses when asked about 'saving' from the IPPK fee

- 'Saving?... just a bit' (respondent in Tanjung Nanga).
- 'I saved some for my children' (respondent in Tanjung Nanga).
- 'All savings were spent on going to the hospital' (respondent in Adiu).
- 'From the US$250 saved, US$125 is already spent' (respondent in Adiu).
- 'Saving was meant as a health insurance. Initially I saved US$375; but now I have already spent US$188 on going to the hospital in Kota Malinau' (respondent in Adiu).
- 'All of the IPPK money was used for a saving for my children' (respondent in Sengayan).
- 'I saved money for travelling to Sulawesi' (respondent in Sengayan).

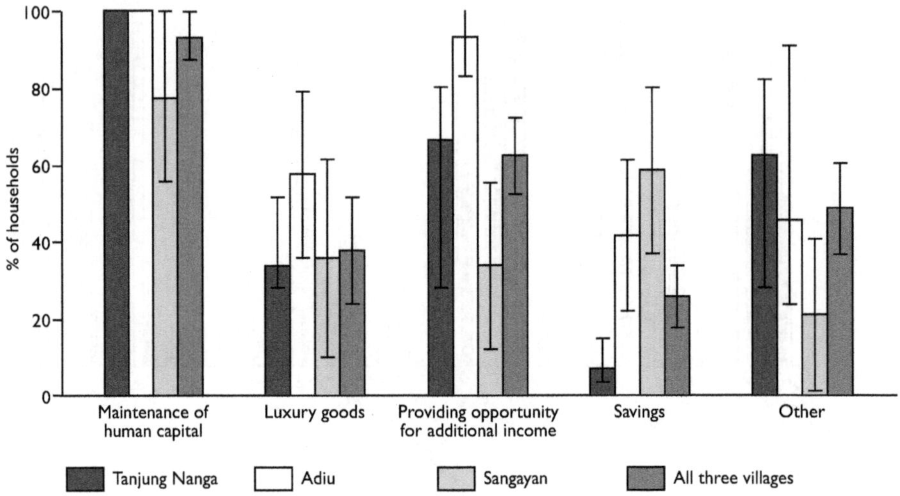

Note: lack of value/error bar indicates that the value/error is zero.

Figure 7.4 *The percentage of households with different investment strategies (disbursements within different categories of commodities and services) by means of the IPPK fee: The percentage is illustrated with an error bar* $\pm$ *2 * SE*

Variations in investment strategies

By grouping the commodities and services into investment categories, it might be possible to uncover the existence of certain trends in investment patterns in the villages. A homogeneity test was applied, and even though no systematic differences appeared between the households' investment strategies (P = 0.99), one result is worth mentioning. Almost all households in Adiu invested part of the fee in commodities that could 'provide opportunities of additional income', which is significantly different from both Tanjung Nanga and Sengayan (see Figure 7.4).

For the total study population, the majority of households spent part of the received fee in 2002 on 'maintaining the household's human capital' by investing in healthcare and education. This is followed by more than three-fifths of the population who invested in 'additional income opportunities'. 'Luxury goods' such as electronic devices and generators are only purchased by approximately two-fifths, while one quarter saved part of the fee; however, the two strategies are not significantly different.

With this knowledge on expenditure trends, it would be interesting to investigate differences in how payments were used and handled. Households generally spent the money from the payments randomly and instantly. Current needs at the time of payment receipt seem to be the main explanatory variable. For some households, subsistence products such as food and medicine have been the main preference at the time of the first payment, while for others, the first choice has been 'luxury goods', such as a bicycle or a television.

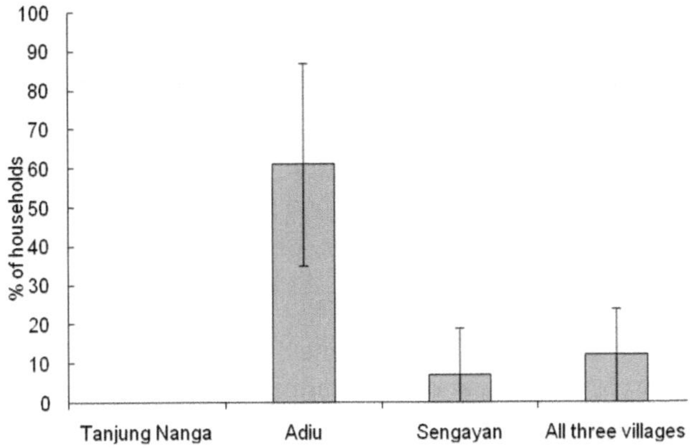

Figure 7.5 *Proportion of households knowing the total amount of fee received by the village: No households in Tanjung Nanga were familiar with the amount of fee received by the village (the average percentage is illustrated with an error bar ± 2 * SE)*

Transparency and participation

To investigate the degree of transparency in the IPPK process, households were asked about their knowledge of the total fee received by the village, familiarity with the names of IPPK operators and with the villagers in charge of fee distribution. More than half of the households in Adiu and less than one tenth in Sengayan were familiar with the total fee amount received by the village (see Figure 7.5). In Tanjung Nanga, none of the households knew about this matter. When we look at households knowing the name of the IPPK operator, a significantly higher share is found in Sengayan (93 per cent) and Adiu (87 per cent) than in Tanjung Nanga (34 per cent) (see Figure 7.6). In terms of familiarity with the villagers who had been in charge of the distribution of the IPPK fee, the

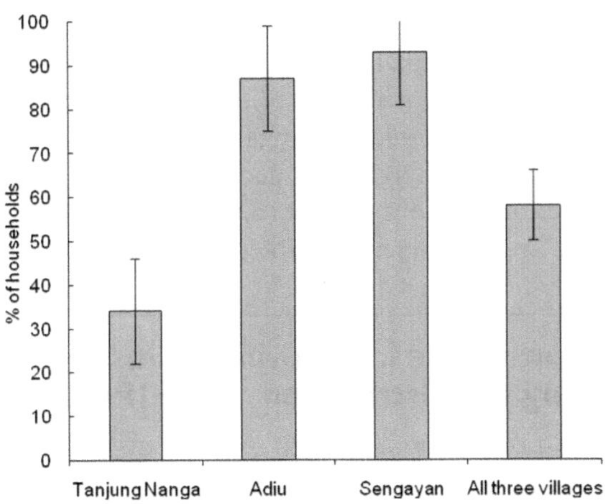

Figure 7.6 *Proportion of households acquainted with the name of the IPPK operator (average percentage is illustrated with an error bar ± 2 * SE)*

Box 7.4 Statements from Adiu and Tanjung Nanga about the possibility of influencing decisions

- 'If the majority of people choose one thing, I just follow that decision.'
- 'I was just listening; only important people speak up.'
- 'I was just listening; I am allowed to speak up, but I don't want to.'
- 'I was only listening; people who are better at speaking do so. '

answers reveal a higher degree of transparency as all households in each of the three villages were well informed about their administrative superiors.

Another indicator of the level of involvement and transparency in the process was the participation of village members in community meetings and actions of collective choice in matters concerning IPPK. Our qualitative data reveal that most of the households were aware of such meetings and were also invited in advance. However, during the community gatherings, it seemed that common villagers were reluctant to express their genuine opinions and left the important decisions to the higher-ranking aristocrats (see Box 7.4). Only two of the nine respondents joining the meetings in Adiu actively expressed their opinions, while in Tanjung Nanga, the proportion was only 4 out of 17 respondents (this aspect was not covered in Sengayan). However, the impression is that many community meetings were held in each of the three villages and many subjects were discussed, which demonstrates some level of participation.

Equitable distribution

For social cohesion within the village, it is essential that appropriators view the allocation of joint benefits as fair (Ostrom, 1990). During focused group interviews in Sengayan, difficulties were experienced in knowing the precise rules for the distribution of IPPK fees. It seemed that rules did exist, but were inconsistent and blurred (see Box 7.5), and were only shared by the *Kepala Desa* and his staff. In Adiu and Tanjung Nanga, the situation was different since rules for distribution of IPPK fees were highly differentiated (see Table 7.4). The administration system seemed to be most advanced in Tanjung Nanga, with a committee consisting of six of the village's aristocrats in charge of all decisions related to IPPK, and sub-committees responsible for different groups (families, widows, youth, students, etc.). However, making any judgements on the fairness of distribution based on

Box 7.5 Comments from a villager in Sengayan regarding the distribution of the IPPK fee

My neighbour went to Kepala Desa *and asked for IPPK fee; he got US$250. Then I went to ask for the fee, but I was told that the fee had not arrived yet. Later on I received US$125.*

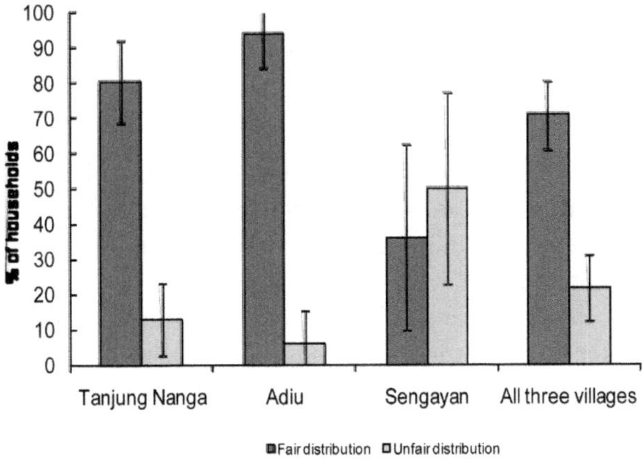

Figure 7.7 *Percentage of households perceiving the distribution of IPPK fee within the village as fair or unfair (average percentage is illustrated with an error bar ± 2 * SE)*

the existence and content of the presented rules would definitely suffer from lack of validity. Instead, the respondents were asked directly about their perception of fee distribution. Once more, Sengayan differs as only approximately one third of the households found the distribution fair, while a majority are satisfied in Tanjung Nanga (80 per cent) and Adiu (94 per cent) (see Figure 7.7). The discontented households in Sengayan were mostly concerned with the lack of community meetings, the considerably lower fee to newcomers and widows, and the relatively higher fee allocated to *Kepala Desa* and *Kepala Adat*. Households in

Table 7.4 *Rules for the distribution of the IPPK fee within the three villages: Allocations are illustrated by percentages*

	Tanjung Nanga[a]	*Adiu*[e]		*Sengayan*[f]
		Long Adiu	*Punan Adiu*	
Adult		100		
Youth	25	35–40	25	25
Students at secondary level and above	10[b]	10	25	
Children (below secondary level)	8 (4)[c]	10		
Household	100		100	100
Widow	50 (25)[d]		33	50
Newcomers	25			
Temporary villagers		33		

Notes: The proportions are based on examples given by the key informants and are listed according to one reference group. In Punan Adiu, the reference is adults, while in the other cases, it is households.

a The rules for distribution in Tanjung Nanga are based on two reports made by the committee for fee distribution.

b In Tanjung Nanga, this category is named scholarship.

c Children living inside the village receive 8 per cent and children living outside the village receive only 4 per cent.

d Widows living alone receive 50 per cent and widows living with relatives receive 25 per cent.

e Rules for distribution in Long Adiu are based on production reports and focused group interviews.

f Rules for distribution in Sengayan are based on focused group interviews.

Adiu and Tanjung Nanga explained their reason for perceiving the distribution as fair: all households received more or less the same amount.

Equal distribution

Except for a few households in Sengayan who state that they have not received any payments, the general situation is that all households have received a share of the IPPK fee. The general explanation for the few excluded households is that they are viewed as 'newcomers' who are not in a position to collect the same amount of benefits as original villagers. They have moved to the village within the last few years and often work in the coal mine or as traders. The entire IPPK fee received by Sengayan was distributed to households and not partially invested in common projects, as in Tanjung Nanga and Adiu (see Tables 7.1 to 7.3).

This might be explained by the fact that many government and private projects had already been realized in Long Loreh and therefore substituted the need for further common projects.

Concerning the number and size of payments received per household, it was revealed that households in Tanjung Nanga, on average, received a significantly lower fee than households in Adiu and Sengayan. Investments in common projects have been relatively highly prioritized by the committee in Tanjung Nanga. Households in Adiu have, on average, received almost three times more than households in Tanjung Nanga (US$1525 compared to US$528). Despite the fact that households in Adiu, on average, received US$500 more than households

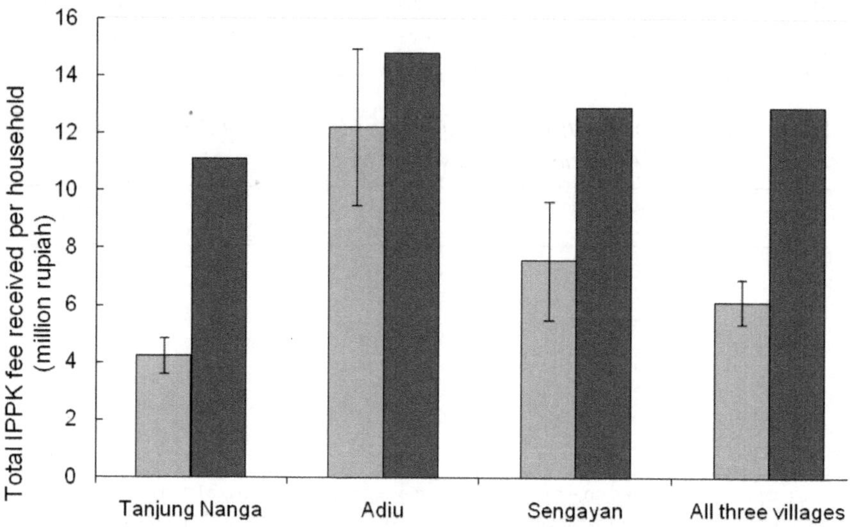

Note: The estimate based on semi-open questionnaires are illustrated with an error bar ± 2 * SE.

Figure 7.8 *Two different estimates of total IPPK fee received per household: One estimate is based on data obtained from semi-open questionnaires at household level (*left light grey column*); another estimate is based on data obtained from focused group interviews and production reports concerning the total number of households receiving the fee and the total fee received at the village level (*right dark grey column*)*

in Sengayan (US$946), the difference was not found to be significant. If considering all three villages as one population, the average total fee received per household was estimated at US$766.The estimates were crosschecked by comparing with calculations based on the amount of total fee received at village level that was distributed to households (see Figure 7.8).

As the household surveys were conducted according to income strata, it was possible to compare the allocation of fee to households in different income groups across the single villages. T-tests revealed that only in Tanjung Nanga were two strata proven to be significantly different (P = 0.04) favouring the better off. Households in the very low-income stratum received only US$243, while households in the medium-income stratum received more than three times as much, at US$785.

Occurrence of conflicts

The number, kind and intensity of conflicts caused by the presence of IPPK could shed light on major obstacles and strengths within social capital. In this respect, conflicts are understood as a negative occurrence with potentially adverse effects on social coherence (Colfer and Wadley, 2001). The proportion of households having experienced conflicts due to IPPK activities did not differ between the three villages. But since more than half of the households in the total study population had come across conflicts due to IPPK, it is important to scrutinize the descriptions and explanations. With the introduction of IPPK in Adiu, some conflicts accompanied the flow of money; not only between Adiu and its neighbouring villages, but also internally. In particular, the relationship between the Merap and the Punan endured many disputes after the introduction of IPPK, in which controversies and misunderstandings caused the former not to share the last payment with the latter. Furthermore, the Punan did not receive the same non-monetary benefits, such as a village office and health clinic. In Tanjung Nanga and Sengayan, the reasons for conflicts were the amount of money received per household; distribution between newcomers and permanent villagers; households with many children compared with households with few; households consisting of one parent compared with households with two parents; elites accused of bribing; and lack of participation and transparency concerning the distribution of the IPPK fee.

Living conditions influenced by IPPK

To allow the respondents to express themselves and thus extract valuable information on changes in the villagers' living conditions due to the presence of IPPK, respondents were asked to compare their current quality of life with the period prior to the advent of IPPK and to elaborate upon causes of change. In spite of the fact that the quantitative ranking was of limited use, since respondents had problems comprehending the applied method, the underlying qualitative explanations turned out to be more useful.

In Adiu, more than half of the respondents perceived their living conditions as having improved due to upgrading of house quality, better tools and better access to medicine. One respondent regarded his household's life situation as harder today due to having more children, while the rest did not feel any difference in this regard. In Tanjung Nanga, two-thirds of the respondents said that their life conditions had become better. Only one claimed his life had become harder, while the rest felt no difference. Here the main causes stated were better health conditions, cleaner houses, better roads, better transportation, better education and higher income for villagers, in general, and therefore more work for craftsmen. In Sengayan, only half of the respondents were able to answer the question. All except one judged their current life situation as better than in the period before IPPK. The respondents did not come up with explanations, whereas the remaining respondent judged his life situation as harder today because of more conflicts, although his economy had become better.

Access to education

The disbursement of IPPK fees (see Figure 7.3) revealed that, on average, almost half of the households in the total study population used part of the IPPK fee on costs related to education. No significant differences appeared between the three villages. Nearly all of the mentioned expenditures were on uniforms, pencils and books; but for one quarter of the responding households, expenses for their children's tuition fees were the most important. Key informants in each of the three villages emphasized the positive implication of IPPK: that the young people now receive a better education.

Access to health facilities

Figure 7.3 shows that more than three quarters of the total study population disbursed part of the IPPK fee on health. Expenses for medicine were mentioned by all of these households, while almost half of them have used part of the IPPK fee on removal to hospital in Malinau or Tarakan. Key informants mentioned that, until now, villagers have not had the financial means to reach Tarakan (one day of travel) and pay for medical help.

Changes in access to employment and income activities

The indirect effects of IPPK on employment and trading opportunities were found mainly to benefit households in Adiu. Three quarters of the households in Adiu had been engaged in selling products, mainly agricultural ones, to IPPK employees (see Figure 7.9). In contrast to Adiu, households in Tanjung Nanga had little opportunity to sell their locally produced products to IPPK workers. In both villages, the potential to trade local products could have been greater, but because many IPPK workers did their shopping in Malinau Kota, where the prices of food and other daily products are normally lower and the range of products larger, this opportunity diminished. With regard to employment opportunities, more than half (all men) of the respondents in Adiu had been employed by the IPPK operator (see Figure 7.10). Employment most often involved practical work as loggers, tractor

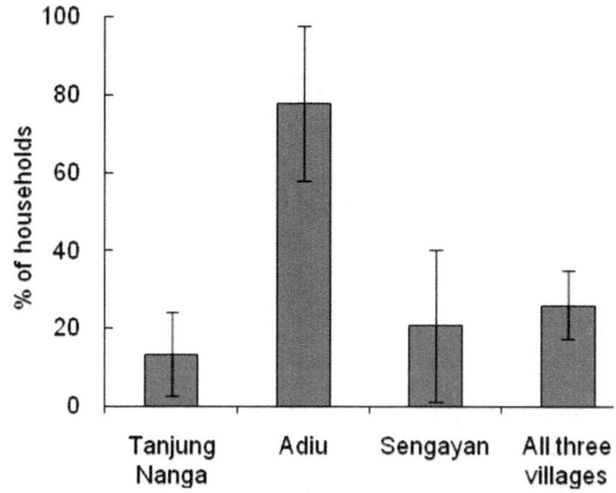

Figure 7.9 *Proportion of households engaged in trading products to IPPK employees (percentage is illustrated with an error bar ± 2 * SE)*

operators, de-barkers, security staff or 'scouts' searching for valuable timber trees (mainly the Punan). Households in Tanjung Nanga were almost never employed in the logging camp. In Sengayan and Adiu, where employment at the IPPK operator seemed to have been a common income option, the impression is that employment has only been for short durations. In Sengayan, key informants stated that even though the village benefited from employment opportunities, villagers could only endure the job for short periods. Only one respondent mentioned that he was employed for a whole year, whereas the normal occupation period was between two and five months. The average wage for an IPPK employee was around US$37 per month, an estimate based on statements from six respondents.

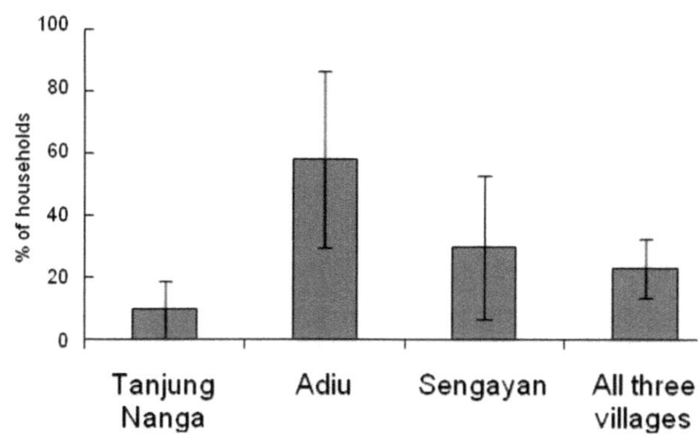

Figure 7.10 *Proportion of households engaged in employment at the IPPK company (percentage is illustrated with an error bar ± 2 * SE)*

Box 7.6 Statement from focus group interview in Long Adiu concerning changed trading behaviour

Malinau is too far away. It's more common that villagers go to Long Loreh where a lot of companies are operating. Villagers sell their forest game and chickens. It is easier to travel because villagers now have a long tail engine which has increased the frequencies of travels to Long Loreh. Even though Malinau is far away, almost every day someone goes there and often returns the same day. The prices are better in Malinau!

In addition to these more or less direct income opportunities, key informants in all study sites stated that the enhanced financial situation in the communities had created a larger local market for vegetables and handicrafts, and improved opportunities for villagers to carry out minor local entrepreneur business. Households' investment in transport facilities was also stated to have a positive income effect as it had opened up the markets in Malinau and Long Loreh (see Box 7.6). Trading behaviour was also affected in a slightly adverse manner: villagers' increased purchasing power was stated to attract more traders who came to sell them clothes, kitchen equipment and other goods.

Farming activities

Initially, IPPK logging activities were assumed to affect the size, location and cultivation of households' individual farming areas. Although around half of the households in all three study locations either increased or decreased their farming areas during IPPK operations, only a minority attributed this to IPPK logging activities. In Tanjung Nanga and Sengayan, the respondents explained that the logging area was too far away to attract any farming area expansion. In Adiu, where the logging area was located very close to the village, only one respondent said that he opened a new field in the logging area due to this proximity and the less demanding clearing effort. Only one respondent (in Adiu) was actually able to reduce his cultivated area due to the increased income that he received from his employment at the IPPK company.

It was emphasized by the key informants in Sengayan that the IPPK fee had made it possible for villagers to cultivate their fields more productively by acquiring better tools and seeds, and by more regularly applying fertilizer and pesticides. This argument was also supported by key informants in Tanjung Nanga, who also stressed that having chainsaws makes the cultivation of fields much quicker and less problematic. The IPPK company provided both Sengayan and Adiu with a rice husk, although the rice husk in the latter was part of the common investments. Only a few or none of the households in each of the three villages adopted new commercial crops by buying seeds or seedlings from the received fee.

Forest activities

Did the communities have any interest in, or exert any influence on, the management of the concession areas? Operational rules established between the IPPK concessionaire and the communities were either vague or absent in all three study locations. It appeared that Sengayan was the only case where an agreement was made on maximum extraction levels per hectare. Communities and IPPK concessionaires seemed to have paid most attention to reaching an agreement on the fee per cubic metre harvested and ignored other management issues (e.g. period of extraction, harvest technology, species restrictions, limitations on diameter of trees logged, etc.). However, the *Kepala Desa* in Tanjung Nanga explained that the IPPK committee had decided to allocate the IPPK blocks at some distance from the village (approximately a two to three hours' drive) in order to avoid any direct impact on the villagers' utilization of the forest resource (see Table 7.1).

Did the villagers experience any changes in the access to forest products due to the logging activities? In general, most of the interviewed households collected forest products as part of their livelihood. The most common activity was fishing carried out by the majority of the total study population (85 per cent). Hunting was the second most common activity (70 per cent), followed by collection of fruit (47 per cent), timber harvesting (46 per cent), rattan collection (35 per cent) and *gaharu* collection (20 per cent). A high proportion of these households also explained that, during the last three years, they had encountered problems in collecting some of these products. However, with the exception of villagers' access to timber, logging activities by IPPK were rarely held responsible for the more difficult harvesting conditions. Almost one third of households stated that IPPK activities were responsible for the shortage in timber supply. These households claimed IPPK activities were responsible for harder harvesting conditions (see Figure 7.11). Although no significant differences among the three villages appeared concerning timber shortages, Adiu showed a higher proportion pointing at IPPK. Demographic pressure (i.e. the growing population) was the main explanation given by the households for the generally poorer harvesting conditions of forest products.

Discussion

Financial capital

Importance of the IPPK fee to the household economy

Cash income is becoming increasingly important to communities in the Malinau area, especially in locations with improved infrastructure (Wollenberg et al, 2001; Levang, 2002). This trend has become even more pronounced in communities influenced by IPPK since the fee is an easy way of obtaining cash. Why make an effort or stay content with second best when it is possible to enjoy the benefits from the forest by striking a deal with the concessionaires (Levang, 2002)?

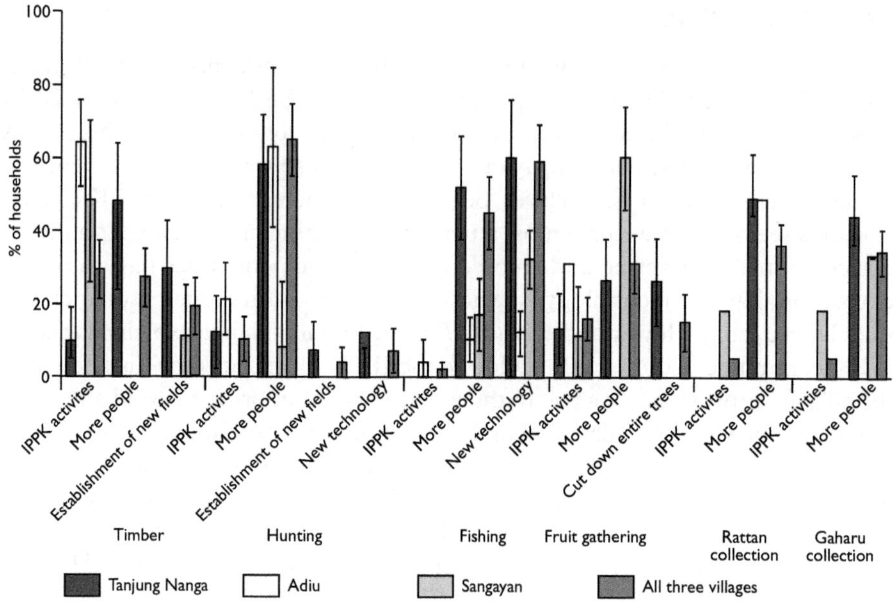

Note: lack of value/error bar indicates that the value/error is zero.

Figure 7.11 *Percentage of people claiming the mentioned reasons as being responsible for decreased access to and/or more difficult harvesting conditions: The percentage is based on the number of households expressing difficulties in obtaining the forest products and not the total number of users (percentage is calculated as the average number of households and is followed by ± 2 * SE)*

According to our results, income from the IPPK fee constituted a main source of cash in each of the three study locations (see Figure 7.1). This was especially pronounced in Adiu and Tanjung Nanga. Results differed in Sengayan, where income from farming activities was greater than the IPPK fee, while income generated from off-farm labour work was considered to be of equal significance. Households in Sengayan have good access to the market and a variety of stable income sources, which provide them with the opportunity to buy other subsistence products and allow them to trade their own farming products. In Tanjung Nanga, the main cash income was provided by the IPPK fee, which might be explained partly by the smaller proportion of households being engaged in other income-generating activities. Although households in Tanjung Nanga received the smallest fee compared to the other two villages, they still perceived this income as greater that other income sources, indicating a slightly lower income level in the village. In Adiu, employment with the IPPK operator and relatively good market access have been options for generating stable and significant income, which facilitate a higher income level. Nevertheless, households in Adiu did not believe that these income opportunities could substitute the IPPK fee. In contrast, households in Sengayan were generally confident that they would have been able to generate the same amount of cash from other activities. Furthermore, households in

Sengayan generally perceived their financial situation to be less affected by the fee than did households in the two other villages. In Tanjung Nanga, households generally perceived their financial state to be greatly influenced, which is in keeping with the findings from pair-wise ranking.

The IPPK fee was generally ranked higher than other income sources in the pair-wise ranking (see Figure 7.1). The reason is due to the combined effect of availability and actual amount (see Figure 7.2). Compared to other cash-generating activities, the IPPK fee was obtainable for most households (only two households in total did not receive it) and is therefore frequently mentioned in the pair-wise rankings. In terms of income size, only wages from employment in Malaysia were found to be more attractive than the IPPK fee. Nevertheless, income from IPPK wage, off-farm labour work or sale of forest products should not be neglected because households considered these activities important in terms of cash income. Households often stated that contrary to most off-farm labour work, income from the IPPK fee was irregular and unstable (see Box 7.2). Not only was there great uncertainty about the IPPK scheme and the prospects of forest dwelling communities to receive monetary benefits derived from forest exploitation (Barr et al, 2001), but also in households' perception of the fee: how much would they receive per payment, how often and when? This lack of income stability can be argued to have influenced the way in which households and communities used the IPPK fee.

It was revealed in Tanjung Nanga that the households belonging to the middle-income stratum had received over three times more than those belonging to the very low-income stratum. The result implied unequal distribution of benefits within the village, which was not only limited to the elite compared with the commoners, but also involved other social levels. Furthermore, this finding endorses the characteristically strong hierarchy in Kenyah villages (Sellato, 2001). Regarding the observed incongruence between estimated fee per household (see Figure 7.8), there seem to be four explanations:

1 Respondents deliberately understated the received amounts and number of payments (as they believed this approach would benefit their intentions of achieving outside assistance) (Sellato, 2001).
2 Respondents forgot to consider payments received by other family members within the household (e.g. children or youth).
3 More households actually received a fee than assumed in the calculations (e.g. an unknown number of households in the village of Long Jalan received fees from the neighbouring Tanjung Nanga).
4 Information was inadequate about expenses and investments at the village level. Realistic estimates of the total fee received at the household level are likely an approximate average of the two estimates.

Short- versus long-term investments

'When you receive money like this, which is not stable, you just use it for fun!' A respondent in Sengayan with a relatively stable income stated this view and implicitly indicated a short-term perspective. Angelsen and Wunder (2003) mention that subjective short-term preferences of local people do not coincide with their goals of long-term welfare improvements or with external values. What we perceive as a short-lived and unwise investment might seem extremely important to the well-being of a local person. Generally, people care not only about their absolute incomes, but also compare their income status to households with whom they would like to 'keep up'. It may be considered more important to buy a television, as their neighbours did, than to improve their basic diet or fix their roof (Angelsen and Wunder, 2003). In cases such as the introduction of IPPK, where sudden large cash inflows appear, poverty in terms of material constraints can be reduced (Angelsen and Wunder, 2003). Unfortunately, however, as Levang (2002) also mentions, modern manufactured goods are not only costly, but are also easily damaged and need maintenance or quick replacement. Furthermore, traditional livelihood strategies may be eroded, which justifies the importance of long-term investments in order to make communities and households less vulnerable (Angelsen and Wunder, 2003).

As part of the household survey, it was intended to reveal whether or not households' disbursement patterns, created by the receipt of the IPPK fee, encompassed investment strategies with long-term perspectives. While the disbursement categories 'maintenance of human capital', 'provision of additional income opportunities' and 'savings' hold prospects for the future, investment in 'luxury goods' is considered a short-term investment. Even though some of the commodities that enable households to generate additional income are only short lived (e.g. long tail engines and agricultural tools), the investments might be stepping stones in the direction of more stable income opportunities. Most households in the study invested part of the received fee in goods 'providing opportunity for additional income' and 'maintaining human capital'. The latter is regarded as a long-term investment because it secures the foundation for future development. Investments that could 'provide opportunities for additional income' have especially been favoured in Adiu. Except for households in Sengayan, many households have invested part of the IPPK fee in tools that could make their farming activities easier and more productive. For example, the fee has enabled an increasing number of households to afford the application of fertilizers and pesticides in their farming fields. Furthermore, in Tanjung Nanga and particularly in Adiu, households purchased chainsaws. The main incentive for more frequent spending on chainsaws in Adiu than in Tanjung Nanga is probably that Adiu's access to timber markets is good due to the short distance to Malinau Kota. The fact that households in Tanjung Nanga have received relatively small fee amounts has, of course, limited their options for disbursements on transport and chainsaws. In Sengayan, the main road and a relatively strong purchase power prior to the introduction of IPPK have most likely influenced the lack of incentive to purchase long tail engines or agricultural tools or chainsaws. Although Levang et al (2005) found that timber harvesting carried out by individual households is an increasingly important source of income to villages that are well connected to the market (greater than US$25 per day for very little risk), the drawback on the forest resource as a safety net function should be kept in mind. Adiu's prevailing geographical and logistical features have also functioned as an incentive for households to buy long tail engines, which was done by almost all households. Although it is possible to go to Malinau in one hour with a four-wheel drive, ground transport is expensive and normally out of reach for most villagers. Consequently, the road provided by the IPPK is seldom used as a short cut to Malinau. It is more convenient to use the waterways, enabling people to reach Malinau in approximately three hours. The marginal benefit of investing in a long tail engine was therefore relatively high. In general, the extra income that has become available with IPPK has increased mobility and therefore also increased villagers' access to markets and information. Villagers in both Adiu and Tanjung Nanga mentioned that they now more frequently travel to neighbouring villages or even to Malinau Kota or Tarakan. We argue that the main effect of improved market access has been on villagers' expenditure rather than on their opportunity to sell their locally produced products. However, trading products to IPPK employees seemed to have affected the villagers more directly. In particular, the distance to the logging camp appeared to have a substantial effect on the possibilities of contributing to the household income by trading products to IPPK employees.

Although households, in general, demonstrated inability or disinclination to save the received fee, it was revealed that about half of the households in Sengayan put aside part of the fee. Income levels and different opportunity costs associated with saving probably comprised part of the underlying explanation for variations in saving patterns. The availability of stable income sources in Sengayan seemed to encourage the households to put aside some of this extraordinary income since the opportunity costs of doing so were low. Another part of the explanation comes from the village's social and ethnic characteristics, as commented on by Carol J. Pierce Colfer (pers comm, June 2003):

> *It's important to remember that in the past (and perhaps still), the emphasis on generosity precludes people from saving their money. If they have money, others know about it and come and request it – requests that people are unlikely to be able to refuse since the person may need medical attention, their children may be forced to quit school without the money or some other very important human concern.*

This social mechanism is particularly pertinent for Kenyah communities (e.g. Tanjung Nanga). Households might also regard saving as a 'waste of cash' because they are convinced that there will be alternatives to IPPK in the future. Why save part of the fee when infinite amounts of timber are available in the backyard and 'outsiders' are willing to pay for it? Contrary to Sengayan, Tanjung Nanga and Adiu have made several and large common investments and savings at the village level. Such approaches could help to secure an equal distribution of the benefits achieved from IPPK, which according to Angelsen and Wunder (2003) is an important issue if the objective is economic growth by means of reducing poverty.

Social capital

Transparency is viewed by Oakerson (1992), Ostrom (1990, 1999) and Wollenberg (1998) as a requirement for trust-building, establishing a sense of community and minimizing opportunistic behaviour. Our study indicates that villagers have only been informed to some extent about the contractual conditions of IPPK. Adiu seemed to possess the highest degree of transparency and Tanjung Nanga, the lowest. The impression is that only a handful of influential villagers have been engaged in negotiation with the companies and then left some of the issues, or rather final agreements, for discussion at community meetings. The negotiations have taken place in a non-transparent manner without the involvement of the rest of the village. However, the well-organized institutional structure that characterizes Kenyah communities and experiences from HPH operations in the area appeared to have shaped the management of the IPPK fee in Tanjung Nanga in a positive manner. The common investments, the well-organized distribution of the fee to villagers (though not equal!) and careful considerations regarding the involvement in IPPK reflect a high degree of institutional development.

Our study in Adiu suggests that the Punan enjoyed less attention as they neither took part in the initial negotiations nor received the last payment and the same collective benefits as the Merap (village office and health clinic). This situation supports what other studies also claim – that the former hunter-gatherers – the Punan – are the most politically marginalized ethnic group in the area (Levang et al, 2005).

Development schemes such as IPPK often suffer from unequal distribution of benefits in the initial phases. It is observed that some actors have assets or skills that put them in a privileged position to take advantage of the new opportunities, while others simply have a more developed entrepreneurial spirit and, consequently, respond more quickly and reap higher benefits, often with the outcome of increased inequality (Angelsen and Wunder, 2003). Parallel circumstances were observed during the fieldwork. In Sengayan, the amount of benefits received by households seemed to depend more upon their social status and relations to the elite than on any fixed set of rules. In the case of Tanjung Nanga, there also seemed to be a tendency towards rent-seeking by the village elite. Many respondents from a household survey in both Tanjung Nanga and Sengayan claim that *Kepala Desa* and, to some extent, *Kepala Adat* have received a disproportionate portion of the fee. In Adiu, the two former *Kepala Desas* had to resign due to rumours and accusations of cheating.

Other studies on the issue of level of participation in forest management have stated that powerful stakeholders such as IPPK companies do not perceive local people to have either the rights or abilities to participate in managing forests (Colfer and Wadley, 2001). Instead, concessionaires prefer the fast and unproblematic track with a low level of participation, which, in the case of the Malinau area, often implies that the *Kepala Desa* is invited by the logging concessionaire for a cautious deal (Sellato, 2001). Our research indicates that villagers have had little opportunity to participate in the formulation of the contracts and only the most important people from the villages have played a role in the negotiation process. A situation like this where only a few stakeholders are taken into account in resource management is argued to result in a low degree of 'sense of ownership' and, therefore, more conflicts, disagreements and less accountability (Ostrom, 1990; Colfer and Wadley, 2001).

Another important aspect in analysing social capital is the level of conflicts. On average, more than half of all households experienced conflicts related to IPPK issues. The qualitative analysis revealed that most conflicts concerned a low level of transparency and participation in the initial phase of IPPK implementation. Many households in each of the three study sites blamed the rules of distribution for causing conflicts. Distributional rules are regarded as a result of the level of transparency and participation within the village since, in order to establish a stake in the management, villagers should have the opportunity to modify and alter operational rules affecting their livelihood (Ostrom 1990, 1999; Colfer and Wadley, 2001). The rules seemed to be less developed in Sengayan, which corresponds to significantly more villagers finding the distribution unfair, compared to Tanjung Nanga and Adiu. In addition, more newcomers are present in Sengayan, which might make the community less homogeneous. The level of cultural homogeneity within communities is identified by many researchers as crucial to the performance of collective management (Ostrom, 1990, 1999; Oakerson, 1992; Wollenberg, 1998).

The communities investigated seemed to have enhanced part of their social capital through social learning. In Tanjung Nanga and Adiu, the rules for distribution of the fee were changed during the IPPK period. The rules that originated from the time of HPH were considered inadequate and in need of adjustment. In Adiu, the criteria were changed so that the fee was allocated to adults, youth and students instead of per household since it was perceived as a more fair system. The two communities seemed to be capable of adjusting and adapting rules to the prevailing conditions, showing signs of institutional robustness (Ostrom, 1990; Wollenberg, 1998; Buck et al, 2001). Social learning in another form was also observed. Other research argues that there has been a growing interest in planning land use in the area, especially as distrust of IPPK permit holders has grown and more forest has disappeared and degraded (Barr et al, 2001; Levang, 2002). In this study, key informants in the three study sites expressed that they wanted to keep part of their forest protected because they want to support conservation of water and game, improve prospects for tourism and keep the door open for the possibility of increasing farming areas. These changes and ideas suggest that communities have experienced forest product shortages and therefore have become more inclined to care about their access to, and the quality of, forest resources. This shift in management focus is seen by other researchers to follow changed economic incentives and resource deficiencies (Ostrom, 1990; Oakerson, 1992; Wollenberg, 1998; Angelsen and Wunder, 2003).

Although conflicts occurred in all three villages, the impression is that Adiu had the most durable and robust social capital. Furthermore, even though

Tanjung Nanga appeared to be the best institutionally structured village, participation and transparency seemed to be generally ignored. Adiu had the highest level of transparency, the highest frequency of participation and the highest acceptance of the procedures for distribution of the IPPK fee. Adiu's strong social capital might be due to the small size of the community compared to the other two study sites. Ostrom (1990) and Wollenberg (1998) argue that cooperative strategies and trust are more likely to emerge and be sustained in smaller rather than larger communities.

Human capital

Educational and health changes

In some studies, it is suggested that the poverty-reducing effect of growth is enhanced by common investments in, for example, health and education (Gillis et al, 1996; Angelsen and Wunder, 2003). Education is viewed as the single most important factor in simultaneously promoting economic growth, poverty reduction and a more equal income distribution (Gillis et al, 1996; Sen, 1999; Angelsen and Wunder, 2003). The impacts of IPPK on education were considered highly significant to the communities. Around half of all households used part of the IPPK fee on costs related to education and one quarter of them used part of the payment on tuition fees for their children attending school above primary level. The impression from focus group interviews in the three villages is that the possibility of studying at higher levels than the primary level were strictly limited prior to the advent of IPPK. The hindrance was mainly households' financial incapability, but also, in the case of Tanjung Nanga and Adiu (Long Adiu), lack of housing facilities in Malinau Kota. Both Tanjung Nanga and Long Adiu decided to spend part of their respective common fee on the establishment of student accommodation in Malinau Kota.

Health is considered an important development target since it forms the basis of a family's or a community's ability to maintain and enhance social wellbeing and to pursue new livelihood strategies (Carney, 1998; Lindenberg, 2002). The presence of IPPK affected the quality of and access to health services, both directly and indirectly. Villagers have been able to more regularly afford to see a doctor in Malinau Kota and Long Loreh, to purchase the necessary medicine and occasionally to leave for Malinau or Tarakan for severe health matters. Additionally, the study reveals that a large proportion of the total number of households spent part of the IPPK fee on daily living and kitchen equipment, which could potentially have a positive impact upon health. Most households in the three villages also judged their living conditions as better today than in the period before IPPK. Although the improvements in quality of life cannot be differentiated according to the impacts of other concessionaires, government programmes, etc., most of the explanations given related to changes induced by the IPPK company, in general, and improved health and education, in particular.

Employment opportunities

It is often stated that wage employment at concessions is one of the most important indirect benefits to local communities in areas affected by timber concessionaires. However, these employment opportunities are also often found to be limited to skilled and educated outsiders (Barr et al, 2001; Levang, 2002; Angelsen and Wunder, 2003). Furthermore, the presence of concessionaires is often stated to alter the communities' portfolio of available income activities since it can create opportunities to trade local products and directly provide the community with cash income that can be reinvested in the means for providing new income sources (Barr et al, 2001; Levang, 2002; Angelsen and Wunder, 2003). In this study, the distance to the logging camp seemed to have a substantial effect on the possibility of being employed. Labour wages have the potential to diversify income opportunities at the household level and make households less risk prone to oscillating economy and climate (Levang, 2002). However, only short-lived and low qualification employment with low wages was offered by the IPPK company in the three study sites. Whether this trend is caused by prejudiced considerations by companies or unavailability of skilled labour in villages has not been investigated in this study. However, other researchers made the same observation that the percentage of the local work force in the companies' personnel is low and of only temporary duration (Sellato, 2001; Levang, 2002; Angelsen and Wunder, 2003).

Natural capital

Around three quarters of households using forests for timber harvesting, hunting or fishing have experienced difficulties or shortages during the last three years. However, IPPK activity was only a partial explanation for the generally more difficult harvesting conditions of forest products. IPPK was held responsible for the severe harvesting conditions of forest products by less than one tenth of all households, with the exception of impacts upon timber availability (28 per cent) and fruit gathering (16 per cent). This result supports what other research from the area have indicated – that local people are experiencing problems with hunting, declining fish populations and gathering of other forest products in areas affected by timber concessions (Barr, 2001; Barr et al, 2001; Colfer and Wadley, 2001; Levang, 2002; Levang et al, 2005). We argue that households more often mentioned IPPK as the reason for decreased timber availability due to visible and immediate timber extraction. It might be more difficult for the respondents to see a causal link between timber removal and a decline in, for instance, fishing and hunting. Furthermore, the consequences of IPPK activities on, for instance, fishing and hunting may need more than two to three years to become visible and are therefore underestimated in this study.

Although our research has not documented the exact economic dependency upon forest products, we found that almost half of the households in each of the three study sites mentioned forest products as an income-generating activity. According to the pair-wise rankings, the economic importance of forest products equals that of income from farming activities (see Figure 7.1). We argue that, from a development and human capital point of view, small timber concessions such as IPPK have the potential to transform natural capital into other assets such as human and physical ones, improving risk buffering and minimizing vulnerability. However, due to the importance of forest products, this transformation has to go together with securing access to an adequate extent of forest so that households have the chance to maintain the forest resource as subsistence as well as an economic gap-filling resource (Byron and Arnold, 1999; Angelsen and Wunder, 2003; Edmunds and Wollenberg, 2003).

The negative impacts upon natural capital that stem from IPPK operations seem to be, *inter alia*, caused by inadequate mapping, boundary demarcation and lack of contractual recognition of monitoring. Furthermore, the lack of legal recognition by the authorities makes the communities vulnerable to political and economic fluctuations and therefore encourages communities to reap the benefits as fast as possible since they risk being pushed out of the area once timber companies realize the opportunities of extracting resources (Barr et al, 2001; Anau et al, 2002; Angelsen and Wunder, 2003). The pronounced insecurity seems to make communities and villagers attribute very little value to benefits that they expect to receive in the future (Ostrom, 1990). These conditions have also affected social capital, possibly to a larger extent since they might cause conflicts and disagreements among and within communities (Barr et al, 2001; Anau et al, 2002; Resosudarmo and Dermawan, 2002; Limberg and Rhee, 2003).

Conclusions

This study has shown that the IPPK fee has been available to all households except for a few newcomers in Sengayan. It has constituted the most important source of cash income to the households in terms of availability and size. This was especially prevalent in Adiu and Tanjung Nanga, which had only poor alternatives for generating cash. Unequal distribution of the fee among income strata was only experienced in the Kenyah community of Tanjung Nanga, where social differences between commoners and aristocrats are more pronounced. Almost all households have disbursed part of the fee on health and education, which could improve their human capital. These disbursements, together with investments that can provide opportunities for additional income, might indicate that part of the IPPK fee has been used in a way that could benefit communities in the long term. However, several households considered the main weakness of the IPPK fee to be instability and irregularity, which impeded the sustainability of household investments. The small number of households putting aside part of the IPPK fee stresses the significance of regularity. Considering the short-term horizons of the IPPK benefits, larger community investments with long-term returns should therefore have higher priority.

The indirect benefits of IPPK were the enhanced possibility of employment at the IPPK company and trading local products to IPPK employees. Communities at a short distance from the logging camp were found to gain larger benefits from these new opportunities than those situated further away. However, the drawbacks of logging undertaken in the village proximity are the higher number of incidences of forest product shortages related to IPPK. Given the short duration of IPPK activities and the importance of forest products to household livelihoods, we consider the impacts of IPPK on forest product availability as severe. Since all three communities have experienced forest product shortages, they tend to be concerned about their access to, and quality of, forest resources.

We argue that institutional robustness and small community size have had a positive influence on the degree of transparency and participation during the process of IPPK. These attributes hold true for the small community of Adiu, which also had the lowest frequency of conflicts and the highest level of acceptance of distribution.

Overall, we argue that small-scale timber concessions such as IPPK can substantially benefit forest dwelling communities with financial, human and physical capital. Increased but short-term cash income is the most direct benefit, with the potential to be converted into other forms of capital, such as human and physical, therefore improving risk buffering and minimizing vulnerability at the household level. However, there is a risk of compromising social capital and reducing households' access to forest products. As a result, clearly defined operational rules concerning the exploitation of the concession area and rules for the community's distribution of benefits deriving from small-scale timber concessions are prerequisites for minimizing conflicts and reducing pressure on natural capital. In addition, a government agency should be in place to assist in contract negotiations, to oversee timber concession activities and to mediate in conflict-solving.

Notes

1 An HPH concessionaire was previously operating in the forest area claimed by Tanjung Nanga; as a result, in 2001, the village received US$12,500 (or 100 million rupiah) in compensation for the timber extracted until 2000. The *Kepala Desa* himself received 20 per cent of this amount as a payment for negotiating the deal. The rest of the money was distributed among the villagers.

References

Anau, N., Iwan, R., van Heist, M., Limberg, G., Sudana, M. and Wollenberg, E. (2002) 'Negotiating more than boundaries: Conflict, power and agreement building in the demarcation of village borders in Malinau', in CIFOR (ed) *Forest, Science and Sustainability: The Bulungan Model Forest. Technical Report Phase I 1997-2001, ITTO Project PD 12/97 Rev. 1 (F)*, CIFOR, Bogor, Indonesia

Angelsen, A. and Wunder, S. (2003) *Exploring the Forest – Poverty Link: Key Concepts, Issues and Research Implications*, Occasional Paper No 40, CIFOR, Bogor, Indonesia

Baland, J. M. and Platteau, J.-P. (1996) *Halting Degradation of Natural Resources: Is there a Role for Rural Communities?* Clarendon Press, Oxford, UK

Barr, C. (2001) *Banking on Sustainability: Structural Adjustments and Forestry Reform in Post-Suharto Indonesia*, CIFOR, Bogor, Indonesia

Barr, C., Wollenberg, E., Limberg, G., Anau, N., Iwan, R., Sudana, I. M., Moeliono, M. and Djogo, T. (2001) *The Impacts of Decentralization on Forests and Forest-Dependent Communities in Malinau District, East Kalimantan, Case studies on Decentralisation and Forests in Indonesia*, CIFOR, Bogor, Indonesia

Buck, L. E., Wollenberg, E. and Edmunds, D. (2001) 'Social learning in the collaborative management of community forests: Lessons from the field', in Wollenberg, E., Edmunds, D., Buck, L., Fox, J. and Brodt, S. (eds) *Social Learning in Community Forests*, CIFOR, Bogor, Indonesia

Byron, N. and Arnold, J. M. (1999) 'What futures for the people of the tropical forests', *World Development*, vol 27, no 5, pp789–805

Carney, D. (ed) (1998) *Sustainable Rural Livelihoods: What Contribution Can We Make?* Department for International Development, London

Casson, A. and Obidzinski, K. (2002) 'From new order to regional autonomy: Shifting dynamics of "illegal" logging in Kalimantan, Indonesia', *World Development*, vol 30, no 12, pp2133–2151

Colfer, C. J. P. (2003) Pers comm, Anthropologist, leader of the CIFOR programme on Local People, Devolution, and Adaptive Collaborative Management, CIFOR, Bogor, Indonesia

Colfer, C. J. P. and Wadley, R. L. (2001) From 'participation' to 'rights' and responsibilities in forest management', in Colfer, C. J. P. and Byron, Y. (eds) *People Managing Forests: The Links Between Human Well-Being and Sustainability*, Resources for the Future, Washington, DC

Edmunds, D. and Wollenberg, E. (eds) (2003) *Local Forest Management: The Impacts of Devolution Policies*, Earthscan, London

FWI (Forest Watch Indonesia)/GFW (Global Forest Watch) (2002) *The State of the Forest: Indonesia*, FWI, Bogor, Indonesia, and GFW, Washington DC

Gillis, M., Perkins, D. H., Roemer, M. and Snodgrass, D. (1996) *Economics of Development*, W. W. Norton and Company, New York, NY

Kaskija, L. (2002) *Claiming the Forest: Punan Local Histories and Recent Developments in Bulungan, East Kalimantan*, CIFOR, Bogor, Indonesia

Levang, P. (2002) 'Peoples dependencies on forests', in CIFOR (ed) *Forest, Science and Sustainability: The Bulungan Model Forest. Technical Report Phase I 1997-2001, ITTO Project PD 12/97 Rev 1 (F)*, CIFOR, Bogor, Indonesia

Levang, P., Dounias, E. and Sitorus, S. (2005) 'Out of the forest, out of poverty?', *Forest, Trees and Livelihoods*, vol 15, pp211–235

Limberg, G. and Rhee, S. (2003) *Highest Profit in the Shortest Time with Maximum Impact*, CIFOR, Bogor, Indonesia

Lindenberg, M. (2002) 'Measuring household livelihood security at the family and community level in the developing world', *World Development*, vol 30, no 2, pp301–318

Oakerson, R. J. (1992) 'Analysing the commons: A framework', in Bromley, D. W. (ed) *Making the Commons Work: Theory, Practice, and Policy*, Institute for Contemporary Studies Press, San Francisco, CA

Ostrom, E. (1990) *Governing the Commons: The Evolution of Institutions for Collective Action*, Cambridge University Press, Cambridge, UK

Ostrom, E. (1999) *Self-Governance and Forest Resources*, Occasional Paper No 20, CIFOR, Bogor, Indonesia

Resosudarmo, I. A. P. and Dermawan, A. (2002) 'Forests and regional autonomy: The challenge of sharing the profits and pains', in Colfer, C. J. P. and Resosudarmo, I. A. P. (eds) *Which Way Forward? People, Forests and Policymaking in Indonesia*, Resources for the Future (RFF)/CIFOR, Washington, DC

Sellato, B. (2001) *Forest, Resources and People in Bulungan: Elements for a History of Settlement, Trade and Social Dynamics in Borneo, 1880–2000*, CIFOR, Bogor, Indonesia

Sen, A. (1999) *Development as Freedom*, Anchor Books Inc, New York, NY

Steins, N. and Edwards, V. (1999) 'Platforms for collective action in multiple use commonpool resources', *Agriculture and Human Values*, vol 16, no 3, pp241–255

Sunderlin, W. D., Angelsen, A. and Wunder, S. (2003) 'Forests and poverty alleviation', in Food and Agriculture Organization (ed) *State of the World's Forests 2003*, FAO, Rome

Warner, K. (2000) 'Forestry and sustainable livelihoods', *Unasylva*, vol 51, no 3, pp1–14

Wollenberg, E. (1998) 'A conceptual framework and typology for explaining the outcomes of local forest management', *Journal of World Forest Resource Management*, vol 9, pp1–35

Wollenberg, E., Adiwinata, A., Uluk, A. and Pramono, H. (2001) *Income is Not Enough: The Effect of Economic Incentives on Forest Product Conservation*, CIFOR, Bogor, Indonesia

Wunder, S. (2001) 'Poverty alleviation and forests – what scope for synergies?', *World Development*, vol 29, pp1817–1833

8

A Comparison of Damage Due to Logging under Different Forms of Resource Access in East Kalimantan, Indonesia

Haris Iskandar, Laura K. Snook, Takeshi Toma, Kenneth G. MacDicken and Markku Kanninen

Introduction

Regional autonomy laws, including the decentralization of responsibilities for the forestry sector, may have important implications for the future extent and quality of natural forest in Indonesia. Until 1999, logging in the country was carried out exclusively under large industrial forestry concessionaires, known as *Hak Pengusahaan Hutan* (HPH). These operators, who were granted 20-year selective logging concessions on national forest lands, were to follow the *Tebang Pilih Tanam Indonesia* (TPTI), the guidelines for Indonesia's selective cutting and replanting system (Departemen Kehutanan, 1993) within a 35-year polycyclic harvested system. A second system of forest resource access, initiated in 2000, is called *Ijin Pemungutan dan Pemanfaatan Kayu* (IPPK), otherwise known as small-scale timber harvesting. These annual permits for commercial logging were intended to provide timber harvesting rights to local communities who lie outside the areas defined by the central government as permanent forest estates, to be granted by district governments under Government Regulation PP No 62/1998 and 6/1999, delegating partial authority in the forest sector to the regions. In practice, however, many have been granted within the boundaries of HPH concessions to private investors – mostly logging companies (Barr et al, 2001).

These permits were granted in almost all districts in Kalimantan, Indonesia. Between April 2000 and May 2001, the district government of Malinau, East Kalimantan, issued 40 IPPKs of 100ha to 3000ha each, covering an aggregate area of 59,850ha (Pemda Malinau, 2001).

The harvesting rules to be followed by these systems were essentially the same, which would sustain timber yield on a long-term basis. In HPH, activities related to forest inventories of commercial and harvestable trees, logging planning, cutting block mapping and logging road construction should be done prior to the logging. However, the IPPK system was not required to work with formal plans due to limited time and area allocated for harvesting. The skid trail network was established during timber extraction according to the position of trees to be felled. In most cases, loggers worked unsupervised; but trees to be felled were marked. Crawler tractors (Caterpillar D7G or Komatsu D60E), with a greater than 4m wide blade, were generally used for both road opening and log extraction. Owing to market conditions where most of the harvested timber is used for plywood, harvesting in HPH was limited to trees larger than 50cm diameter at breast height (dbh) to ensure that the future harvesting cycle remains viable. On the other hand, in IPPK, the diameter limit is much more flexible and can be lower because the local community also used sawn timber for local sawmills. However, our observation prior to the study showed that harvesting practices implemented by these two systems, as well as the damage to the residual stand, were essentially the same due to TPTI violations by HPH. Moreover, after the IPPK system was put in place, we also observed that several HPH operations were operating in a similar fashion to IPPK. There were several reasons – social, cultural, political and economic – to identify why and how these practices happen (Barr et al, 2001; Resosudarmo, 2002). Furthermore, after decentralization, there was a significant increase in the percentage of the community who received financial and in-kind benefits from the new system. However, the conflicts between communities over land claims and borders also increased (Palmer and Engel, 2005).

The main objective of this study was to evaluate post-harvest damage from road construction, felling activities and skidding operations in areas harvested under short-term IPPK permits compared with longer-term HPH selective logging. Furthermore, we wanted to identify the influence of this IPPK new system to the earlier HPH system granted by the Ministry of Forestry in terms of harvesting practices and logging damage to the forest stands. We hypothesized that harvesting practices under the selective logging of the HPH system, which was more regulated, resulted in similar logging damage of practices by IPPK permit holders that incorporate no logging guidelines.

Study sites

The study was carried out in Malinau District, East Kalimantan (see Figure 8.1), located between 116° 29' and 116° 44' E longitude and 2° 59' and 3° 28' N

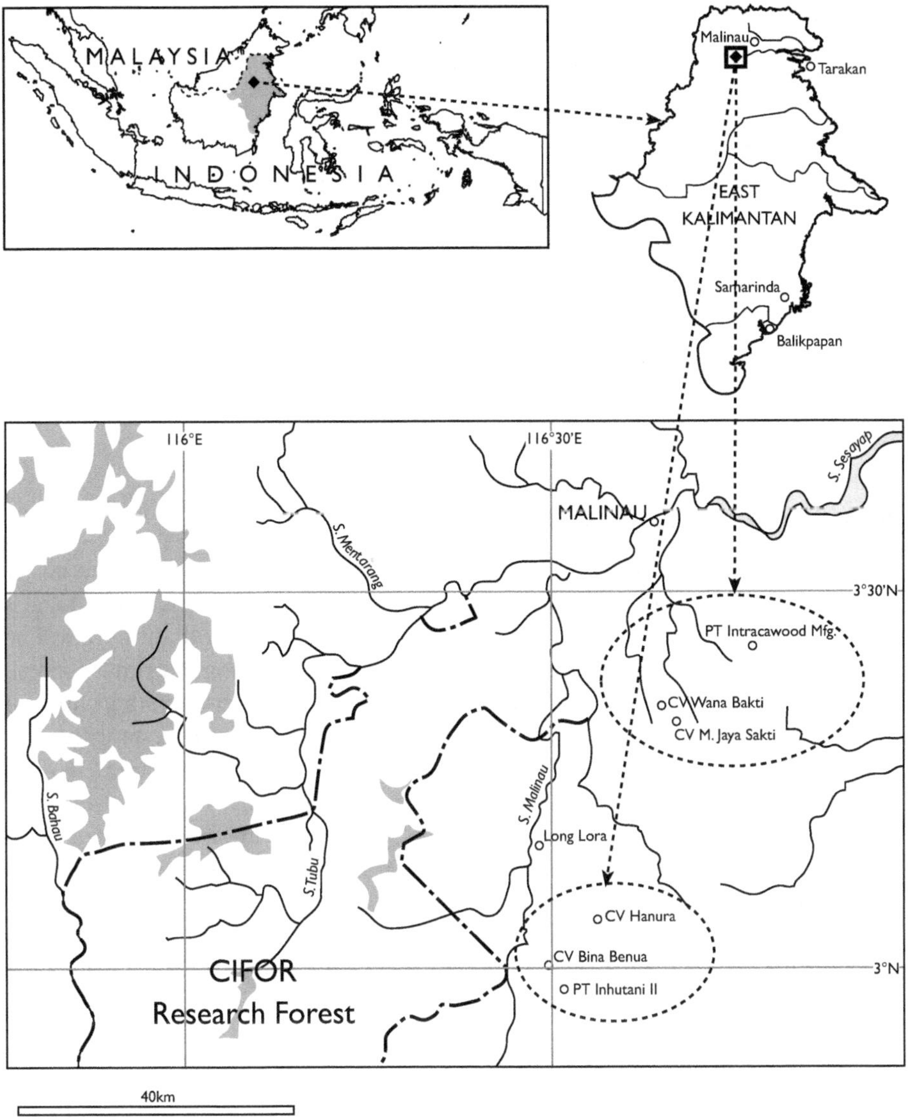

Source: Iskandar et al, 2006

Figure 8.1 *Map of Indonesia archipelago (*top left*), East Kalimantan province (*top right*), and map of the study area in Malinau (*bottom*)*

latitude. The forest type is predominantly lowland and sub-montane tropical forest, dominated by trees of the *Dipterocarpaceae* family. In most study areas, altitude ranges from 100 to 200m above sea level (asl), with flat to undulating terrain (0 to 8 per cent slopes), although some areas are steeper (8 to 15 per cent slopes). The monthly rainfall varies between 200mm and 400mm, totalling approximately 4000mm annually (Machfudh, 2001).

The sample areas were selected based on information regarding accessibility (i.e. road conditions and equipment) obtained from concessionaire management and government office staff. Our preliminary evaluation of companies operating in Malinau revealed that several selective logging concessionaires' licences were terminated for various reasons: expiration, topography constraints or conflict with local communities. We were therefore only able to evaluate two selective logging concessionaires: one belonging to a state-owned company and the other to a privately owned company, with annual cutting areas of 700ha and 5000ha, respectively. Three short-term permit holder operators and a subcontractor of selective logging concessionaires were also evaluated. Logging activities for all operators took place mostly in early 2001.

Methods

Sample plot establishment

Two forest types were evaluated in this study: primary forest, which had larger trees and had never been logged, and logged-over forest, which had been logged under prior selective logging by concessionaires and had fewer mature trees. The state-owned HPH holder and one IPPK holder were harvesting in primary forest, while the privately owned HPH and the remaining IPPK holders had harvest permits for logged-over forest. Six 20ha sample plots were established in November 2001 and January 2002 in forests harvested by each operator (36 plots in total). For most IPPK holder operators, who lack maps, we established our plots within the boundaries of the cutting blocks based on information from camp managers and village staff. Imaginary boundaries for each plot were established by mapping roads and skid trails using a global positioning system (GPS) Trimble Geo Explorer-3. Data on harvest intensity and damage to residual trees were collected inside the imaginary plot boundaries along logging roads and skid trails, as well as within gaps produced by felling.

Measuring harvest intensity

Tree stump counts and measurements were used to estimate harvests, as has been done elsewhere (Chhetri and Fowler, 1996). In each sample plot, stumps were counted and measured along the logging roads and skid trails, and within felling gaps. Diameters were measured at the top of the stump using diameter tape. We converted the diameter at the estimated stump height (dsh) to dbh using an allometric equation derived from tree measurements in the same area (Iskandar et al, 2003):

$$dbh = 0.518 + 0.966 \times dsh \ (SE = 0.850; R^2 = 0.998).$$

Logging damage assessment

In each sample plot, skid trails and logging roads were mapped and measured using GPS. To inventory damage in areas affected by harvest operations, we measured the canopy opening and felling impact zones along roads and skid trails using a laser range finder (Impulse 200), a hand-held laser ranging instrument that allows one to quickly and conveniently measure distances, height and vertical angles (Anon, 1998). To estimate the width of the open areas, we measured the distance in eight directions from the centre of the plot to each edge.

Damage to residual trees from logging operations was assessed along skid trails and logging roads using a system applied in a previous CIFOR experiment (Sist and Nguyen-The, 2001). All injured trees greater than 10cm dbh were measured and recorded and classified according to six main classes of damage:

1 bark-wood damage;
2 leaning;
3 crown injury;
4 broken trunk alive;
5 broken trunk dead; and
6 uprooted.

Stratification and data analysis

To compare rates of post-harvest damage, we stratified the 36 plots according to the operator type and type of forest as follows:

- in primary forest, HPH n = 6 plots (PT Inhutani II); subcontractor n = 6 plots (CV Bina Benua); IPPK n = 3 plots (CV Hanura); and
- in logged-over forest, HPH n = 6 plots (PT Intracawood) and IPPK n = 15 plots (CV Hanura, CV Malinau Jaya Sakti and CV Wanabakti).

Data on tree damage were analysed per unit area or length of skid trail or road, and per number and volume (cubic metres) of harvested trees using SPSS (1999, Version 9.0.1). A comparison of damage levels used one-way analyses of variance (ANOVAs). Characteristics of damage and frequency of stumps by diameter class were compared using a non-parametric test for two independent samples (Mann-Whitney). A confidence level of P = 0.05 was used for all statistical comparisons.

Results

Harvesting practices by HPH

Harvesting practices by selective logging HPH in primary forests were operated in two systems, first by HPH logging staff and then by HPH's subcontractor, which was equipped with better machinery. Initial analyses, however, revealed that harvests by the subcontractor of the selective logging HPH were significantly higher than by HPH logging staff (three stumps per hectare compared to one stump per hectare). Of these stumps, 10 per cent and 3 per cent, respectively, were below the TPTI diameter limit. There were no significant differences between HPH's subcontractor and HPH in skid trail length (52m per hectare compared to 34m per hectare) and road length (27m per hectare compared to 25m per hectare). However, HPH's subcontractor logging resulted in higher rates of damage compared to HPH. Significant differences were found in the number of damaged trees (32 trees per 100m compared to 20 trees per 100m, respectively) and total canopy openings (0.3ha gap per hectare of forest compared to 0.1ha gap per hectare of forest, respectively). The sum of harvested and damaged trees on HPH selective logging was also significantly higher for the HPH's subcontractor among all diameter classes, and twice as high for trees less than 50cm, as on selective logging by HPH logging staff (19 trees per hectare compared to 7 trees per hectare, respectively).

When harvesting practices in primary forest by HPH's subcontractor were compared to IPPKs, no significant differences were found in harvest intensities (3 stumps per hectare compared to 4 stumps per hectare, respectively), skid trail length (52m per hectare compared to 49m per hectare), logging road length (27m per hectare versus 25m per hectare), number of damaged trees (32 trees per 100m compared to 38 trees per 100m) and total canopy openings (0.3ha gap per hectare of forest, respectively). In addition, when HPH logging staff harvesting practices were compared to the IPPK, there were also no significant difference in terms of skid trail and logging road length. However, significant differences were found in terms of harvest intensities (1 stump per hectare compared to 4 stumps per hectare), number of damaged trees (20 trees per 100m compared to 38 trees per 100m) and total canopy openings (0.1ha gap per hectare of forest compared to 0.3ha gap per hectare of forest, respectively). These differences led us to exclude areas logged by the HPH's subcontractor from the comparison between HPH and IPPK permit holder logging operations in primary forest. A more detailed description of the harvesting practices by HPH and IPPK in primary forest is presented in the 'Discussion' section below.

Harvest intensity

Harvest intensities in primary forest were one and four stumps per hectare, respectively, for selective logging HPH and IPPK permit holders. In logged-over forest, harvest intensities were relatively similar for logging by HPH and IPPK permit holders (three stumps per hectare compared to four stumps per hectare, respectively). In primary forest, IPPK permit holders left more stumps per hectare for all diameter classes and more stumps in diameter classes 30cm–50cm, 50cm–70cm and 70cm–90cm than logging HPH (Mann-Whitney test; see Figure 8.2(a)). IPPK logging in logged-over forest resulted in more stumps per hectare than HPH selective logging for all diameter classes (see Figure 8.2(b)) and significantly more stumps at class 10cm–30cm.

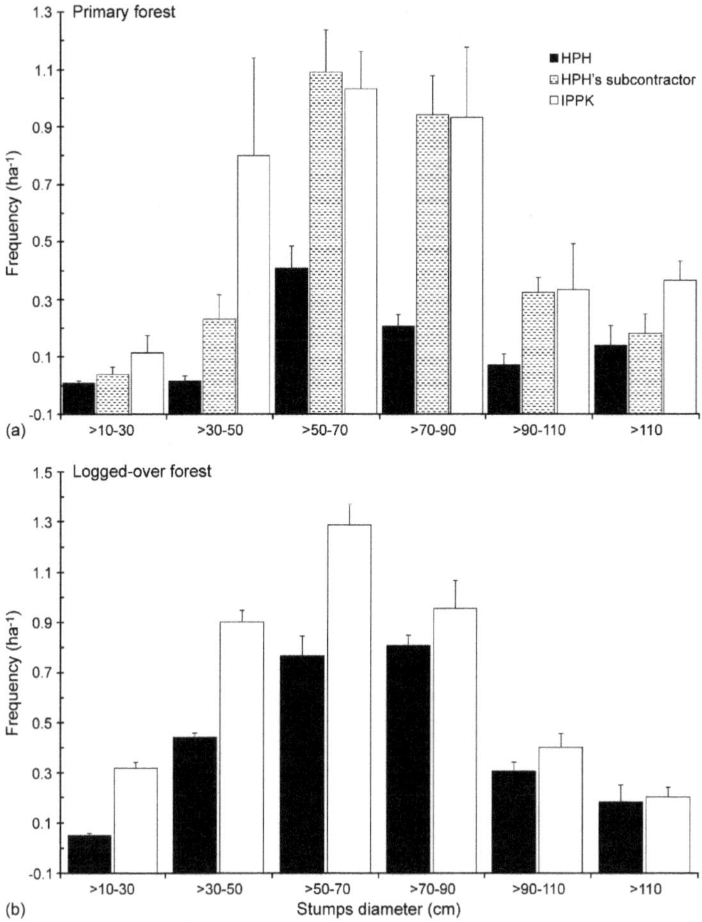

Source: Iskandar et al, 2006

Figure 8.2 *Frequency and distribution of stump diameters of trees harvested (stumps per hectare) by logging concessionaires (HPH), HPH's subcontractor, and short-term permit holders (IPPK) in both forest types*

In both forest types, IPPK permit holder operations cut more trees with diameters smaller than the TPTI diameter limit (50cm) than did HPH selective logging. In primary forest, the percentages for HPH and IPPK logging were 3 and 26 per cent, while in logged-over forest they were 19 and 30 per cent, respectively.

Logging road and skid trail and length

IPPK permit holder operations resulted in higher lengths of skid trails and roads per hectare than HPH logging in primary forest, and higher skid trail length in logged-over forest (see Table 8.1A). However, differences in road length per hectare between IPPK permit holders and HPH logging were not significant in

Table 8.1 *Comparison of skid trail and logging road length between HPH, HPH's subcontractor and IPPK*

A Per unit area (ha^{-1})	*HPH Mean ± SE*	*IPPK Mean ± SE*	*HPH's subcontractor Mean ± SE*
Primary forest			
Skid trails length (m/ha^{-1})	33.6 ± 5.9 a	48.9 ± 13.5 a	51.6 + 8.8 a
Logging road length (m/ha^{-1})	24.9 ± 4.4 a	27.0 ± 6.4 a	27.3 + 2.8 a
Logged-over forest			
Skid trails length (m/ha^{-1})	43.4 ± 5.9 a	62.7 ± 5.2 b	n.d.
Logging road length (m/ha^{-1})	34.2 ± 1.9 a	29.8 ± 1.8 a	n.d.
B Per number of trees harvested	*Mean ± SE*	*Mean ± SE*	*Mean ± SE*
Primary forest			
Skid trails length (m)	41.4 ± 8.2 a	14.1 ± 4.5 b	18.3 ± 1.7 b
Logging road length (m)	31.5 ± 7.1 a	7.4 ± 1.4 b	11.1 ± 2.1 b
Logged-over forest			
Skid trails length (m)	17.4 ± 2.7 a	18.6 ± 2.4 a	n.d.
Logging road length (m)	13.5 ± 0.5 a	8.9 ± 1.3 b	n.d.
C Per volume (m^3) *of trees harvested*	*Mean ± SE*	*Mean ± SE*	*Mean ± SE*
Primary forest			
Skid trails length (m/m^{-3})	3.4 ± 0.7 a	1.2 ± 0.2 b	1.7 ± 0.1 b
Logging road length (m/m^{-3})	2.5 ± 0.4 a	0.7 ± 0.2 b	1.1 ± 0.2 b
Logged-over forest			
Skid trails length (m/m^{-3})	1.6 ± 0.2 a	2.3 ± 0.4 a	n.d.
Logging road length (m/m^{-3})	1.3 ± 0.1 a	1.1 ± 0.2 a	n.d.

Notes: Plots for HPH and IPPK in primary forest (6 and 3 plots, respectively) and in logged-over forest (6 and 15 plots, respectively).
n.d. = no data available.
Means followed by the same letter are not significantly different by ANOVA at $P < 0.05$.
Source: Iskandar et al, 2006

either forest type. The only significant difference between the two types of operators was for skid trail length in logged-over forest (63m per hectare for IPPK compared to 43m per hectare for HPH).

When analysed per harvested tree, IPPK operations produced lower lengths of skid trails and roads per tree harvested than HPH logging in both forest types (see Table 8.1B). However, the difference was significant only for road length in primary (7m compared to 32m per stump, respectively) and in logged-over forest (9m compared to 14m per stump, respectively). IPPK permit holders also built lower lengths of road per unit volume (cubic metres) harvested than HPH logging in both forest types, and lower lengths of skid trails in primary forest (see Table 8.1C). However, the only significant difference was in logging road length in primary forest (1m compared to 3m per cubic metres for IPPK and HPH logging, respectively).

Logging damage to residual trees

In both forest types, IPPK permit holders damaged more trees per hectare than HPH logging. There were significant differences in primary forest for diameter

Table 8.2 *Comparison of logging damage between HPH, HPH's subcontractor and IPPK logging operations*

A Per unit area (ha^{-1})	*HPH Mean ± SE*	*IPPK Mean ± SE*	*HPH's subcontractor Mean ± SE*
Primary forest			
Trees damaged per length (100m^{-1})	19.9 ± 4.1 a	38.0 ± 6.1 b	32.2 + 2.6 b
Total canopy opening (ha ha^{-1})	0.11 ± 0.02 a, *	0.31 ± 0.01 b	0.30 + 0.04 b
Logged-over forest			
Trees damaged per length (100m^{-1})	32.0 ± 2.0 a	43.3 ± 2.2 b	n.d.
Total canopy opening (ha ha^{-1})	0.19 ± 0.14 a	0.32 ± 0.03 b	n.d.
B Per number of trees harvested	*Mean ± SE*	*Mean ± SE*	*Mean ± SE*
Primary forest			
Number of trees damaged	13.6 ± 3.3 a	7.8 ± 0.8 a	9.5 ± 1.2 b
Total canopy opening (m^2)	1506 ± 387 a	865 ± 59 a	1099 ± 84 b
Logged-over forest			
Number of trees damaged	9.9 ± 1.1 a	11.2 ± 1.1 a	n.d.
Total canopy opening (m^2)	757 ± 77 a	904 ± 105 a	n.d.
C Per volume (m^3) of trees harvested	*Mean ± SE*	*Mean ± SE*	*Mean ± SE*
Primary forest			
Number of trees damaged	1.2 ± 0.3 a	0.7 ± 0.1 a	0.9 ± 0.1 b
Total canopy opening (m^2 m^{-3})	120.4 ± 18.6 a	77.1 ± 7.6 a	102.2 ± 8.0 b
Logged-over forest			
Number of trees damaged	0.9 ± 0.1 a	1.3 ± 0.2 a	n.d.
Total canopy opening (m^2 m^{-3})	72.2 ± 7.9 a	114.2 ± 22.7 a	n.d.

Notes: Plots for HPH and IPPK in primary forest (6 and 3 plots, respectively) and in logged-over forest (6 and 15 plots, respectively).
n.d. = no data available.
Means followed by the same letter are not significantly different by ANOVA at $P < 0.05$, and followed by * are significant difference at $P < 0.01$.
Source: Iskandar et al, 2006

classes 10cm–30cm (17 trees per hectare compared to 4 trees per hectare, respectively), 30cm–50cm (8 trees per hectare compared to 4 trees per hectare, respectively) and 50cm–70cm (2 trees per hectare compared to 1 tree per hectare, respectively). Moreover, 82 per cent of trees damaged by IPPK operations were less than 50cm dbh.

In logged-over forest, most damaged trees were in diameter classes 10cm–30cm (24 trees per hectare compared to 13 trees per hectare, respectively) and 30cm–50cm (12 trees per hectare compared to 8 trees per hectare, respectively), which represented 86 per cent of all damaged trees.

Significant differences were found between IPPK permit holders and HPH logging operations in both forest types with regard to the number of damaged trees and area of canopy opening per unit length or area (Table 8.2A). However, there were no significant differences between HPH's subcontractor and IPPK in those two categories of damage in primary forest. On the other hand, a significant difference was found between HPH's subcontractor and HPH logging staff. In primary forest, values for IPPK, HPH's subcontractor and HPH were 38, 32 and 20 trees per 100m for damaged trees, with 0.3ha, 0.3ha and 0.1ha per hectare for

Table 8.3 *Trees damaged per 100m length of skid trail or logging road (trees 100m^{-1}) due to HPH and IPPK logging in primary forest*

Types of damage/locations	*Skid trail and felling gap*		*Logging road*	
	Mean ± S. E.	*Range*	*Mean ± S. E.*	*Range*
HPH (n = 6 plots)				
Bark and wood damage	2.3 ± 0.5	0.1–3.5	0.4 ± 0.2	0.0–1.3
Leaning	3.9 ± 1.5	0.1–10.4	0.6 ± 0.3	0.0–1.7
Crown injury	8.2 ± 1.7	2.8–14.6	3.4 ± 1.5	0.0–9.3
Broken trunk – alive	0.6 ± 2.4	1.3–4.9	1.1 ± 0.5*	0.0–3.4
Broken trunk – dead	5.5 ± 1.4	1.3–11.1	1.8 ± 0.8*	0.0–5.4
Uprooted	7.6 ± 3.2	1.0–22.3	1.4 ±0.6*	0.0–4.1
IPPK (n = 3 plots)				
Bark and wood damage	2.4 ± 0.5	1.5–3.0	0.5 ± 0.2	0.0–0.9
Leaning	1.7 ± 0.2	1.5–2.1	0.7 ± 0.3	0.0–1.1
Crown injury	8.2 ± 2.5	3.2–11.2	9.1 ± 1.1	7.1–11.2
Broken trunk – alive	2.8 ± 1.1	0.7–4.0	4.9 ± 0.9*	3.3–6.2
Broken trunk – dead	4.3 ± 0.8	2.9–5.8	11.8 ±2.0*	8.5–15.5
Uprooted	15.4 ± 4.1	7.3–21.1	25.7 ± 4.9*	17.0–34.1

Note: Statistically significant difference by Mann–Whitney test – SPSS, between HPH and IPPK at P < 0.05
P-value: 2* (1-tailed sig) – not corrected for ties.

canopy opening, respectively. In logged-over forest, values were 43 trees per 100m compared to 32 trees per 100m for damaged trees, and 0.3ha per hectare compared to 0.2ha per hectare for canopy opening, respectively.

When damage was analysed per number and volume of trees harvested (see Table 8.2B and C), no significant differences were found between types of operators in either forest type. However, HPH logging operations produced higher levels of damage than IPPK in primary forest, while in logged-over forest, IPPK logging produced higher levels of damage than HPH.

Table 8.4 *Trees damaged per 100m length of skid trail or logging road (trees 100m^{-1}) due to HPH and IPPK logging in logged-over forest*

Types of damage/locations	*Skid trail and felling gap*		*Logging road*	
	Mean ± S. E.	*Range*	*Mean ± S. E.*	*Range*
HPH (n = 6 plots)				
Bark and wood damage	1.4 ± 0.3	0.5–2.7	0.1 ± 0.0	0.0–1.1
Leaning	2.1 ± 0.4	1.1–3.0	0.4 ± 0.1	0.0–0.7
Crown injury	15.3 ± 1.5	10.4–19.8	11.3 ± 1.3	7.9–15.1
Broken trunk – alive	4.4 ± 0.6	2.7–6.7	4.5 ± 0.5	2.9–6.1
Broken trunk – dead	4.7 ± 0.5	2.6–5.6	4.6 ± 0.6	2.4–6.5
Uprooted	9.5 ± 1.1	5.0–13.7	5.3 ± 0.8*	2.8–8.0
IPPK (n = 3 plots)				
Bark and wood damage	1.3 ± 0.2	0.0–3.1	0.4 ± 0.1	0.0–1.2
Leaning	1.6 ± 0.2	0.0–2.9	1.2 ± 0.3	0.0–4.4
Crown injury	16.8 ± 1.2	8.4–22.2	17.2 ± 2.5	6.9–47.5
Broken trunk – alive	4.3 ± 0.3	2.7–6.1	5.2 ± 0.4	3.1–8.3
Broken trunk – dead	4.2 ± 0.3	2.3–6.5	7.5 ± 0.9	1.4–14.2
Uprooted	14.3 ± 1.6	6.1–27.2	15.0 ± 2.9*	1.6–38.7

Note: Statistically significant difference by Mann–Whitney test – SPSS, between HPH and IPPK at P < 0.05
P-value: 2* (1-tailed sig) – not corrected for ties.

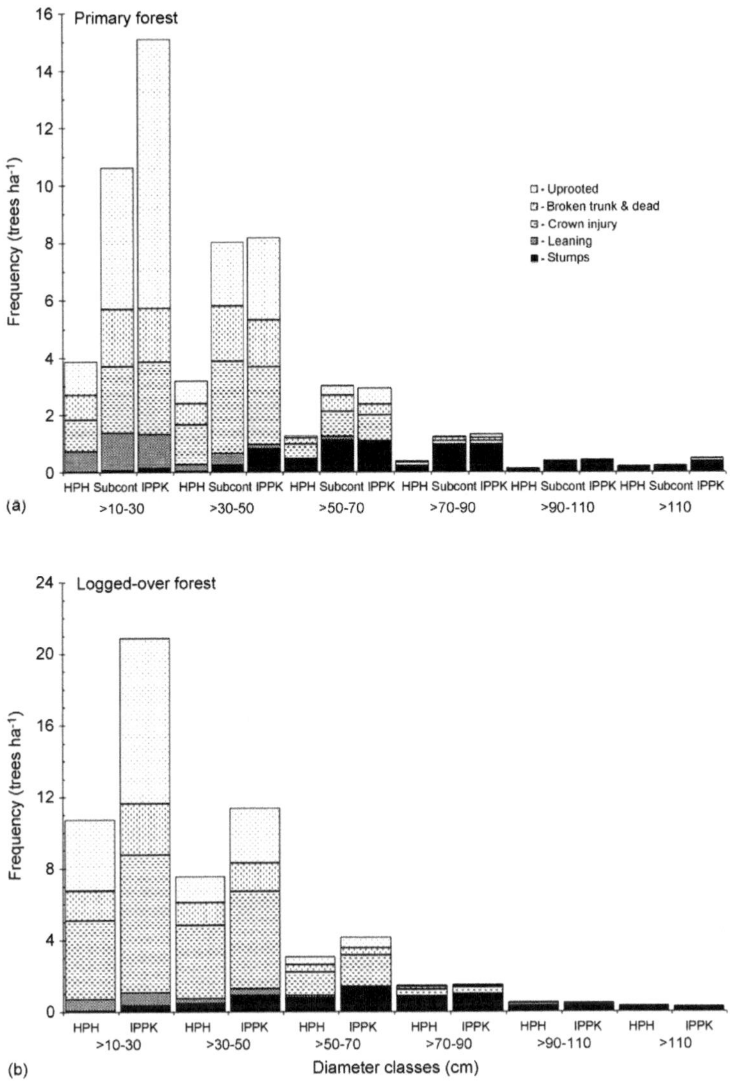

Source: Iskandar et al, 2006

Figure 8.3 *Frequency and distribution of potential loss of trees (trees per hectare) by logging concessionaires (HPH), HPH's subcontractor and short-term permit holders (IPPK) in both forest types*

Category of damage to residual trees

In primary forest, for both operators, the most common types of harvesting damage to residual trees are crown injury, broken trunk (dead) and uprooted (see Table 8.3). Along skid trails and within felling gaps, the number of damaged trees was similar between HPH and IPPK logging. However, along roads, IPPK damaged significantly more trees than HPH logging in the categories 'broken trunk (alive)' (4.9 trees per 100m compared to 1.1 trees per 100m, respectively), 'broken trunk (dead)' (11.8 trees per 100m compared to 1.8 trees per 100m, respectively) and 'uprooted' (25.7 trees per 100m compared to 1.4 trees per 100m, respectively). The common categories of damage in logged-over forest were 'crown injury' and 'uprooted' (see Table 8.4). However, the only significant difference was in the number of uprooted trees along roads (15 trees per 100m compared to 5 trees per 100m for IPPK and HPH logging, respectively).

We hypothesized that trees suffering four of the categories of damage (leaning, crown (injured), broken trunk and uprooted) would either rot or die, and thus be lost from the population of future harvest trees. In primary forest, IPPK permit holders would thus lose significantly more trees from this population than HPH logging from all diameter classes (28 trees per hectare compared to 9 trees per hectare, respectively; see Figure 8.3(a)). More than 80 per cent of these trees were in diameter classes 10cm–30cm (15 trees per hectare compared to 4 trees per hectare, respectively) and 30cm–50cm (8 trees per hectare compared to 3 trees per hectare, respectively). IPPK harvests would also result in higher losses than HPH selective logging from the population of future harvest trees in logged-over forest (39 trees per hectare compared to 28 trees per hectare, respectively; see Figure 8.3(b)). Significant differences were in diameter classes 10cm–30 cm (21 trees per hectare compared to 11 trees per hectare, respectively) and 30cm–50cm (11 trees per hectare compared to 7 trees per hectare, respectively). However, HPH's subcontractor harvesting in primary forest resulted in significantly higher losses than HPH logging staff in all diameter classes (23 trees per hectare compared to 9 trees per hectare), with similar losses to IPPK (28 trees per hectare). The difference was only found in diameter class 10cm–30 cm dbh (11 trees per hectare compared to 15 trees per hectare, respectively).

Discussion

Harvest intensity

Harvest intensities documented in this study (1 to 4 trees per hectare) can be classified as 'low' compared to other selective logging studies in Borneo (5 to 15 stumps per hectare: Bertault and Sist, 1997; 6 to 14 stumps per hectare: Pinard et al, 2000a; 7 to 11 stumps per hectare: Sist et al, 1998). Removals might be described as under-harvesting, being lower than permissible harvest levels, which range from 8 to 10 trees per hectare, depending upon tree diameter and the annual allowable cut (AAC) calculated from inventories. Our low harvest estimates may reflect the different methodology used for our study from others. Estimates in other studies were based on permanent sample plots ranging in size from 1ha to 4 ha, which were established before logging operations (Suhartana and Idris, 1998; Pinard et al, 2000a; Sist et al, 2003). We used sample plots of 20ha, established in areas previously harvested. We may have underestimated harvest levels because we could not be sure that all stumps of harvested trees in cleared areas (i.e. log landings and logging roads) had been counted. However, low levels of harvest by HPH logging crews in primary forest might have been deliberate, reflecting a recognized lack of machinery or resources for harvesting on the part of this state-owned company. It is not uncommon for low harvests by HPH logging crews to be followed by subsequent harvests, usually a few months later, by a HPH's subcontractor that is better equipped to fulfil their AAC. Strictly speaking, this qualifies as repeat-entry harvesting, which is a violation of TPTI regulations.

The different stump size distribution and harvest intensities between HPH and IPPK logging may reflect differences in their respective markets for logs. A study by Obidzinski and Suramenggala (2000) in East Kalimantan revealed that many logs from IPPK harvests were sold to local sawmills for local consumption. These small- and medium-sized sawmills process smaller logs than those that HPH logging operators send to plywood industries in East Kalimantan. The differences in stump size and harvesting intensities between IPPK and HPH logging may also reflect the fact that IPPK permits do not incorporate exact guidelines, whereas HPH selective logging operators are required to follow TPTI guidelines, with their minimum diameter limits of 50cm. However, considering the result from HPH's subcontractor, after the IPPK system was put in place, several HPH operations were operating in a similar fashion to IPPK. This is evidenced in the similarity in harvest intensity and damage from their logging operations. FWI/GFW (2002) found that IPPK holders' permits, lacking guidelines, favour maximizing current harvests, including trees below the TPTI diameter limit. In the case of HPH selective logging, many of the stumps less than 50cm in diameter were found along logging roads, where trees had been cut to clear land for road maintenance and building materials, rather than for timber.

Logging road and skid trail and length

Several authors have reported that, at low harvest levels (less than four trees per hectare), most logging damage results from road construction rather than the felling of trees (Gullison and Hardner, 1993; Sist et al, 2003). Similarity between operators in road length per hectare reflected that many IPPK harvesting operations were using logging roads that had been previously constructed by HPH. They also used the same type of machinery. This similarity in road length may also reflect the topography in the study area. A study by Johns et al (1996) in Brazil reported that planned and unplanned logging operations in even topography constructed similar lengths of roads.

Logging road length per hectare in this study (from 25m to 34m per hectare) was relatively similar to those described by Johns et al (1996) (i.e. 23m per hectare in a 105ha area to 27m per hectare in a 75ha area, with a harvest intensity of 5 and 6 stems per hectare, respectively). In a larger study area in Bolivia, Gullison and Hardner (1993) reported that the 8m per hectare of roads were built for a harvest intensity of 0.1 stems per hectare within a 602ha felling area. These differences

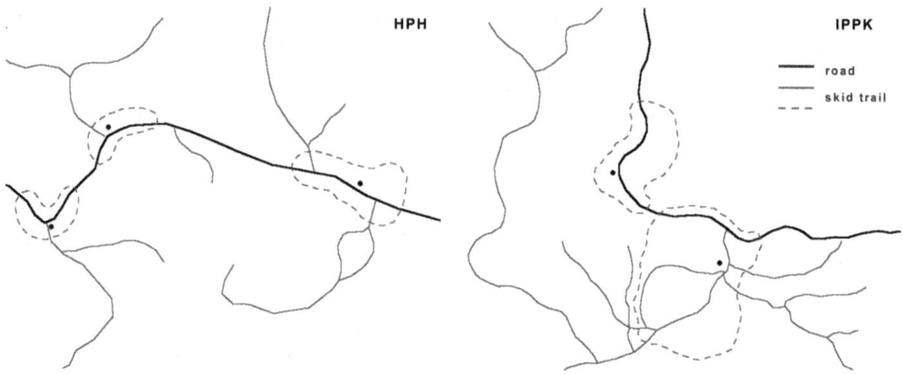

Source: Iskandar et al, 2006

Figure 8.4 *Differences of road, skid trails and log landings constructed in logged-over forest between concessionaires (HPH) and short-term permit holders (IPPK) logging operations*

may reflect the different methodology used for these studies (i.e. the average logging road length per hectare declines as the sample area increases).

A reason for extensive skid trail length per hectare in IPPK and HPH's subcontractor logging in primary forest (49m and 52m per hectare, respectively), as well as IPPK logging in logged-over forest (63m per hectare) was revealed by visual checks of operators' logging areas. The choice of where to create a skid trail was determined by the crawler tractor drivers trying to get access to areas where trees had been cut. This resulted in a network of unplanned skid trails to pull the logs from stumps to the log landing (see Figure 8.4). Many of these skid trails were not utilized for log transport but as shortcuts to felling gaps, leading to many dead ends that resulted in damage. It is widely acknowledged that the key to reducing impact and ground area disturbance is to reduce unnecessary skidding (Suhartana and Idris, 1998; Jackson et al, 2002).

The total length of skid trails and road length per number or volume of harvested trees was generally shorter for IPPK than for HPH logging (except for skid trail length in logged-over forest). This reflected the higher number of trees harvested by IPPK and HPH's subcontractor, which reduced the length of skid trails and roads per unit harvested. Accordingly, IPPK and HPH's subcontractor are more efficient in terms of utilizing machinery than for road and skid trail construction than HPH logging staff operations.

Skid trail lengths measured in our study (34m to 63m per hectare in a 20ha plot) were much lower than those reported by Johns et al (1996) in Brazil (158m per hectare in a 105ha plot to 171m per hectare in a 75ha plot with a harvest intensity of 5 and 6 stems per hectare, respectively) and Whitman et al (1997) in Belize (104m per hectare of skid trail with a harvest intensity of 0.5 stems per hectare in a 92ha plot). Moreover, Pinard et al (2000a) in Malaysia reported that higher harvest intensity of 14 stems per hectare resulted in 168m per hectare of skid trails in a 50ha plot. However, in a larger study area in Bolivia, Gullison and

Hardner (1993) reported that 14m per hectare of skid trails resulted from a harvest intensity of 0.1 stems per hectare in a 602ha plot. These differences may reflect the different methodology used and different harvest intensities among the studies. Average length of skid trails per hectare is reduced as study areas increase and harvest intensity declines.

Logging damage to residual trees

The direct impact of logging upon forests is obviously a net reduction in trees. However, the indirect impacts from damage, mortality and rot to residual trees also affect the future health of forests (Pinard and Putz, 1996). Logging activities by IPPK permit holders resulted in higher numbers of damaged trees per hectare than logging by HPH in both forest types. In primary forest, the higher number of trees greater than 90cm dbh harvested by IPPK and HPH's subcontractor would be expected to have affected the size of canopy openings and the number of damaged trees. Other studies in Kalimantan suggested that felling intensity was an important feature in the damage caused by logging (Bertault and Sist, 1997; Sist et al, 1998). In logged-over forest, the differences in damage between the two types of operators reflected better harvesting techniques of the private HPH logging operator. IPPK permit holders' harvests resulted in excessive skid trail networks and log landings. However, the high levels of harvest by IPPK reduced the incidence of damage per unit and volume (cubic metres) of trees harvested, compared to the HPHs.

Overall, IPPK permit holders' operations resulted in a higher number of damaged trees per 100m unit length compared to HPH logging. The number of trees damaged per 100m length of skid trail and road (from 20 to 43 trees per 100m; see Table 8.2) was much lower in this study than in that by Johns et al (1996), which revealed 67 to 135 damaged trees per 100m in length in planned and unplanned logging areas, respectively. However, our results were similar to those reported by Jackson et al (2002) in Bolivia (42 to 56 trees per 100m road length). These differences probably reflected higher levels of harvests (six stems per hectare; Johns et al, 1996) than in our study, which ranged from one to four stems per hectare. Most damaged trees were in diameter classes 10cm–30cm, paralleling observations by Sist et al (2003) in East Kalimantan that the highest number of damaged trees were in diameter classes 20cm–30cm and 10cm–40cm dbh reported by Pinard and Putz (1996) in Malaysia.

Total canopy opening per hectare, ranging from 1106 square metres to 3226 square metres per hectare (see Table 8.2A) was relatively similar to rates reported in other studies of conventional logging (2276 square metres per hectare: Johns et al, 1996; 2575 square metres per hectare: Jackson et al, 2002). A study by Pinard et al (2000a) in Malaysia reported that a harvest intensity of 14 to 11 stems per hectare resulted in 1660 square metres to 1020 square metres per hectare of canopy opening in conventional logging areas of 50ha and 70ha, respectively. In Brazil, Pereira et al (2002) found that, at a harvest intensity of 6 trees per hectare, conventional logging operations in a 14ha area produced 1120 square metres per hectare of canopy openings, whereas in a 112ha area with a harvest intensity of 4 trees per hectare, 890 square metres per hectare of canopy openings resulted. The high levels of harvest by IPPK and HPH's subcontractor reduced the incidence of logging damage per number and volume of trees harvested compared to HPH logging staff. The number of trees damaged and total canopy opening were not found to be significantly different among operators in both forest types. This reflected that harvest intensity would affect damage by logging, which reduced the degree of disturbance from logging operation per hectare or unit area. With all details collected on harvesting practices, the low damage by IPPK and HPH's subcontractor indicate that they perform harvesting operations more efficiently than HPH logging staff. However, when considering sustainability, timber harvesting on a long-term basis and the future health of forests, logging practices by IPPK and HPH's subcontractor were classified as a low-grade system. Large canopy openings can promote the regeneration of pioneer species and may limit the establishment of Dipterocarp regeneration due to competition (Pinard et al, 2000b). However, several authors have reported that regeneration and growth of commercial trees tended to be greater in the most severely damaged stands more than a year after the canopy was opened (Sist and Nguyen-The, 2001; Fredericksen and Pariona, 2002).

Observations in logged-over forest areas in East Kalimantan show that many Dipterocarp species (*Shorea laevis*, *S. leprosula*, *S. parvifolia* and *Dryobalanops aromatica*) regenerate on skid trails, in felling gaps and along logging roads (Kartawinata, pers comm, 2003).

Category of damage to residual trees

In both forest types, significant differences between operators in types of tree damage were found only for uprooted trees along logging roads. This reflects the higher number of smashed trees that had been pushed by the crawler tractor to the roadside during road construction on short-term permit holder areas. Many of these uprooted trees were of small diameter (less than 30cm dbh) or lower-value timber species, which encouraged the crawler tractor drivers to clear them. Another reason is that when the temporary log landings were full, logs were piled along the roads, damaging smaller trees along the edges. Trees with crown injury and broken trunk (dead) dominated the damage category along logging roads in both forest types. Most were in small diameter classes (less than 50cm dbh), which reflects the higher level of secondary damage to residual trees during road construction. The secondary damage due to road construction is typically greater than that caused by the felling of commercial trees (Gullison and Hardner, 1993). Trees pushed over fall into adjoining forest, injuring more trees and creating a corridor of damage considerably wider than the road itself. Although the length of logging road per plot was lower than skid trail length, there was substantially more damage to the forest along roads.

This study quantified the total number of harvested and damaged trees; but data were not collected on the number or species of remaining undamaged trees.

Another study in this area recorded the density of 210 trees per hectare less than 50cm dbh (Sist et al, 2003). This would mean that the rates of damage that we recorded, which ranged from 18 to 32 trees per hectare less than 50cm dbh (in both forest types), would correspond to 9 to 15 per cent of pre-harvest tree density. These estimates are lower than those described by Pinard and Putz (1996), which ranged from 17 to 57 per cent, and Sist et al (2003), which ranged from 19 to 56 per cent of initial tree density. The distinction may reflect the different methodology used as well as the different harvest intensity in our study. It is hard to know whether levels of damage of 9 to 15 per cent of pre-harvest trees less than 50cm would have a significant impact upon the future harvest potential of these stands. Furthermore, Pinard and Putz (1996) found that 10 to 43 per cent of trees with moderate types of damage (stem and bark damage) and trees snapped off in diameter classes less than 60cm died during the first year after logging.

Other studies have reported that the mean annual mortality rate in logged-over forest during the two years following logging operations was 2.6 per cent per year, reflecting higher mortality of trees damaged during logging (4.9 per cent); this declined in year four after logging to 1.5 per cent per year (Sist and Nguyen-The, 2001). This suggests that between now and the time of the next polycyclic harvest, in 35 years, the number of dead trees resulting from damaged trees documented in this study would be low compared to the number of dead trees that would result from average annual mortality rate. In another study, Bertault and Sist (1997) found that 23 per cent of pre-harvest trees (115 trees per hectare) died during conventional logging operations.

Conclusions

IPPK and HPH's subcontractors certainly perform harvesting operations more efficiently in terms of total number and volume of trees harvested. However, these practices were classified as a low-grade system for sustainably managed timber harvesting and the future health of forests. These high levels of harvest combined with large numbers of damaged trees by IPPK and HPH's subcontractor will reduce future harvest options and timber potential in these areas. Increases in total canopy openings beyond natural conditions in unplanned logging by IPPK and HPH operations may favour regeneration of pioneer trees and limit the natural establishment and growth of Dipterocarp species due to competition. However, the long-term density of trees growing on areas with high levels of disturbance during logging remains uncertain (Pinard et al, 2000b).

This suggests that the lack of long-term licences and an almost total lack of government regulations encourage higher harvest and harvest-related damage rates. However, without proper harvesting regulations and support by the well-established control institution, harvesting practices under the HPH system resulted in similar logging damage to harvesting under the IPPK system. Both types of operators could reduce damage to the soil and residual trees, which may contribute to the sustainability of timber harvesting in the future for both the community and the district government by switching from the use of heavy and destructive crawler tractor to lighter and more environmentally friendly machinery and following reduced-impact harvesting practices. Previous studies in Kalimantan have quantified the difference in networks of skid trails constructed with different machinery (Sist et al, 2003; Klassen, pers comm, 2004). Skidding with lighter and smaller machinery, such as the pulldozer tractor CAT 527, with narrow blades (3m wide) and enhanced manoeuvrability, resulted in lower total skid trail length and impact to the soil than skidding with the CAT D7G crawler tractor.

References

Anon. (1998) *Impulse User's Manual*, 8th edition, Laser Technology Inc, Englewood, CO

Barr, C., Wollenberg, E., Limberg, G., Anau, N., Iwan, R., Sudana, I. M., Moeliono, M. and Djogo, T. (2001) *The Impacts of Decentralization on Forests and Forest-Dependent Communities in Malinau District, East Kalimantan, Case studies on decentralisation and forests in Indonesia*, CIFOR, Bogor, Indonesia

Bertault, J.-G. and Sist, P. (1997) 'An experimental comparison of different harvesting intensities with reduced-impact and conventional logging in East Kalimantan, Indonesia', *Forest Ecology and Management*, vol 94, pp209–218

Chhetri, D. B. K. and Fowler, G. W. (1996) 'Estimating diameter at breast height and basal diameter of trees from stump measurements in Nepal's lower temperate broad-leaved forests', *Forest Ecology and Management*, vol 81, pp75–84

Departemen Kehutanan (1993) *Pedoman dan Petunjuk Teknis Tebang Pilih Tanam Indonesia (TPTI) pada Hutan Alam Daratan*, Direktur Jenderal Pengusahaan Hutan, Departemen Kehutanan, Jakarta, Indonesia

Fredericksen, T. S. and Pariona, W. (2002) 'Effect of skidder disturbance on commercial tree regeneration in logging gaps in a Bolivian tropical forest', *Forest Ecology and Management*, vol 171, pp223–230

FWI (Forest Watch Indonesia)/GFW (Global Forest Watch) (2002) *The State of the Forest: Indonesia*, FWI, Bogor, Indonesia, and GFW, Washington, DC

Gullison, R. E. and Hardner, J. J. (1993) 'The effect of road design and harvest intensity on forest damage caused by selective logging: Empirical results and simulation model from the Bosque Chimanes, Bolivia', *Forest Ecology and Management*, vol 59, pp1–14

Iskandar, H., Hubble, D. L. and MacDicken, K. G. (2003) 'Estimating dbh of commercial trees from stump measurements in Malinau, East Kalimantan', *Journal of Tropical Forest Science*, vol 15, no 3, pp502–504

Iskandar, H., Snook, L. K., Toma, T., MacDicken, K. G. and Kanninen, M. (2006) 'A comparison of damage due to logging under different forms of resource access in East Kalimantan, Indonesia', *Forest Ecology and Management*, vol 237, no 1–3, pp83–93

Jackson, S. M., Fredericksen, T. S. and Malcolm, J. R. (2002) 'Area disturbed and residual stand damage following logging in a Bolivian tropical forest', *Forest Ecology and Management*, vol 166, pp271–283

Johns, J. S., Barreto, P. and Uhl, C. (1996) 'Logging damage during planned and unplanned logging operations in the eastern Amazon', *Forest Ecology and Management*, vol 89, pp59–77

Machfudh (2002) 'General description of the Bulungan Research Forest', in CIFOR (ed) *Forest, Science and Sustainability: The Bulungan Model Forest. Technical Report Phase I 1997-2001, ITTO Project PD 12/97 Rev 1 (F)*, CIFOR, Bogor, Indonesia

Obidzinski, K. and Suramenggala, I. (2000) *Illegal Logging in East Kalimantan: Papers on Social, Economic and Political Implications*, Illegal Logging Project Report, CIFOR, Bogor, Indonesia

Palmer, C. and Engel, S. (2007) 'For better or for worse? Local impacts of the decentralization of Indonesia's forest sector', *World Development*, vol 35, pp2131–2149

Pemda Malinau (2001) *Keputusan Bupati Malinau tentang Ijin Pemanfaatan Kayu (IPK) Kabupaten Malinau tahun 2000/(2001)*, Bagian Sosial dan Ekonomi, Malinau, Indonesia

Pereira Jr., R., Zweede, J., Asner, G. P. and Keller, M. (2002) 'Forest canopy damage and recovery in reduced-impact and conventional selective logging in eastern Para, Brazil', *Forest Ecology and Management*, vol 168, pp77–89

Pinard, M. A. and Putz, F. E. (1996) 'Retaining forest biomass by reducing logging damage', *Biotropica*, vol 28, no 3, pp278–295

Pinard, M. A., Putz, F. E. and Tay, J. (2000a) 'Lessons learned from the implementation of reduced-impact logging in hilly terrain in Sabah, Malaysia', *International Forest Review*, vol 2, no 1, pp33–39

Pinard, M. A., Barker, M. G. and Tay, J. (2000b) 'Soil disturbance and post-logging forest recovery on bulldozer paths in Sabah, Malaysia', *Forest Ecology and Management*, vol 130, pp213–225

Resosudarmo, I. A. P. (2002) 'Timber management and related policies: a review', in Colfer, C. J. P. and Resosudarmo, I. A. P. (eds) *Which Way Forward? People, Forests and Policymaking in Indonesia*, Resources for the Future (RFF)/CIFOR, Washington, DC

Sist, P. and Nguyen-The, N. (2001) 'Logging damage and the subsequent dynamics of a dipterocarp forest in East Kalimantan (1990–1996)', *Forest Ecology and Management*, vol 165, pp85–103

Sist, P., Nolan, T., Bertault, J. G. and Dykstra, D. (1998) 'Harvesting intensity compared to sustainability in Indonesia', *Forest Ecology and Management*, vol 108, pp251–260

Sist, P., Sheil, D., Kartawinata, K. and Priyadi, H. (2003) 'Reduced-impact logging in Indonesian Borneo: Some results confirming the need for new silvicultural prescriptions', *Forest Ecology and Management*, vol 179, pp415–427

Suhartana, S. and Idris, M. M. (1998) 'Controlled skidding to minimize forest damages: case study at two forest companies in Central Kalimantan', *Buletin Penelitian Hasil Hutan*, vol 16, no 2, pp69–78

Whitman, A. A., Brokaw, N. V. L. and Hagan, J. M. (1997) 'Forest damage caused by selection logging of mahogany (*Swietenia macrophylla*) in northern Belize', *Forest Ecology and Management*, vol 92, pp87–96

9

Tane' Olen as an Alternative for Forest Management: Further Developments in Setulang village, East Kalimantan

Ramses Iwan and Godwin Limberg

This chapter describes how a conservation movement came to the village of Setulang, East Kalimantan, Indonesia. Despite having settled in the area since 1968, the people are perceived as immigrants with fewer rights to resources than the 'original' people. As a result, different parties made claims on the forest region with its potential for timber in the hope of gaining the attention of timber entrepreneurs willing to make offers to the local people. The people of Setulang rejected the small-scale timber harvesting (IPPKs) introduced during the decentralization era (2000 to 2002). In contrast to other villages in the region, which invited investors to clear fell their forest lands, Setulang's fight to protect its forests brought it the 2003 *Kalpataru* award[1] from the president of Indonesia.

A conservation movement had come about in the village.

Setulang village

Setulang Village, with a population of 860 (208 households) Dayak Kenyah Oma' Lung people, is situated on the lower reaches of the Malinau River. It is a two-hour trip by long boat from the capital of Malinau District. More recently, in 2006, a road was constructed to connect Setulang to Malinau Town.

According to oral history, the Kenyah Oma' Lung people lived in Long Sa'an, a village located on the upper reaches of the Pujungan River, a tributary of Bahau River, for 13 generations. The place was so remote that the people decided to look for somewhere to settle nearer the town, closer to markets and the government centre. In 1963, a scouting group went to look for a suitable destination in which to settle. The group had a meeting in Langap with Alang Impang, the chief *adat* leader of the Malinau River, to ask for his permission. Alang Impang then showed them several places, one of which was Long Setulang. They also had meetings with *adat* leaders in Gong Solok and Sentaban, and finally with the sub-district administrator in Malinau. In 1968, the whole community settled in the area that is now Setulang. A written decree of 1974 by the Bulungan Regent officially acknowledged Setulang's existence as a village, determining its upstream boundary with Setarap and downstream boundary with Sentaban.

At that time, Malinau was still a remote sub-district in the hinterland of East Kalimantan. But processes to integrate the margins of the state also reached these parts. One such process was introduced by Law 5, 1979, on village government, which imposed a uniform structure on all villages in Indonesia and called for the public election of a village leader. With reforms and decentralization, Malinau became an autonomous district and, at the same time, was finally completely integrated into the Indonesian state. Then the IPPKs arrived.

Since settling in Long Setulang, the people have developed crop rotation patterns and tree crops. They use their produce to fulfil their everyday requirements. Some is also sold, particularly the popular upland rice. With sufficient produce from their fields and gardens, the Setulang villagers have no difficulty in fulfilling their basic needs, even though they do not receive IPPK fees as do other villages in the Malinau River catchment area.

Many of the young people travel to Malaysia mainly to work in timber companies. Their experiences there make them aware of the impact that felling forests has upon local people and their environment. The money they earn there is used to buy more expensive goods, such as generators or outboard motors, and also to build houses.

Development of the *Tane' Olen*

The forest in the upper reaches of the Setulang River is virgin lowland Dipterocarp forest dominated by large trees, including *kapur* (*Dryobalanops* spp), *meranti* (*Shorea* spp), *keruing* (*Dipterocarpus* spp), *tengkawang* (*Shorea macrophylla*) and *ulin* (*Eusideroxylon zwageri*). The forests are a vital source of game (wild boar and deer), frogs, fish, fruits, medicinal plants and building materials for the Setulang villagers, who also use the river as a source of clean water for drinking, washing and bathing, and as a means of transport to their fields.

In 1974, two logging companies were operating near Setulang: PT Trang Jaya Nugraha in Setarap and PT Sanjung Makmur in Sentaban. The impact of the companies' activities resulted in murky water in the Setulang River and damaged forests, forcing villagers to look further afield for their building materials. Consequently, the people of Setulang rejected offers of cooperation from IPPK businessmen, who came promising many kinds of aid.

These changes gave rise to a desire to protect the 5300ha of forest in the upstream part of the Setulang River. The forest is protected as a buffer to ensure a clean water source for the village and provides a place for collecting building materials. According to traditional customs, the forest was declared *Tane' Olen* (protected land).

In the old days, *Tane' Olen* was part of the forest personally owned by the customary leader and his family. Its use was controlled by the *Paren* (aristocrats), and no one was allowed to use it or even enter. Exceptions are made, however, for special occasions to fulfil a public need. For example, in the case of disasters, such as the longhouse burning down, the necessary building material, such as poles and rattan, could be taken from the *Tane' Olen*. Also, in communal ceremonies, a hunt for bushmeat would be allowed in the *Tane' Olen*. Normally, the *Paren* would accompany people into the *Tane' Olen* as a form of direct control and require part of the harvest for their personal use.

Over time, with the erosion of *adat* and its institutions, people's understanding of the *Tane' Olen* changed. Today in Setulang, it is perceived as communally owned forest to be used for the common interest. Similarly, there have been changing perceptions of the role of *Tane' Olen* in some other villages in Malinau District.

With the changing perception of *Tane' Olen*, some community members also changed their opinion on its use. Initially in Setulang, certain factions still wanted control over the *Tane' Olen* for personal and family benefit. However, since community decisions are made through general village meetings, the opinion of the majority was followed. The villagers felt that *Tane' Olen* was so important that they needed their own organization to handle its management. A meeting was held by the village assembly – *Badan Perwakilan Desa* (BPD) – to discuss a suitable form and function for the organization. Villagers decided to found the Setulang Village *Tane' Olen* Management Body. The *Tane' Olen* in Setulang is currently managed by a body known as *Badan Pengelola* (management body) *Tane' Olen*. This body arranges all activities and enforces customary rules relating to *Tane' Olen*. Nevertheless, while the institution controlling the *Tane' Olen* has changed, the function and use of the *Tane' Olen* are still focused on forest conservation.

How the villagers fought for *Tane' Olen*

The people of Setulang are proud of their *Tane' Olen* and consider it an asset to the village. The management body is slowly growing into its role and learning to plan for the future.

In 2001, relevant customary rules were first written to be used as the basis for making decisions and, most importantly, so that the younger generation would know about customary rules. The rules relate to the removal and use of timber and non-timber forest products (NTFPs) found in the *Tane'Olen*, but also include decision-making processes. These rules have been a great help in facing conflicts with timber companies.

The first time that a timber company approached Setulang was in 1995, when Elisar Ipui was village head. A timber businessman came to Setulang offering cooperation to fell timber in the upper reaches of the Setulang River (Banjir Kap). But after a village meeting, in which the people considered the damage that would be caused, they rejected his offer.

From 2000 to 2002, the local government in Malinau District issued 46 IPPK small-scale timber harvesting permits to timber businessmen. During this period, eight timber businessmen came to Setulang officially, offering joint ventures with the people of Setulang to log the upper reaches of the Setulang River. Meetings were held in which only a minority of community figures were in favour of accepting the offers. Fearing any cooperation with these companies would lead to the destruction of their forests and water source, most of the people rejected all of the offers. A few of these cases are highlighted below.

Problems with CV Gading Indah[2]

CV Gading Indah had an IPPK small-scale timber harvest permit to operate in the area around Setarap village. Villagers in Setulang discovered that timber was being taken across their border with the Punan Setarap village when the water in the Setulang River started to change colour.

At the beginning of August 2002, Abuk Ipui and Yusten Sarun from Setulang visited CV Gading Indah's camp in Kuala Semiling. They saw many large logs that they suspected could not have been left behind by the company which once operated in the Semiling River area. Their suspicions were reinforced on 7 September 2002, when Krimson Nyepan and Pebing Lihai told the Setulang village head that they had found evidence of company timber felling while out hunting in the upper reaches of the Setulang River.

On 14 September 2002, following up on the report, the Setulang village head sent five people out to the location to investigate. Their inspections revealed that 33 meranti trees, 720m into Setulang village territory, had been felled and their trunks removed.

These findings were discussed in a village meeting on 16 September 2002, in which the villagers decided to put a stop to CV Gading Indah's activities until the situation was solved. On 17 September 2002, 116 Setulang villagers approached the CV Gading Indah camp and seized all the heavy machinery working in the felling block, gathered it all at the logging camp at the confluence of the Semiling River with the Malinau River, and symbolically fenced it all in using tree roots.

Their seizure of heavy machinery brought company operations to a standstill, and the company promised to go to Setulang. Initial discussions were held in Setulang on 20 September 2002, attended by the manager of CV Gading Indah, the South Malinau sub-district secretary, members of the Malinau Police Force and the Malinau Regional Military Command, and the chairman of the Malinau District Kenyah Dayak Community.

In the meeting, the villagers submitted a demand for 5 billion rupiah, stating that they had suffered losses from damage to the forest and that timber had been removed without their permission. The company admitted its wrongdoing, saying that it was the result of careless workers and was not done intentionally. The company asked for concessions from the Setulang villagers; yet by the end of the meeting, no agreement had been reached.

Following the discussions in Setulang, the manager of CV Gading Indah and the South Malinau sub-district secretary sent reports to the Malinau regent on the problem. On 27 September 2002, a Setulang village delegation was called to make a statement to the regional secretary in the Regent's Office in Malinau. Villagers sent seven people to report on what had happened in the forest.

A second discussion was held in Setulang on 29 September 2002 with the manager of CV Gading Indah, the South Malinau sub-district secretary, the chief of the Malinau Police Force with one of his officers, members of the Malinau Regional Military Command and the chairman of the Malinau District Kenyah Dayak Community.

After the company offered 100 million rupiah, the villagers lowered their demand from 5 billion rupiah to 3.5 billion rupiah. The company then offered 150 million rupiah and the villagers went lower still, to 2.5 billion rupiah. Because no agreement was reached in the discussion and the representatives of the company were not willing to make any decision, affirming that there were four other partners to consider, the villagers asked that all partners attend the next meeting.

At the third discussion held in Setulang on 4 October 2002, with the South Malinau sub-district administrator, the South Malinau sub-district secretary, members of the Malinau Regional Police Force and the Malinau Regional Military Command, the chairman of the Malinau District Kenyah Dayak Community, and five people from CV Gading Indah, the company made an offer of 200 million rupiah. Villagers stuck to their demand for 2.5 billion rupiah. Because the two parties could not reach an agreement, the South Malinau sub-district administrator suggested that they take the problem to the regency level.

A meeting was then held on 15 October 2002 in the meeting room of the Malinau Regent's Office. The meeting, led by the secretary of Malinau District, was attended by the head of the Forestry Office, the chief of police, the district military commander, two staff members from the Prosecutor's Office, one from the government administration, one legal staff member, the manager of CV Gading Indah, and 15 representatives from Setulang village. The meeting got under way at 2.00pm; but because no agreement had been reached after 2 hours, it was adjourned for 15 minutes to allow each party the opportunity to speak in private. During the negotiations, the villagers left the government with the responsibility of finding a good solution.

When they reconvened, the regent suggested that the head of general affairs and the head of programme planning, who were both from Setulang, take part in order to prevent conflict, or worse. The meeting ended at 8.30pm, with an agreement that the company should pay 400 million rupiah in damages to the people of Setulang.

The company could not manage this amount, arguing that their permit would expire on 3 December 2002. The regional government thus gave CV Gading Indah until January 2003 to work on four logging blocks that had not yet been felled and to pay their dues.

Finally, on 16 October 2002, a ceremony took place in which the agreement was signed where 200 million rupiah in cash were handed over to the people of Setulang and the company was to pay the remaining sum in installments.

Problems with PT Inhutani II

In May 2003, PT Inhutani II ordered villagers who were working in the fields to leave the area that they claimed belonged to a farmers' group of Tanjung Lapang with whom they had negotiated a deal. In response, the Setulang villagers went to the production camp on 13 May 2003 to protest. There has been no follow-up to the problem as Setulang villagers are still waiting for a response from the Malinau Barat sub-district administrator.

Problems with PT Lestari Timur Indonusa (LTI)

When heavy rains in September 2002 caused the Setulang River to flood, villagers became suspicious when they noticed that the river water was cloudier than usual. They believed the changes were the result of company operations in the neighbouring village area.

From 2 to 6 December 2002, several villagers from Setulang were sent out to investigate. It turned out that PT Lestari Timur Indonusa (LTI), which was operating in the Paking village area, had encroached on Setulang village territory and removed export quality timber from 117 trees.

On 7 December 2002, Setulang villagers descended on the PT LTI camp in Paking and seized the heavy machinery operating in the forest. The villagers then sealed the area and took the ignition keys of the heavy equipment home. A letter was made on the spot and signed by the PT LTI manager acknowledging the confiscation.

On 10 December 2002, a meeting was held in Setulang between the villagers and PT LTI. Staff from the Mentarang Sub-district Office also attended; but nothing was solved because PT LTI refused to believe the Setulang villagers' findings. An agreement was reached to conduct another field inspection.

This second inspection took place on 13 and 14 December 2002. The participants included a staff member from the Mentarang Sub-district Office, one staff member from the Malinau District Forestry and Plantations Office, and villagers from both Paking and Setulang. The villagers, upset that their findings had not been believed, conducted a stricter count this second time, including all stumps over 10cm in diameter. They also counted damage to NTFPs and the number of landslides. They counted 709 timber tree stumps less than 10cm in diameter, 78 fruit and palm trees, landslides in 14 places with a combined length of 1400m, three junctions with logging trails totalling 400m in length, and a logging road 4720m long and 25m wide leading into the Setulang village area.

Results of the second inspection were discussed at the next meeting held on 19 December 2002 in the Mentarang Sub-district Office. Taking part were PT LTI, the Mentarang Sub-district, the local police, the Mentarang Sub-district military commander, Malinau District Forestry and Plantations Office staff, and villagers from Paking, Semolon, Sentaban and Setulang. Discussions focused on village borders; the company stated that they were unclear. No agreement was reached, so the Mentarang Sub-district concluded the meeting, suggesting that another be held at a later date.

The next meeting was held in Setulang on 19 March 2003; but the topic of discussion was no longer PT LTI taking timber – rather, it focused on village borders. At the meeting were the South Malinau Sub-district head and his staff, the Malinau Barat sub-district head, staff from the Forestry and Plantations Office, the head of legal affairs, officers from the Malinau Regional Police and the Regional Military Command, and villagers from Sentaban and Setulang. In this meeting, staff from the Forestry and Plantations Office said that the forests in the upper reaches of the Setulang River were designated as protected forest area as determined in the 1971 RTRWP (*Provincial Land-Use Plan*) for East Kalimantan. The Setulang villagers were disappointed with the announcement; to them, conservation areas meant government-owned areas where villagers had no rights to make demands on a company working in the area. They questioned why they had not been told before. The situation began to get out of control, so the meeting was brought to a close.

Villagers returning from a hunting trip in the upper reaches of the Setulang River said that they had heard the sound of tractor engines working in the area. Consequently, on 2 May 2003, several youths from Setulang went to the PT LTI work site. It turned out that the company was still logging despite the ignition keys from its heavy machinery being seized. The youths called all the heavy machinery operators, pulled out the dynamo starters from five machines and took them home to Setulang.

Not long after the dynamo starters were seized, operations had begun again; so on 20 May 2003, ten villagers with experience as operators and mechanics were sent to remove the heavy machinery. They took two tractors back with them to Setulang.

PT LTI reported the seizure of equipment to the Malinau deputy chief of police, saying that the Setulang villagers had been abusive and made threats. Consequently, on 26 May 2003, the deputy chief of police called the Setulang village chief and customary chief to testify to these actions.

After taking testimony from the village head, the police called two youths who were involved when the heavy machinery was seized. When a letter arrived from the deputy chief of police calling the two as witnesses, a village meeting was held in which a statement was made that all the Setulang villagers felt responsible and that those who had seized the heavy machinery had done so in the interests of all the people in the village.

At that time, the villagers hoped they would get an opportunity to meet with the regent, who was planning a visit to Setulang on the 29 May 2003. He did not come, however; so on 2 June 2003, 112 villagers decided to go to the Malinau Police Station to testify. The chief of the Regional Police Force said that the problem would be handed over to the Malinau regional government; but there has not yet been any follow-up.

How did Setulang manage to keep their *Tane' Olen*? There are several factors that have made it possible:

- *Peoples' cohesion.* The unity and cohesion of the Setulang people comprise its main asset. Collective action is helping to implement most village development activities and to deal with outside threats.
- *Village institutions.* The presence of *Tane' Olen* in Setulang has stimulated thinking within the village on how to divide tasks. The *Tane' Olen* management body is in charge of organizing activities and the implementation of customary regulations and sanctions.
- *Fulfilled basic needs.* Not only are subsistence needs fulfilled, but most people also have a little surplus, which provides cash for other needs such as clothing or costs of education. There has been no need to sell the forest for cash.
- *Attention from outsiders.* The development of *Tane'Olen* and its *de facto* recognition by the government and other villages has been much aided by the high level of outside interest. Assistance in the form of cross visits, training in forest inventory, mushroom cultivation, charcoal production and handicrafts has helped to raise the community's awareness of the importance of conservation and to open their minds.
- *Support from other parties.* Being different and more visible than people from other villages around them, the efforts of Setulang villagers in protecting their forest have attracted the attention of other parties. This attention has been shown in a number of ways. The people of Setulang village have received two awards since they began managing their *Tane' Olen* protected land: a grant to participate in the World Water Forum Kyoto in 2003, and the *Kalpataru* (the Indonesian environmental award), also in 2003.

Several organizations have provided input in support of the villagers' efforts:

- *CIFOR:* ongoing research and visits from various researchers kept discussions focused on conservation, conservation concessions and management of conservation areas, as well as potential benefits to be derived from conservation areas.
- *Seoul National University (Republic of Korea)* sent students to hold discussions with villagers on Dayak communities' dependence upon the forest.
- *The Ministry of Forestry* planned some research in the area.
- *The Indonesian Academy of Sciences (LIPI)* set up a research plot to investigate the potential and diversity of plant species in the region. The results will provide extra information and serve as a guide for the *Tane' Olen* Management Body set up by the people of Setulang.

Visits from tourists and researchers provided moral support to the people. During these visits, villagers are often invited to discuss various aspects and possibilities of forest management. These discussions convinced the people of the importance of taking into consideration sustainability in management planning for the benefit of present and future generations in Setulang and its neighbouring villages.

In addition, the *Tane' Olen* of Setulang has become the subject of several media reports on national and international television.

Until now, *Tane' Olen* has been widely supported, and many people have been involved in its management. However, a new generation is growing up with different needs and ideas for the future. Basic needs are also expanding. Cash, for example, is increasingly becoming a basic need, and it may not be completely supplied by the rice surplus. In the past, young people would work for some years in other places, such as Malaysia, earning cash. Today, administrative requirements to work in other countries are more complicated.

Although the establishment of *Tane' Olen* was a joint decision, there are some individuals who hope to obtain personal gain from its management. This raised a discussion on whether to push for stricter implementation of the rules related to managing the *Tane' Olen*. To date, this has not yet occurred; but when people become serious about maintaining the resources, this might become a real need.

New developments around the *Tane' Olen*

The strengths of the Setulang community have enabled them to maintain their protected forest. While still disputed by some neighbouring communities and not legally recognized by the government, their *Tane' Olen* has been recognized *de facto*. Are these strengths sufficient to face new challenges? Can the Setulang community withstand the pressure to fragment? Will the management body be strong enough to adjust to changing situations?

Over the last few years, the people of Setulang have started to experience a land shortage due to the delineation of village territories, logging operations, reforestation programmes, plans to establish oil palm plantations and population growth. This situation prompted them to consider how to manage land and resources in the village territory. The village land-use plan was developed in village meetings, facilitated by CIFOR staff. After the village meetings, a small committee was formed, including the village head, customary leader and village leaders, to translate these ideas into a village land-use sketch map.

In the process, the people of Setulang reached an agreement on the division of their village territory. They designated areas for agricultural tree crops, reserve forest, *Tane' Olen*, settlement expansion and area for the development of oil palm plantation. These uses have been visualized with a village land-use sketch map (see Figure 9.1), and customary regulations have been drafted for part of it. Some villagers consider customary regulations sufficient to regulate land use. Others think that the village plans should be linked to the district land-use plan to avoid

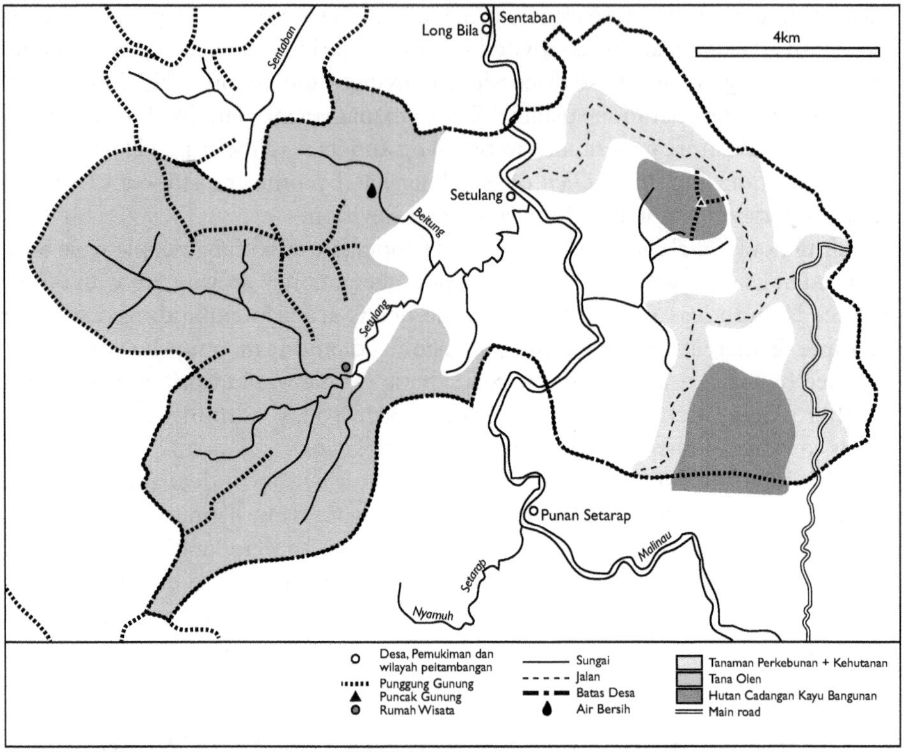

Source: Sidiyasa et al, 2005

Figure 9.1 *Land-use map of Setulang*

overlap in land use and to allow village planning to be accommodated at the district level.

Although the land-use planning exercise applied to the general village, it raised questions on management in the future. The *Tane' Olen* is not seen as independent from the rest of the village, and discussions on future needs resulted in changes in village land use. For example, if *Tane' Olen* is to be managed for ecotourism, not much logging can be allowed. At the same time, population increase necessitates the expansion of the settlement and the establishment of new houses. Where is the construction timber to come from? The village decided that the remaining patches of forest in the area designated for plantations should be kept in reserve as production forest. These patches are not large and will probably only suffice for the immediate expansion; but the people have already been planting trees in their abandoned fields for the longer term.

Meanwhile, district government development planning is still top down. Official land-use plans are said to include plans for oil palm and acacia plantation establishment. Since the people were not involved in the planning process and were not informed, they are perplexed and worried that these programmes are suddenly being implemented.

One great concern of local people is, therefore, the expansion of oil palm. The district government is planning the development of oil palm plantations, ostensibly to increase people's well-being. This plan has been debated among villagers because there is no clarity on the status of the land that will be planted. Also, it is not clear how benefits will be shared and/or distributed. Meetings to explain the plan have been conducted; but information was only presented verbally and people did not dare to put forward proposals to reconsider the plan. Many people feel that they lack the experience and knowledge about oil palm cultivation, particularly maintenance, fertilization, harvesting and marketing of produce. Another concern is environmental impact. Because the plantations will include agricultural lands, the plan also threatens the availability of farmland for the villagers.

If the income from the oil palm is not sufficient to fulfil basic needs, and there is no other land for rice cultivation, the *Tane' Olen* will come under threat.

Outsiders have labelled the forest protection activities undertaken to date as conservation activities. This was initially discussed by CIFOR staff with the people of Setulang. At that time, the possibility of compensation payments by conservation organizations was mentioned. Although no compensation payments have yet been made, the forest is still protected. However, the increasing need for cash keeps the issue relevant. One open question is whether people will continue to protect the forest if no compensation payments are received. The management body is trying to develop options where conservation activities can provide cash benefits to support the villagers' economy.

Although the people of Setulang disagree to some extent, they all agree:

- that conservation at the community level works if the people are united and if social cohesion is strong;
- that the conservation area will not be disturbed if basic food needs are fulfilled;
- on how community-based conservation can be recognized at higher levels;
- on how village land-use planning can be integrated within higher-level planning processes;
- that although the concept of *Tane' Olen* has changed, the community needs recognition and rights for the protected area;
- that, in the case of Setulang, the concept of *Tane' Olen* has been adjusted, but that the aim is still the management and conservation of natural resources, particularly forests;
- on supporting factors: unity, fulfillment of basic food needs from other areas or sources, and rules and institution that are to be obeyed;
- on threats: legal uncertainty, conflict over village boundaries, lack of government support, lack of integration of community plans with district government plans.

Notes

1 The *Kalpataru* award is given annually by the Ministry of Environment for those who are concerned with, conserve or protect the environment.

2 CV, or *Commanditaire Vennootschap*, is Dutch for 'limited partnership'.

Reference

Sidayasa, K., Zakaria, A. and Iwan, R. (2005) *Hutan Desa Setulang dan Sengayan Malinau, Kalimantan Timur – Potensi Dan Indentifikasi Langkah-Langkah Perlindungan Dalam Rangka Pengelolaannya Secara Lestari*, CIFOR, Bogor, Indonesia

Part III

Other Developments Driven by Decentralization

10

Winners Take All: Understanding Forest Conflict in the Era of Decentralization in Indonesia

Made Sudana

Introduction

Changes brought about by reforms and decentralization have also brought increased levels of conflict. Conflicts lying dormant for years have re-emerged and new conflicts are more open and explicit. Protests, sometimes violent, have become a new phenomenon among local communities.

Many of these conflicts are directly related to the increased value of natural resources. With reform and decentralization, local governments are required to raise part of their own revenue. This is most commonly done by granting as many permits as possible to small-scale logging companies. At the same time, reforms have given local communities an opportunity to demand their share in benefits. The resulting competition for compensation paid by mining and logging companies has triggered boundary disputes between communities and conflicts within them. Conflicting claims over land and territories, using ethnicity or history, have exacerbated the problem. Uncertain government policy, the lack of appropriate conflict resolution approaches and low capacity of both communities and local government in conflict management resulted in prolonged and increasingly bitter rivalry among ethnic groups.

Based on observations of conflict in villages along the Malinau River in the district of Malinau, East Kalimantan, this chapter describes the characteristics of local conflict, identifies causes and describes its impacts in order to better understand how conflict can stimulate change.

Methodology

From 2000 to 2003, observations and interviews were carried out in 27 villages along the Malinau River, comprising approximately 10 ethnic groups, each with its own history, customs and land-use systems. The year 2000 was taken as the cut-off date to compare the incidence of conflicts before and after decentralization.

We limited our observations to conflicts linked to the exploitation of timber and non-timber forest products (NTFPs), the use of agricultural lands and village boundaries. In order to gain an understanding of conflict, we tried to identify causes, factors leading to conflict escalation and actors involved. We observed the frequency of contact between actors, the settlement process (negotiation and mediation) and its outcome (agreement and contracts). An incident was used as the unit of analysis, where an incident is defined as the presence of conflict indicated by demonstrations, fights, discussions, letters, complaints or threats. Each conflict was observed independently and analysed in relation to one another.

Conflict before and after decentralization

Change in frequency

For the purposes of analysis, three periods were distinguished: the New Order period from 1967 to 1996; the period of reforms, from 1997 to 1999; and the period of early decentralization, from 2000 to 2002. Over the whole period, from 1967 to 2002, 94 cases of conflict were observed: 8 cases (8.5 per cent) occurring during the New Order; 17 (18.1 per cent) during the reform period; and 69 (73.4 per cent) during the third period. Some conflicts are recurrent due to not having been satisfactorily resolved. During the New Order, in particular, many conflicts were repressed or settled by force. The new government of the early years of the decentralization provided opportunities for communities to reclaim resources taken away during the earlier period. Therefore, of the 69 conflicts erupting during this period, 25 are continuations of previous conflicts (see Figure 10.1).

Although it is clear that the root of the problem originated during the New Order, reform and decentralization have had significant effects on the escalation of conflict. This is shown by the increased incidence of demonstrations, visits to government offices and debates. The question is, however, who is responsible for managing conflicts over the natural resources?

Factors that cause conflicts

Conflicts along the Malinau River take many forms depending upon whether they concern village boundaries, agricultural lands or competition in sharing benefits of timber and NTFPs. However, the issues are not always simple or easy to solve. In fact, the apparent simplicity of the problem during the New Order, where

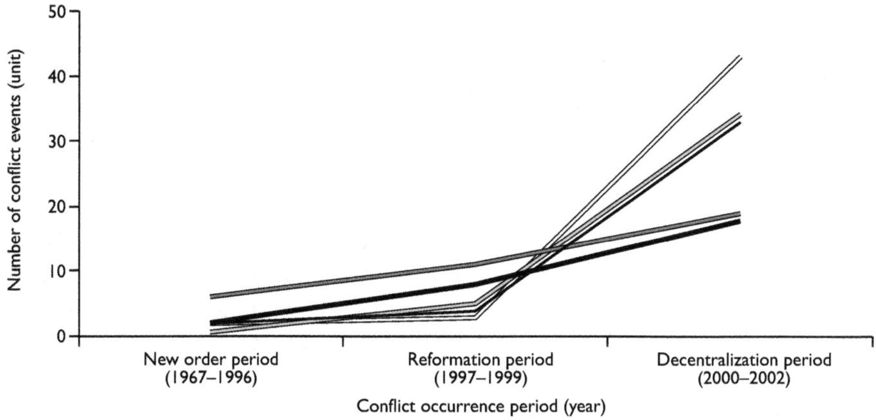

Source: Sudana, 2004

Figure 10.1 *Conflict frequency before and after decentralization*

conflict was often suppressed through intimidation of the powerful, is a main cause of latent conflict. Reform and decentralization changed this. Simple solutions are no longer acceptable. A main source of conflict, for example, is due to encroachment on community territory by logging companies. A certain amount of money is usually demanded in compensation. Arguments are brought to bear on the value of the timber, boundaries, money and, at times, the environment. The course of the conflict is dynamic as it changes over time in response to counterclaims.

As shown in Figure 10.2, before decentralization, the most frequently occurring conflicts, in 17 out of 27 villages, were arguments over agricultural land – farms, rice fields and plantations. Other conflicts occurred over rights to forest products such as edible birds' nests and eaglewood (*gaharu*); compensation by a coal mining company (in 6 villages); competition over timber (3 villages) and conflicting land claims based on different histories (in 13 villages) (Anau et al, 2001). During decentralization, conflicts arose from encroachment by logging activities into village territories causing loss of forest products, such as eaglewood, rattan or birds' nests, or damage to agricultural fields. In compensation, villagers demanded money and/or public utilities.

Figure 10.1 shows sources of conflict and the changes before and after decentralization, while Figure 10.2 shows the different triggers or factors causing conflict.

Agricultural land

Agricultural land has been a source of conflict ever since the Kenyah migrated to the area. Conflicts include the Kenyah Uma' Long in the village of Batu Kajang with the Merap in Gong Solok, in 1967; the Kenyah Uma' Long in Setulang with the Abay in Sentaban, in 1969; and the Kenyah Lepo Ke' in Long Loreh with the

Merap in Langap, in 1974. All of these conflicts concerned the acquisition of fertile land and were usually settled amicably. Solutions were sought through *musyawarah* (village meetings) to come to an agreement, usually with regard to village boundaries. Agreements were, unfortunately, often short lived as one or the other parties would stake another claim. Claims on land were eventually linked to village boundaries and became complicated long-lasting conflicts.

The increase in population added to the problem. Between 1980 and 1999, some Punan people moved from upstream to the lower part of the Malinau River as part of a government-promoted resettlement scheme. In the later period, many other people moved in search of better economic opportunities, health facilities or schools. During this same period, logging companies entered the area, leading to conflict between companies and individuals whose land was used in building logging roads. In most such cases, the conflict was solved by a compensation settlement.

Conflict over agricultural lands increased significantly during the reform period of 1997 to 1999, when a mining company started operations in the area of Long Loreh. The company offered compensation for land and crops affected. Typically, people tried opportunistically to claim fallow fields, thereby causing additional conflicts among villagers. During the decentralization period from 2000 to 2002, individual conflict decreased, but conflict between ethnic groups increased.

Territorial boundaries

During the New Order, village boundaries were not a significant source of conflict. One example concerns the boundary between the villages of Langap and Loreh. The original agreement of 27 February 1975 was broken in 1977, when the Merap people of Sengayan shifted the boundary upstream and moved to Loreh.

The coal mine that started operations in Loreh in 1995 triggered a renewed set of conflicts. Compensations paid for the use of land for mining changed perceptions on the value of land. Villages started to delineate their territories more explicitly. In 1998, the villages of Langap and Loreh clashed: Langap demanded that the boundaries be returned to the 1975 agreement. During the same year, boundary conflicts erupted between the villages of Langap and Tanjung Nanga, and between Langap and Laban Nyarit, disputing rights over land with coal deposits for which compensation payments were expected. For the same reason, Langap was in another conflict with Nunuk Tanah Kibang.

Further downstream, the villages of Setulang and Sentaban had disputed land rights since the Kenyah had settled in this area. In 1992, the two villages disputed rights to rice fields and forests along the Senarau and Sebatiung rivers. At that time, the conflict was settled by the government of the sub-district of Malinau. On 21 April 1999, boundary conflict between Setulang and Sentaban recurred. While not directly due to reform and decentralization, the spirit of the times empowered communities to articulate their demands.

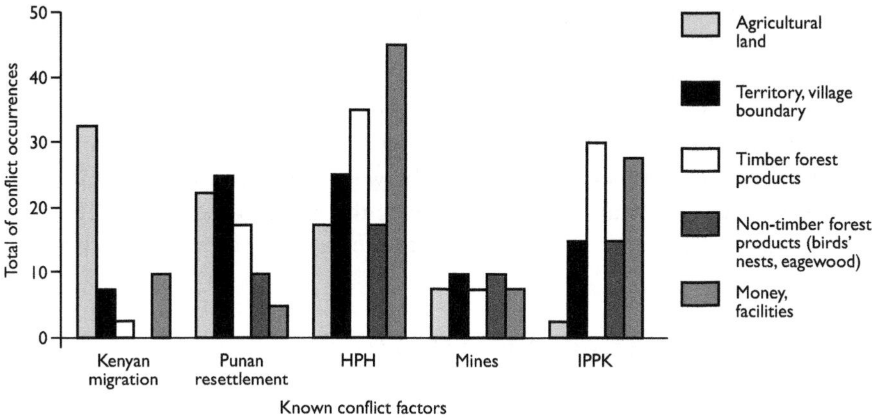

Source: Sudana, 2004

Figure 10.2 *Conflict events and triggers*

Boundary disputes in the Malinau Basin have still not been settled satisfactorily. Langap and Loreh are still in dispute, as are Batu Kajang and Gong Solok.

Timber forest products

Although previously little known, timber harvesting has been a source of conflict between communities and logging companies since the New Order. At that time, such conflicts were normally settled by the military, as experienced by people in Loreh and Langap with the state-owned company, PT Inhutani II. Villagers could only harvest timber for the construction of their houses and public buildings through permission from the concessionaire.

After decentralization, conflict over timber harvesting increased sharply. This was initially triggered by Governor Decree No 20 of 2000, concerning the payment of compensation by forest concessionaires to communities in the territory in which they operated. The granting of the Timber Utilization and Harvesting Permit (small-scale timber harvesting known as IPPK) by the regional government since 2000 stimulated more conflict: communities used all kinds of arguments for claiming shares in the fees and compensation payments. This period saw the rise of multidimensional conflicts: communities with large-scale forest concessionaires (HPH), with IPPK holders and with owners of birds' nest caves; between villages; HPH with IPPK and with cave owners; and between different IPPKs.

Conflicts between local communities and HPH holders occurred most frequently in the villages of Langap (against PT Inhutani II), Metut (against PT Meranti Sakti Indonesia) and in Laban Nyarit (against PT Tribudi Wisnu). When the local government started to issue IPPKs in cooperation with local communities, HPH holders were increasingly excluded, leading to more conflict.

With regional autonomy, district governments took control over the forests in the region. Meanwhile, HPHs received their mandate from the central government and, in the first confusing years, saw their control over their concessions disappearing and being replaced by IPPKs. As described in the previous chapters, the relationship between communities and IPPK evolved. The former learned to demand that promises be fulfilled or else communities would resort to other actions, such as demonstrating, blocking logging roads and even seizing equipment. IPPK-related conflicts were not only between companies and communities, but also between communities over who had the right to receive the fees and compensation, and within communities over the sharing of fees and compensation.

Non-timber forest products (NTFPs)

Conflicts related to NTFPs were few during the New Order. Only two cases are recorded affecting people in Long Jalan and Adiu. On 20 August 1992, villagers of Long Jalan prohibited eaglewood harvesters from entering their territory through a letter signed by sub-district military command (Koramil). On 17 November 1988, Adiu villagers stood up against Datuk Husein, the owner of a rattan stand who claimed ownership of the forest on Gunung Bintang Mountain.

After decentralization, conflict over birds' nest caves occurred more frequently, although most used the pretext to obtain IPPK benefits. Such conflicts occurred mainly in the Gong Solok and Langap territories where the caves are located. It should be noted that these cave owners are descendants of the Merap, who were the first settlers, or Tidung people, who obtained legal rights from the Dutch colonial government. There were frequent conflicts in Gong Solok and Langap villages: cave owners with local communities and with HPHs or IPPKs, as well as IPPK holders with the coal mine operator. In these conflicts, the Punan, who are heavily dependent upon the harvesting of NTFPs, have the most to lose, but are the least able to defend their claims.

Money and facilities

The most significant effect of reforms and decentralization has been the empowerment of local communities. No longer silent observers, communities now forcefully demand their perceived rights, unfortunately most often translated in monetary terms. Payments for timber harvested or compensation for land used for mining have greatly contributed to this attitude. Today, people will use any reason to collect money. Encroachment of village boundaries, damage to land, dust from the road covering their crops, water pollution and harvest failure are among the many reasons to ask for financial compensation. Usually, companies will oblige, as it is simpler to pay than sort out the case. The root of the problem (i.e. who has the right) is not solved, however.

General patterns

Demonstrations and protests

Decentralization has changed the patterns by which conflicts are expressed. In the period before decentralization (the New Order and the reform period), conflicts were expressed through complaints most commonly settled through community or village meetings. Conflict rarely erupted in physical fights.

During the decentralization era, conflict became more complicated. A common pattern was as follows:

1. The village would send a messenger to gain information and convey the demands of the claimant.
2. If the messenger failed in getting the demands fulfilled, after a meeting to determine actions, timing and division of tasks, mass mobilization would follow.
3. Some actions would be taken, such as demonstrations, blocking access roads, raising banners or gathering in front of the opponent's office.
4. Negotiations were held.
5. Solutions were proposed and accepted, which usually included the payment of a certain amount of cash.
6. When the solution was not satisfactory or demands not met, a follow-up demonstration was staged, repeating the process from phase 2, sometimes including confiscation of equipment to increase pressure on the other party to fulfil demands.

Along the Malinau River, 26.1 per cent of (18 out of 69) conflict incidents during the decentralization era involved demonstrations with mass mobilization by local people, as occurred in Loreh, Langap, Gong Solok, Setarap and Setulang. In some places, such as in Loreh and Langap, demonstrations are repeatedly carried out. Most participants of the demonstration are men. Only in one case did women participate.

Conflict is also used by third parties, most often local businessmen, to advance their business interests. Bribing and manipulation of village and political elite are common. Another complication is that many of the small companies causing conflict are, in reality, smokescreens for larger companies. For example, many IPPKs are subsidiaries of, or sometimes directly operated by, large concessionaires. For the HPH, this is lucrative since IPPKs operate on lower costs than HPHs.

Negotiations

Prior to decentralization, negotiation only involved the village head, *adat* leaders and community leaders. After decentralization, community members also became involved, either through representatives or by as many people as possible. Successful negotiations today tend to involve mass mobilization ('people power') to apply political pressure. In such cases, demonstrations are a part of the negotiation process.

Today, in areas along the Malinau River, conflicts involve many parties, including the local and the national government, local companies with international networks and the community. Therefore, negotiations are difficult and seldom achieve a stable agreement. For example, conflict between the Langap villagers and the state-run PT Inhutani II has yet to be fully settled, although the demands have been met. PT Inhutani II, whose shares are held by the state and whose management consists of state-appointed staff, is helpless to face local politicians and entrepreneurs who want to take over forest management and replace the position of PT Inhutani II. Through negotiation after seizing heavy equipment of PT Inhutani II's contractor, Langap villagers have already succeeded in being paid a fine according to *adat* law and compensation of 2.3 billion rupiah; but PT Inhutani II is still not able to operate.

The role of local government

Before decentralization, conflict was usually handled by the Malinau sub-district head. The involved parties or their representatives were called to the sub-district office for a meeting. The meeting had to resolve the problem and the parties were required to make peace. In this way, the conflict between Setulang and Sentaban was settled in 1999. Minor problems were usually resolved in a village meeting, after which the resulting agreement would be sent to the sub-district head for acknowledgement and signature.

During the decentralization era, both the Malinau District and the Malinau sub-district governments were actively involved in resolving conflicts. The sub-district government is responsible for settling conflicts within its area and the district government is called in only when a solution cannot be found. During this period, 13 cases were resolved in the South Malinau sub-district and seven cases settled in Malinau District with the local consultative assembly's involvement. The sub-district government usually invites the conflicting parties after receiving reports from the claimant. At the sub-district office, a meeting is held to discuss solutions acceptable to both parties. Unresolved conflicts will be handed to the district government; for example, the conflict between PT Inhutani II and Langap village required several discussions at the district level. Conflict between Tanjung Nanga and Langap and the IPPK Meranti Wana Lestari was also settled at the district level. Conflict between PT Wana Yasa Kahuripan Indonesia (PT WYKI), who opened an access road from Tanjung Nanga to Long Alango, and the people of Laban Nyarit could not be resolved at the sub-district level because PT WYKI did not respond to the invitation. The people of Laban Nyarit then staged a demonstration, blocked the access road and seized the company's heavy equipment. This case was subsequently handled by the district government, who sent representatives (assistant I and II, the district secretary and his staff, the sub-district head, the chief of the sub-district military command, and police personnel) to the location. This conflict was settled in a meeting at the District Head Office.

Settled conflicts (fee distribution) and unsettled conflicts (land claims)

Conflicts between companies and communities involving monetary payments are usually settled quickly. On reaching an agreement over payment and distribution of fees and compensation, the conflict was resolved between Metut and PT Meranti Sakti Indonesia; between Laban Nyarit and PT Tribudi Wisnu; between Halanga and PT Wahana Stagen Lestari; and between Pelancau and PT Wana Yasa Kahuripan Indonesia. Furthermore, several conflicts were resolved between the people and IPPK holders, such as Bila Bekayuk and CV Sebuku Lestari; Setarap and CV Gading Indah; and Langap and CV Hanura.

Conflicts over the use of natural resources are more difficult to resolve. There are still unresolved conflicts over agricultural land between the villages of Langap and Loreh, Gong Solok and Batu Kajang, and Setulang and Sentaban. Conflicts relating to forests or mines occurred between the villages of Halanga and Laban Nyarit and between Paya Seturan and Langap, as well as between cave owners and the surrounding villages, among others. Most of these conflicts have not been resolved and may even have worsened.

Government interests in increasing local revenues play an important role in the takeover of natural resources management from central government. Companies who have obtained licences from the central government will therefore receive less attention from the district government with regard to conflict settlement. The district government will support people if they are involved in a conflict against these companies. Therefore, conflicts will not be resolved until the regional government manages to take over the affected company.

Inadequate institutions

Malinau District is rich in natural resources and home to at least 18 ethnic groups. Various mining and logging companies own concessions in the district, several of which operate in the Malinau River watershed, where concessions overlap with community territories. The entire forest in this region, including people's land, has been divided into concessions. Villages often have no clear boundaries and rights to the forest are unclear. It is therefore no surprise that land and forests are a main source of conflict between villages, and between communities and other parties, such as companies and birds' nest cave owners. Reform and decentralization have encouraged people to boldly and freely express their wishes, thus leading to open conflict.

Aware of the situation, the district government of Malinau established a conflict mediating team to cope with the rampant conflicts relating to lands, forest and other natural resources. On 27 February 2003, the district head of Malinau issued Decree No 26 of 2003 Concerning the Establishment of a Mediating Team to Settle Disputes/Claims between the People and Companies in the

Malinau District. Malinau does not, however, have a district regulation addressing the settlement of natural resources-related conflicts within communities. There is still no agreement on how to utilize or distribute the available natural resources fairly and proportionately. There are also few people with sufficient knowledge and experience in managing conflicts. The parties involved in conflict, especially the villagers, have little understanding of how to channel them and how to participate in the settlement process. The strong will repress the weaker, resulting in unfair solutions and recurrent claims and conflicts. There have been no serious efforts by the government to address conflicts in villages.

In the face of such constraints, the mediation team is not fully operational. At present, the team only acts in the event of urgent and incidental conflicts that cannot be settled at the village and sub-district levels. In the future, when the requirements are fulfilled and capable manpower is available, we believe that conflicts between many parties will be better resolved.

Conflict analysis

Conflicts, at present, usually relate to the value of natural resources. Their monetary value has increased the conflict and pressure on them. In the past, exploitation of natural resources was limited to the harvesting of NTFPs, such as resin and birds' nests. Conflict over rights to collect these products was common. When logs became a high-priced commodity, conflicts over rights to timber soon followed.

During the New Order, the most dominant conflict in the Malinau River watershed was related to agricultural land. This conflict stemmed from the fact that the Kenyah have always depended more upon rice fields. They tend to make larger fields and work these more intensively. As a result, they view land as an important resource and require the certainty of long-term tenure. It is not surprising, therefore, that they soon used all of the land allocated when they arrived and began to encroach upon neighbouring areas. Rice fields are still an object of dispute, although not always involving the community as a whole.

When HPH holders operate in this region, other forms of conflict, albeit minor in influence, began to appear. It was the coal mining company that, by paying compensation for land used, triggered an increase in conflict. To be eligible for compensation payment, people use many arguments, including claiming previous ownership, claiming rights to fallow land and knowledge on the location of village boundaries. Among the prominent conflicts were those between the villages of Loreh and Langap; between Langap and Nunuk Tanah Kibang; between Langap and Paya Seturan and Long Lake; between Langap and Tanjung Nanga; and between Laban Nyarit and Tanjung Nanga. Most of these conflicts occurred between 1996 and 1999, partly coinciding with the reform movement. While reforms cannot be considered a cause of conflict, they certainly empowered local people to raise their voices and to demand their rights.

The sharp increase in conflict during the reform and early years of decentralization is the result of a combination of events over several years. The arrival of

HPHs allowed by the central government, the policy of the mining company to pay compensation, reforms and decentralization, and the opening of forests to small-scale logging enterprises by the district government all contributed to the creation of conflict. Provincial Decree No 20 of 2000 exacerbated conflicts by requiring companies to pay a stumpage fee to local communities.

Conflict related to coal mining also intensified with the release of a new mine concession permit by the district government. With decentralization, district governments have the right to administer and manage their own natural resources. District governments are also required to increase their local revenue. In Malinau, this is done through increasing taxes, locally called '*retribusi*', from forest exploitation, as well as from mining. This has created conflict between the different levels of governments as the central government still claims control over both forests and mining. In fact, there are still a number of companies operating in the Malinau River watershed with permits issued by the central government. These companies are suspected of bringing a great deal of profit to Jakarta, but less to the district government. While, from an administrative point of view, there could be a transfer of control from the central to the district level, when companies refuse to follow the legal procedures, the district government might not be able to enforce it.

In view of the above facts, the district government of Malinau chose an 'ignore, wait and see' policy. When conflict occurred between these companies and local people, it was ignored by the district government. The district government clearly preferred companies operating based on permits issued by the district, which surely would profit their own region. This was shown by the conflict between PT Inhutani II and the Langap people, and between a coal mining company in Loreh and the Langap people. The district head felt that PT Inhutani II did not respect the district government. On another occasion, in 2003 in Respen, the district head said: 'For 30 years, the people of Malinau shed tears seeing timber hauled by people from the centre, then their eyes became dry seeing coal passing, while no benefit ever reached them.' This statement indicates the local government's resentment against companies operating with central government-issued permits.

Uncertain claims

One important problem underlying many conflicts is the lack of clarity on village boundaries and rights to agricultural land. Village boundaries are a particularly complicated issue because of the many ethnic groups residing in the area, each with a different basis for their claims. As mentioned previously, the many different ethnic groups – each with different backgrounds and culture – arrived at different times to settle in the area.

The Merap are acknowledged as the oldest community and their *adat* leader is recognized as being the chief *adat* leader of Malinau (*Kepala Adat Besar Sungai Malinau*). As a result, the Merap try to regulate tribes who came after them. In the meantime, communities have also become organized in administrative villages not always coinciding with customary territories. Exact boundaries between villages have not been decided and accepted by all parties. The problem is exacerbated by different ethnicities residing in one village and the fact that, during the forced resettlement, administrative villages settled as a whole, thus resulting in two to four separate villages within one location. Customary or *adat* institutions do not have the authority to administer village boundaries, nor are the traditional rights to allocate and regulate the use of natural resources acceptable to the state.

Reforms and decentralization have promoted stronger group identities through the revival of *adat* (customs and traditions). These *adat* identities, however, sometimes comprise only small groups – for example, the Kenyah from the Bahau, although all Kenyah identify themselves in seven sub-groups. Each group has their own leader whom they respect. Similarly, the nine groups of the Punan, while living together with other ethnic groups, have their respective leaders. Even so, distrust of the leader is one of the factors that pose difficulties in administering agricultural land and village boundaries. The leaders are frequently part of the conflict. The increasing population within ethnic groups complicates the conflict over natural resources in the Malinau River watershed.

Conclusions

Conflict over natural resources in the Malinau River watershed is influenced by a number of factors:

- increase in the value of natural resources, forest and lands because of market, money and compensation;
- reform-stimulated freedom of expression among the people;
- the decentralized government granting small-scale logging permits (IPPK);
- uncertain village boundaries and rights to lands and forest;
- lack of agreement on how benefits from natural resources could be fairly and proportionately divided;
- hesitance and lack of capacity in government to address tenure issues; and
- hesitance and lack of capacity by government or civil society to manage conflict.

Conflict is also deeply rooted in historical relationships among local settlements and ethnic groups in Malinau. These relationships traditionally defined access to land and forest resources, as well as relations of power, conflict and cooperation among groups. Reform has strengthened the identity of these groups through revival of *adat*, the presence of more local representation in district government and freedom of expression. In order to fully understand conflicts and their underlying causes, in-depth comprehensive studies are necessary.

References

Anau, N., van Heist, M., Iwan, R., Limberg, G., Sudana, M. and Wollenberg, E. (2001) *Pemetaan Desa Partisipatif dan Penyelesaian Konflik Batas. Studi Kasus di Desa-Desa Daerah Aliran Sungai Malinau, Januari s/d Juli (2000) Laporan Pengelolaan Hutan Bersama*, CIFOR, Bogor, Indonesia

Fisher, S. and Kartikasari, S. N. (2001) *Mengelola Konflik: Ketrampilan dan Strategi untuk Bertindak*, British Council, Responding to Conflict (RTC), Jakarta

Kaskija, L. (2000) *Punan Malinau and the Bulungan Research Forest*, Report submitted to CIFOR, Bogor, Indonesia

Sellato, B. (2001) *Forest, Resources and People in Bulungan: Elements for a History of Settlement, Trade and Social Dynamics in Borneo, 1880–2000*, CIFOR, Bogor, Indonesia

Sudana, M. (2004) 'Winners take all: Understanding forest conflict in the era of decentralization in Indonesia', paper presented at the 10th meeting of the International Association for the Study of Common Property, 9–13 August 2004, Oaxaca, Mexico

Yasmi, Y. (2002) *Conflict in Forest Management: A Study for Collaborative Forest Management in Indonesia*, Wageningen University, The Netherlands

11

Tenure and *Adat* in Malinau[1]

Moira Moeliono and Godwin Limberg

One important aspect of state–society relations is the way in which control over access and use of land and resources or resource tenure is arranged. While the state has a role to play in regulating and protecting the rights of citizens, too often this role is translated as taking control, exercising authority over people by controlling resources upon which they depend. In its attempt to control, the state put into place an often centralized legal system, with uniform rules, thereby ignoring the varied local systems of regulating property rights (Scott, 1998).

In Indonesia, this took the form of one agrarian law, Law 5, 1960, based on concepts introduced by the Dutch, where land is either private individual property or belongs to the state. Tenure, however, was inexorably linked to *adat*, understood as 'cultural beliefs, rights and responsibilities, customary laws and courts, customary practices and self-governance institutions' (Alcorn, 2000), including customary rules of regulating rights of access, use and ownership of resources. When, following independence, the state took over the regulating of tenure, it was believed that the official legal system based on Western law would gradually take over the *adat* systems. Because of the inability of the state to enforce legal regulations effectively through time, however, traditional informal systems have persisted. The result of this encounter of the traditional and the official legal systems is a complex mix of the informal and formal through which local people have to navigate to survive.

Decentralization and reform in Indonesia should have provided the opportunity for districts to accommodate these local systems and to establish a locally appropriate system of formal tenure. Few, if any, districts took this opportunity. Instead, the inability of either central or regional government to control and monitor the use of natural resources has led to a situation of almost open access and increased competition for resources and political power. As shown in the preceding chapters, this situation has been used by local elites to raise capital to strengthen their position or to raise even more capital. In the process, they used *adat* selectively in their own favour, often marginalizing the weaker groups such as the Punan.

This chapter will discuss tenure in Malinau where the increased monetary value of resources has led to increased competition among a wide range of stakeholders and ethnic groups. In the absence of a formal system that protects their rights and is acceptable and easy to implement, local people have adjusted to and modified rules in sometimes unexpected ways to produce unique, locally specific and highly resilient systems of tenure.

Particular attention is given to the Punan, one of the many ethnic groups in Malinau with special land-use requirements. Because their way of life is still partly nomadic, their values differ from those of the more sedentary groups. With a government that perceives traditional culture to be backward and primitive, the Punan are perceived as a problem (Dove, 1993), especially because they are now claiming ancestral rights to large tracts of forests where they used to range.

The national and legal context

One source of local problems can be found in the ambiguity and complexity of national land policies in which land is administered by the national land agency – *Badan Pertanahan Nasional* (BPN) – while forest lands are administered by the Ministry of Forestry. Although the legality of the jurisdiction of the Ministry of Forestry (MoF) over the 60 to 70 per cent of land designated as forest land (*kawasan hutan*) can be debated (Fay and Sirait, 2002), for the past four decades, the problems of tenure in the forest-rich areas of Kalimantan, Sumatera and Papua have been – *de jure* although not always *de facto* – administered by the MoF, whereby local communities or even local administration have often been ignored.

Between the two offices, all land was administered through branch offices representing MoF and BPN at the province and district levels. After decentralization, most of these offices were liquidated or taken over by autonomous districts. In Malinau, the government then established a forestry office that also administers plantations (*Dinas Kehutanan dan Perkebunan*) and a land office (*Dinas Pertanahan*) as part of the district government structure.

Decentralization of these two departments did not occur smoothly. Both departments tried to keep control over these strategic resources. MoF claimed that forestry, as a strategic national resource, was not included in the general decentralized system, and has continued attempts to regain control.

With regard to land issues, the formal process of decentralization[2] was put on hold less than a month after the implementation of the decentralization and autonomy policies. Presidential Decree No 62 of 2001 stipulated that a number of the administrative functions of BPN in the regions are to remain under the authority of the central government – in fact, returning to the de-concentration mode of the New Order until all legislation on land administration has been ratified (Toyamah, 2002). This would take about two years and a date was set, 31 May 2003, by which land affairs would be handed over to the districts. In 2003,[3] the government decentralized land administration again, but retained part of the authority at the centre. The BPN would remain as a national institution, setting guidelines and standards, whereas the districts would be in charge of nine aspects of land administration,[4] including the resolution of conflicts over agricultural land and the determination and administrative processing of *adat* lands.

Confusion arose, however, when the branch office of BPN in Bulungan established a sub-branch in Malinau (Gelora Mahardhika, 1–15 November 2002) to administer land. The BPN sub-branch office claimed it was responsible for collecting the land tax regulated by the national government, although almost no one in Malinau, except large concessions and town dwellers, has registered land or pays land tax. *Dinas Pertanahan* claimed that it was responsible for land registration (which is also the task of BPN) and regulating use (which is also the responsibility of the regional planning agency – BAPPEDA). On the other hand, *Dinas Pertanahan* said its work was limited to land procurement for housing for government staff and public use such as graveyards, while BPN claims authority over all land.[5] Lack of understanding of legal rules and their mandate is also illustrated by the fact that both *Dinas Pertanahan* and BPN claimed that the rights of communities to their fields and settlements were automatically recognized and were even eligible for certification, although within formally designated state forest areas.[6] Such areas might be assigned as 'enclaves' or 'in-holdings' within the forest estate, although there is no actual legal basis for their establishment.

To date, neither office has been effective. In addition to overlapping mandates and lack of understanding of land laws and regulation, they have to cope with a serious lack of manpower. Only one of the four sections of *Dinas Pertanahan*, the administrative section, has a qualified head, and there is general lack of technical personnel. For land mensuration, the two skilled individuals hired by the district are often loaned to BPN, which makes the actual land mensuration. The obvious solution (i.e. to pool resources and merge) is considered impossible since the *Dinas* is responsible to the local government, while BPN is responsible to the central government.

Meanwhile, the district forestry office gained importance as the office that decided on permits for small-scale logging (IPPKs). In accordance with the spirit of local autonomy, district policy recognized local community rights to share the benefits of logging. As illustrated in Part II and discussed below, the IPPK led to a race to claim rights over territories in order to be eligible for fees and compensation. To date, these processes have not resulted in formal recognition of property rights within the state forest.

Meanwhile, although reforms and decentralization resulted in the formal recognition of customary rights in both the constitution and the 1999 Forestry Law,[7] the districts continued to manage resources in the same manner as the central government during the New Order.[8] Land and forest resources are mainly seen as a source of revenue despite functions such as protection and conservation. District governments, such as that of Malinau in which more than 90 per cent of the area is legally designated state forest, have not disputed traditional rights and the presence of local communities in these areas, but have also not been willing to formally recognize these rights. Confused by the contradictory and complex legal system, and the changing value of the resources, people have consequently continued to apply customary rules or *adat* to sort out their property rights.

De facto land tenure and the role of *adat*

Tenure in Malinau is *de facto* regulated by practical needs and traditional norms and institutions. On the ground, occupation, history of use and customary rules (*adat*) are the dominant basis for claims. Although not free from dispute, claims to areas used and settled by local communities were usually, although sometimes reluctantly, accepted and recognized by neighbouring communities, founded on agreements sealed through customary rituals. These claims were mostly based on the need to make a living and covered agricultural fields and forest areas used for hunting and the collection of NTFPs (Anau et al, 2001). People were often free to use each other's lands, sometimes with, but more often without, explicit agreements. Boundaries were fuzzy, although village agreements specified identifiable features of the landscape such as roads and rivers. Most of the forest remained open access and was used for hunting and collecting forest products. These open access areas were often claimed by the first group to arrive as communal property, although individuals could claim rights by being the first to clear land for agricultural fields.

The way in which people of Loreh on the Malinau arranged their space shows the flexibility of the system. In 1972, a group of Kenyah Lepo' Ke from Pujungan settled in this location with permission of the customary leader (*Ketua Adat*) in the neighbouring village of Langap, which had first claim to the land. During the 1980s, the government resettled two Punan communities, Pelancau and Bila Bekayuk, in the same area. Together with a Merap group who moved from the Sengayan River, the four communities arranged and negotiated ways of living together, while each kept their village status. Except for individual fields, most of the area remained open access. Expansion of fields was made possible by an agreement with Langap on the use of additional land.

Before decentralization and the advent of small-scale logging operations, the four villages had probably thought little about territories and village boundaries. For a while, it was accepted that they would share one territory. However, the start of coal mining during 1994 to 1995 (Anau et al, 2001) and the opening of forest for small-scale exploitation changed this, and each village is now claiming its own territory and disputing its boundaries with the neighbours. In the competition, the two Punan groups are losing out. The Merap of Sengayan claim status as the oldest settlers and the Kenyah claims are based on the agreement with Langap. The Punan have no such basis and, as a result, feel threatened. Bila Bekayuk planned to remove the whole settlement a few kilometres down the road because they felt that even their housing space was being claimed by others. At present, some villagers of Bila Bekayuk have moved about 15km away to some lands, which include a fish pond that they claim.

The changes brought by reform, decentralization and the following wave of small-scale logging activities played a significant role in shifting the implicit rules of *adat* to more formal explicit ones. The compensation payments and fees allocated to local communities increased the value of the forest, resulting in a race to stake claims to forest resources by claiming territories based on *adat*.

The use of *adat*, however, is problematic in several ways. The *adat* rights legally recognized in law[9] are assumed to be always communal. In fact, *adat* rights can be either communal or individual, or often both. It is not at all unusual to have individual rights within a communal rights regime (Gönner, 2002). Furthermore, *adat* can mean many different things and has been given even more meanings with the reforms and decentralization. *Adat* is also exclusive, very localized with social and cultural variations (Drake, 1989), and always dynamic and fluid, changing over time. The flexibility inherent in *adat* leads to a redefining of rules according to need. Thus, in Malinau, the interpretation of rights to land now includes a group's original traditional lands (Barr et al, 2001).

The formal regulations on *adat* issues are drafted within the framework of bureaucratic national laws. For *adat* rights to be recognized, several criteria have to be fulfilled and verified by an independent third party of experts. The criteria are that *adat* is practised according to its rules; *adat* leadership/institutions still exist; territory is clearly established; and the people still depend upon the forest. While the last two items might be easy to prove, the existence of *adat* rules, especially those concerning land tenure, are difficult to explain in terms comprehensible to an outsider, but can be shown through practices. When claims are secure and there is little dispute over individual holdings, people see no real need to make claims formally or involve legal offices. Most disputes are settled informally through *adat* or village leaders who presumably use *adat* rules to settle the dispute. With increasing competition, claims tended to be more visibly expressed (e.g. by fencing) and disputes can become intense. When *adat* leaders are then unable to settle the dispute, the sub-district leader (*xamat*) or other district officials are asked to mediate. Interestingly, formal legitimization through certification by BPN is most frequently requested when claims are insecure, as in the case of Chinese traders and Javanese transmigrants.

In addition, the rules that have survived generally concern relationships among people, rather than the governing of resource use. One exception is the *Tane' Olen*, which was originally an area set aside for the exclusive use of the nobility of the Kenyah (see Chapter 10). Today, it is used to denote an area set aside for special purposes, such as the conservation area in Setulang village, and has developed into a local concept of conservation.

Moreover, *adat* communities are always in flux, forming alliances or breaking apart (Drake, 1989), reinventing themselves in contrast to other *and* non-*adat* groups. The different faction of the Punan of Malinau, for example, tried to create one Punan *adat* community (Cesard, 2001) in relation to the other groups who consider them 'clients' under their patronage. The Tidung, who are descendants of the Tidung Kingdom, have redefined themselves as Dayaks.[10] The Merap claim ascendancy over other groups because of their position as *Kepala Adat Besar* (chief *adat* leader), but have invented the term *pewaris* (inheritor) to claim exclusive rights to birds' nest caves and the forest surrounding them. To a large degree, the revival of *adat* institutions in Malinau was linked to the hope of reclaiming and asserting ownership of territories based on *adat* rights.

In order to fulfil government requirements of formal documentation to substantiate their claims, *adat* communities are writing down their *adat* and *adat* law (Anau et al, 2001; Wollenberg, 2003), framed in their group history as evidence of their claims. The Kenyah and the Merap had already written down their *adat* law in 1968, interestingly, using the official format of government regulations. In so doing, they neglected the fact that actual rulings of *adat* have always been typically based on negotiations, precedents and the rules of necessity. More recently, the Punan have also tried to write down their *adat* and history.

Constructing history, however, has been shown to become a contest over time and place (Erb, 1997), where the different interpretations of historical events and agreements have not been conducive to solving conflicts (Moeliono, 2000; Anau, et al, 2002). Some prominent Tidung families, for example, claiming to be descendants from the sultanate that encompassed the Malinau region, have tried to claim large areas covering several smaller village territories. Descendants of the Sultan of Bulungan claimed the area of Gunung Bintang (Bintang Mount), but added that the family had released its use to local people.[11]

As mentioned, in theory, local autonomy allows a return to *adat* governance. Even villages, where a return to *adat* would be relatively simple, have hesitated. The only changes have been in naming – for example, in Kutai Barat, villages are called *kampung* instead of *desa*[12] in the New Order terminology, but its governance remains unchanged.

The government, to date, has remained ambivalent about *adat* and claims based on it. Nevertheless, one of the first district regulations issued by the new district of Malinau was Regulation No 4, 2001, concerning the empowerment, preservation, protection and development of *adat* customs and institutions. Although copied almost verbatim from regulations of other districts, which, in turn, have adopted it from central government regulations,[13] it can be taken as a commitment of the district to acknowledge *adat* as a social convention. The role of *adat* in regulating tenure and property, however, is not recognized. The regulation lacks clear definitions of rights and procedures for claiming and recognizing rights, as well as responsibilities and mechanisms for conflict resolution. On the other hand, it does recognize *adat*'s role in regulating community management of *wilayah adat* (territory) (Patlis et al, 2001); but there is no agreement yet as to how these rights might be recognized. As the *Bupati* of Malinau told us in August 2001, we should be careful in discussing *adat* forests, *adat* land and *adat* rights since 'at present, *adat* communities are too eager to claim rights, although they themselves do not exactly know what these rights entail'. Indeed, communities are often unclear about what *adat* rights involve. As Colchester (2001) has noted, *adat* communities, in general, are unsure how these rights translate into legal definitions as it can mean a range of rights between full ownership and use rights. In an attempt to clarify this, government staff participating in the two legal drafting trainings organized by CIFOR proposed to legally regulate *adat* land. Both the Lundaye and Punan *lembaga adat* have organized workshops to discuss recognition and certification of *adat* land in which the issue of certification was hotly debated. Certification was seen by some as weakening the *adat* institution, while others feared the consequences (i.e. having to pay taxes). This debate is significant in that it shows that formal recognition by *adat* communities and their institutions by itself is not sufficient (ACM-CIFOR, 2001), but requires further negotiations and agreements.

In Malinau, *adat* claims are further complicated by a history of migration and government-sponsored settlements, resulting in a mixing of ethnic groups in one location. Each ethnic group, in turn, brought its particular *adat* and often its legal village status.[14] As a consequence, there were multiple villages and multiple ethnic groups mixed in one location. In the resettlement area of Sembuak, for example, ten villages co-exist in the same location, and, as indicated above, Long Loreh is one of four villages, each of a different ethnic group, situated in one location.

Formal administration is based on villages designated by decree of the *Bupati*. In Malinau, the last time 'new' villages were recognized was during the early 1980s, when Lio Mahan was established in Long Loreh, which itself was set up around 1974. Legal status, however, did not include identification of village territory or the designation of boundaries. All of the villages along the Malinau River are legally on state forest land; but it is unclear if the forests should be considered within village territory or whether villages should be seen to be in the forest area.

The differences and overlap of *adat* territories and administrative divisions in these multiple villages have never been considered or explained. Until recently, this blur has not seemed to matter. When small-scale logging included a condition of sharing benefits with communities who held rights to the forest resources, however, the difference became important and, hence, a source of conflict. Communities claimed rights based on their *adat* but associated them only with their village territory. The inability to distinguish between the two has often led to seemingly unsolvable conflicts: neither have been clearly defined or delineated in the field. Although the interior and stratified Dayak groups such as the Kenyah were strongly territorial, their territories were not always permanent. The coastal polities that were built on trade and the more nomadic Punan groups had less defined territories (Sellato, 2001). Migration was common, and with a small population, land seemed endlessly available. A settlement might define its territory based on actual use, bound by rivers or other physical markers. Boundaries were generally unclear and permeable. When groups migrated, the old settlement was usually abandoned, free to be settled by another group or used as old gardens. With increased competition over resources, boundaries and their rules have been made more explicit (Li, 1999; Wollenberg, 2003) with claims to territories including abandoned sites.

Claims based on historical presence deny the rights of people who arrived during the 1960s and 1970s in the resettlement era, giving up their original lands (Barr et al, 2001). To defend themselves, people now invoke the new definition of territories and claim rights over the lands left upstream, or even in other watersheds, where, in some cases, small settlements might have remained. Examples are the villages of Bila Bekayuk, Nunuk Tanah Kibang, Pelancau, Metut and Liu Mahan (Anau et al, 2002).

In an attempt to solve the problem, the current government has merged multiple villages at one location into single administrative units and further ruled that people have to live within the territory they claim. This decision has created much anxiety, especially among the Punan, who have not lived for some time in the area they now claim as *adat* territories (see Chapter 4) and who fear being subsumed under more powerful groups in their current settlements. In response, some Punan groups, such as Long Rat, Mirau, Halanga and Metut, have moved back upstream to their ancestral territories.

Conflicts over claims are often solved using ethnic and personal family connections. Thus, during a CIFOR-sponsored participatory mapping process during 1999 to 2000, six villages (Long Jalan, Long Lake, Pelancau, Metut, Laban Nyarit and Langap) sought agreements based on compromises due to family relations with another village (Anau et al, 2001). Political connections and strength of numbers also often help. The Punan, who live in small scattered groups, are therefore doubly disadvantaged. Traditionally, these groups moved around frequently, rarely occupying clearly defined fixed territories (Barr et al, 2001). Unlike other groups, they have been late in claiming territories and lack access to the political alliances of other groups. As a result, the Punan to this day remain the weakest group in Malinau, politically as well as economically.

In contrast to the interior part, where land is considered a common resource with individual use rights, landownership around Malinau Town is based on the national law and recognized as individual property. Ownership over these lands is recognized based on history, usually because an individual or their ancestors were the first to settle on a particular site, but also because a community was the most powerful at the time of government recognition or because the land was given to them by the government at the time of resettlement. Large pieces of land in and around Malinau Town were claimed by a few of the most powerful families at the time and are now recognized as *de facto* private property.

Much of this *de facto* privately owned land might legally be within a state forest area, but buying and selling is now common, especially around town. Privatization of landownership is becoming the norm, even though the land is not legally registered. In remote districts such as Malinau, few people, if any, register their land; most people outside the town do not even know that they should. This is not much different from the situation two decades ago across most of Indonesia (Moniaga, 1993). Furthermore, as in many developing countries (De Soto, 2000), the procedure for registering land has not been made easy or cheap. The presence of two land regulating offices adds further confusion.

In summary, the current *de facto* tenure system in Malinau has developed through the collusion of the official legal and traditional systems. Recent changes in the value of land and its resources have also altered people's views on tenure, making it more explicit with fixed boundaries.

The story of Malinau also shows the difficulty in deciding who has what rights and how conflicting rights can be equitably resolved. In the absence of strong government interference to protect the rights of weaker communities, people will arrange to provide for their own economic security. The basic right to make a living should remain the guiding principle to protect their way of life.

Land-use planning

With decentralization, new hope arose for communities to have more secure rights over land and to be able to influence decisions over land use. Law 24, 1992,[15] on spatial planning, together with its implementing regulations,[16] includes rules on the rights and responsibilities of citizens to participate in planning, use and monitoring of land use. In addition, community participation is legally part of the development planning process that follows annual cycles of public meetings from village to *kabupaten* (district). Unfortunately, development planning does not refer to the land-use planning.

In theory, the district government provides improved channels for local people to express their views and exercise influence. Potential channels include:

- New opportunities for local people to make government officials accountable and to have them represent the people's interests through elections of district house of representative – *Dewan Perwakilan Rakyat Daerah* (DPRD) – officials; the district leader; and village assembly – *Badan Perwakilan Desa* (BPD) – representatives.
- Technical requirements in policies for government consultations with local communities (e.g. in providing input to land-use plans). Many of these requirements existed before *Otonomi Daerah* (regional autonomy), but were rarely implemented. Since decentralization, there has been more pressure to demonstrate compliance with them.
- Legal provisions through district regulations (*Peraturan Daerah* or *Perdas*) and decrees. In Malinau, *Perdas* provide for the registration and recognition of customary organizations and the creation of the village council *(Badan Perwakilan Desa)*. Decrees for small-scale logging permits required community consent and a letter of agreement between the community and the logging company.
- The establishment of a more accessible local government.

In reality, community participation is practised differently (see Chapter 13). High stakes, vested interests and vast uncertainties have led to politics of greed and resource exploitation where local government has ignored its accountability to its rural constituents. Those holding power are also engaged in fierce competition among themselves to claim land, forests and related resources such as minerals and birds' nests.

In a discussion on spatial planning with the *Bupati* (district head) in February 2002, CIFOR proposed to designate sufficient village forest to be managed by the villages, with the aim of resolving the problem of overlap between village territories and state forest. Local communities would have certainty of tenure, and the government would be able to retain control over large areas of forest for their revenue. The *Bupati* noted that such a step, however, where districts claim state forest land and give villages the right over large parts of the forest, would probably be unacceptable to the central government. He also noted that villages should be entitled to only 2ha of land per family as that was BPN's policy.

Malinau hired consultants for its land-use planning. Community participation was minimal, with the consultant producing maps based on information from the district of Bulungan, of which Malinau was a part until 1999, and through discussions with district officials. Available information and data offered by CIFOR on the ecology, forest and social conditions of the district were ignored. Public consultation mandated by law was limited to a one-time presentation of the final draft a few weeks before the deadline. No inputs were solicited. Although legalized in 2003 by district regulation, the plan was not actually accepted or used. The atlas with maps was not made available to the public because the district government wanted to revise them.

Actual developments bear no relation to the land-use plan. Although the government had banned the IPPKs in 2002, it then issued 11 medium-scale concessions in total, covering 360,000ha (Andrianto, 2006). The government contracted companies to construct important arteries paid through permits to log a 1km wide strip along the proposed trajectory and is actively seeking investors for oil palm plantations. Malinau also plans to convert 400,000ha[17] to 600,000ha

Box 11.1 Results of land-use planning discussions with 12 focus groups in Long Loreh

- Swidden fields (*ladang*) were the most important land type for 10 out of 12 groups. In addition, problems with making swidden fields and in finding work were listed by four groups as their major economic concern although the profile of problems mentioned was different for each group.
- Most important uses of the forest were rattan (nine groups); *daun sang* – a palm – (eight); *gaharu* (eaglewood) (eight); wildlife (eight); and fruits (seven).
- Land use suitable for their village lands was reported as large plantations, tree crops, protection forest and wet rice cultivation (or *sawah*).
- The predominant hopes (of five groups) were to manage land use, ensure sustained forest and obtain development from CIFOR and the regional government (*Pemda*).
- The case studies made it clear that land rights issues have to be settled before land use can be discussed.

Box 11.2 Local people's visions of village land use

In a workshop sponsored by Pemda Malinau and CIFOR, representatives of 27 villages produced a drawing of their vision of village land use, which included different ways in which they perceived the use of forests.

The visions

Most pictures showed a clear sense of spatial arrangements of different land uses. People imagined dryland farming (*ladang*) and *sawah*, coffee gardens and teak plantations. Most viewed their forest as a source of additional income; two villages mentioned re-investment. One Kenyah community began its presentation by saying that its forest was already gone; but it planned to conduct replanting and logging for village income to be spent on a clean water supply, a hospital, a school and as a safety net for future use.

Punan villages, on the other hand, emphasized the need to continue hunting and to follow their traditional life of looking for fruits, animals and *gaharu*. One Punan village said they wanted a forest concession that would practise selective cutting and replanting, and provide the village with coffee and rattan plantations. Another village further upstream provided areas for cultivation and an area of forest for each of the building materials, *gaharu* collection, sago collection, birds' nests and hunting. In addition, they pointed out areas of the forest where *tengkawang*, sago, pigs, fish, rattan and resin were to be found. Although their map was very detailed, it did not show any boundaries, unlike some of the others that divided the area into clearly bounded areas for certain uses.

No village saw commercially oriented forest management as an activity that they themselves would be able to do. One village hoped to implement a conservation concession, but expected CIFOR to provide support for starting it. Several villages designated areas for replanting of forest, especially teak and mulberry, which might be the result of recent government programmes.

Discussion on how to achieve the vision

This discussion focused on constraints, prerequisites, skills and support needed to achieve each village's vision. There was a general agreement that finalizing village boundaries was a prerequisite for further efforts in forest management. The lack of clarity on the extent of rights on birds' nest caves was also considered an important constraint. In addition, the villagers wanted clarity on, and explanations of, government regulations pertaining to forest management, community rights and spatial planning, as well as capital, education funds and access to markets.

People mentioned skills needed, including technical skills in forest management, agriculture and cultivation of tree crops, but also skills in community organizing, accessing capital, communications, village governance and legal knowledge. In order to develop or obtain these skills, they would need the support of the local government as well as from outsiders such as investors, CIFOR and non-governmental organizations (NGOs).

Other issues that emerged from the discussion were the lack of community cohesion; the lack of government support; the lack of representation in village activities and in government projects; confusion about current regulations; and a lack of regulations on forest management.

of forest to non-forestry use (Andrianto, 2006). There is also talk that the central government wants to designate a 5km strip along the border with Malaysia, consisting of around 425,000ha, as a state area of strategic importance – *Kawasan Strategis Negara* (KSN) – and convert this land to oil palm plantation.

Meanwhile, a number of villages drafted their own land-use priorities (see Box 11.1) and visions for land use in the future (see Box 11.2), expecting CIFOR to play the role of broker in linking these plans to the district-level processes.

The district did not seriously accept or consider village land-use plans, while only a few villages actually apply their plans in a formal sense. Thus, the potential of a participatory land-use planning (*tata ruang*) process is not realized and the formal plans remain as usual, stored and unused by the government. Land use is still regulated by the non-formal and extra legal (de Soto, 2000) rules of need and opportunity.

A new spatial plan using the old official delineation, where land outside the forest estate would amount to only 17 per cent of the area, was accepted by the government in December 2003. Thus, most of Malinau's population has been relegated to illegal squatters on forest land.

The Punan of Malinau and claims to land

The Punan in Malinau have a special position in the power constellation of the district. Until the 1970s, many of them were largely nomadic, basing their livelihoods on hunting and gathering. They are less organized than the other groups and have almost no representation in the government except for a small number of Punan teachers. They have therefore been especially disadvantaged in negotiations over boundaries and benefits. An example is the case of Mirau during mid 2000, when the inhabitants returned from a forest product collecting trip and found their boundary with Laban Nyarit being marked by a joint Mirau–Laban Nyarit mapping team of which they were not previously informed.

Even when the Punan have shared decision-making with other ethnic groups, they face the additional burden of prejudice since they are fewer in number and lack political clout. Where negotiations have occurred and agreements have been stable, villages have had limited success in enforcing the latter. IPPK activities that involved cutting roads through the territory of other villages have, in at least four cases (Setulang, Gong Solok, Laban Nyarit and Long Loreh), removed timber or destroyed valued fruit trees, generating anger among Punan villagers with acknowledged claims to these lands (Anau et al, 2002).

With valuable resources at stake, a highly competitive atmosphere and a lack of trust, the threat of merging territories has caused new anxieties. Most of these are concerned with questions of people's rights. Do they now have to share their rights with everyone in the new village, even those outside their ethnic group or clan? Do they lose their rights to individual property located in a village, other than that at which they reside? What happens to claims to abandoned territories?

These anxieties were expressed during CIFOR's annual community workshop in Setarap in April 2003. When asked their opinion on the merging of villages, 31 of the 51 participants expressed disagreement. In addition to a lack of trust in the government (11 responses), the main reason was the fear of losing rights to resources to which they claimed title (9 responses). Although not all participants in this workshop were Punan, the Punan were clearly the most opposed to village mergers. In a follow-up meeting in Long Loreh, the four villages in this location also rejected the merger. In this case, the Punan of Pelancau and Bila Bekayuk were most decisive, with 41 out of 45 opposing it. Again, the Punan feared marginalization in an ethnically mixed village. They are now discussing the possibility of establishing a separate territory upstream, and some have already moved upstream to claim and settle territories. In fact, CIFOR suggested that the government establish a special-purpose forest area upstream as territory for the Punan, following the precedent of the Krui area in Lampung; but the government's reaction, to date, is unclear. It appears committed to its decision to merge villages despite local opposition.

Conclusions: Disconnects of policy

The inherent contradiction and disconnects between legal and customary tenure are typical of the situation throughout Indonesia. On the ground, where the two systems face each other, there can be conflict, rejection, encapsulation or assimilation. In the absence of clear and strongly enforced government regulations, the property rights regime remains largely informal and, generally, in contradiction to national law.

Only when the contradictions become explicit, as when government development projects require large areas of land, does the formal law overrule the informal – usually in favour of government interests. During the New Order, this occurred frequently, with huge numbers of people being dispossessed and marginalized. Malinau was fortunate because the difficult terrain limited the activities of the logging companies in the area. When logging concessionaires occupied the forest, villagers usually tolerated their presence, although often because of intimidation (Anau et al, 2002). The concessionaires, in return, accepted the presence of communities and their activities, often making deals directly with the communities to share resources (Barr et al, 2001). Even so, many of the communities resented the presence of these companies and their non-delivery of promises for village development.

Why didn't the district use the opportunity of decentralization to regulate tenure? Why is tenure such a taboo? Law 22, 1999, which decentralized administrative authority to the district level, has been applied uniformly throughout Indonesia without considering the diversity of circumstances or the varying capabilities of the district governments. Unlike earlier laws, no criteria were ever developed to see how districts were able to handle autonomy (Wignjosoebroto, 2005). More importantly, the autonomy given did not really allow the freedom envisioned. Subsequent regulations, allegedly designed to implement the law, simply imposed limitations, uniform structures or, in some cases, recentralized authority. One example is village government. According to Law 22, 1999, villages are autonomous and are free to revert to *adat*. GR No 76, 2001, on village government, however, contains guidelines for village government structure and management that, while improving the old structures, are also expected to be applied uniformly throughout the country. Several regulations (i.e. Government Regulation No 34 of 2002 on Forestry; Presidential Decree No 62 of 2001 on Returning Authority to Administer Land Matters to BPN) simply re-centralized authority.

At the same time, regional autonomy strengthened the position of *adat*. The high diversity of local *adat* systems, however, makes the tenure issue even more complex and prone to conflicts. Although conflicts due to unclear or overlapping property rights were frequent in the past, the government had always managed to suppress them, with or without force. Most importantly, however, there is little incentive for districts to regulate tenure, especially in favour of local people.

Regional autonomy also meant that those districts had to raise a large part of the revenue to finance district administration and development. Lacking a tax base, and with few district-run enterprises, administrations have had to resort to licensing (license fees, concession fees, charges, etc.) as their main source of revenue. Ideally, of course, the state and district governments should not be involved in doing business and should leave it to private enterprises to generate wealth. But this can only happen if:

- the laws and standards are in place to regulate business behaviour within socially acceptable norms;
- There is an adequate, fair and efficient taxation system (including a system for collecting royalties, if necessary, by which the state or district can appropriate a share of the wealth created; and
- there is a transparent and accountable system in place to ensure equitable and effective use of these revenues for the public good.

The problem is that most of these provisions are not yet properly in place in Indonesia, at least at the district level. Together with resolving issues of tenure, establishing these basic elements of governance is the big challenge.

Rethinking tenure, therefore, remains taboo because it implies a rethinking of territorial control by the state through the government. In this case, it would require the Department of Forestry to devolve control over huge amounts of land. The department and the district government are reluctant to do so as the land and forests on the land are still an important source of revenue and power.

Given the disconnects, it is unclear as to who, then, is responsible for security of tenure. While it is commonly accepted and believed that the government is responsible, as indicated above, it has not been able or shown any desire to regulate or provide security of tenure. It may be in the district administration's interest to maintain multiple levels of ambiguity to allocate land in ways that they judge most important and which meet its own interests. Since formalization is perceived as resulting in winners and losers, the district is wisely wary of the possibilities of increased ethnic strife and a decline in political support from key groups. As such, keeping the status quo might be a purposeful strategy similar to the way of old-style Javanese kings, which allowed for the establishment of numerous *kabupaten* (districts) to prevent a concentration of power able to challenge the state (Onghokham, 2003).

The confusion over who has authority to regulate tenure, together with inconsistent and un-enforced regulations, has resulted in a free-for-all intense competition over resources. Coupled with the lack of long-term vision, decisions are made based on short-term benefits rather than long-term strategies. No rules mean people do as they wish, resulting in the marginalization of weaker parties such as the Punan and benefits taken by the village elites. Following decentralization, local communities' right to obtain a share in benefits has been recognized; but their weak bargaining power and unclear legal standing allow other parties to reap more.

What, then, are the options? First, the many disconnections and roots of the problem need to be recognized and understood. The government needs to be convinced that it is in their interest to regulate property rights in ways that are fair to all. According to de Soto (2000), where assets are not described or organized, development opportunities are limited. Security of tenure is therefore a prerequisite for development.

Will the government, then, be responsible for the security of tenure and the protection of the rights of the weak? As Thiesenhusen (2003) notes, many governments are not willing to fulfil this responsibility for both political and economic reasons. Other than the costs involved, assisting local communities is low on the priority scale of most governments. If communities are given responsibilities, however, their scope needs to be clearly defined and coupled with appropriate rights. Government policy needs to support this.

If the government is then forced to recognize the existence of *adat* communities and their rights, together with the rights of other local communities, a participatory approach will be needed by which it can start discussions and negotiations with communities. A government regulation is ultimately required to give *masyarakat adat* legal status and have their rights legally recognized. Although there is no legal obligation by the government to have public participation in producing legal documents, the laws will only have meaning and respect on the ground if there is cooperation and collaboration with the people whose tenure is to be secured and whose rights should be protected.

Second, there is a need to rationalize the designation and ownership of forest areas. With 95 per cent of the land being designated state forest land and with a growing population, communities must be involved. Much of this land has been deforested, degraded or is already occupied by local people. The occupied parts are often still covered by natural forest, agroforests or planted forests, showing how inherently responsible local people are when the care of the land is in their hands.

As previously mentioned, there is a growing movement of local people demanding that their rights be recognized. Although these rights are acknowledged in Law 41, 1999, no steps have been taken to release control over these lands. The closest the government has come is to consider the possibility of leasing logged-over concessions for community forestry (*Hutan Kemasyarakatan*), although it would still retain control over these denuded lands. One option, therefore, would be to continue the community forestry programme, but with improvements. It is probably more effective to create enabling policies that allow communities to exert their rights and responsibilities than by promulgating regulating policies that limit the options for management.

Several other legal options are available. The government could recognize *adat* rights and designate *adat* forests, giving rights and management responsibility to *adat* communities. Alternatively, where *adat* is problematic, communities could be given rights and responsibilities to clearly delineated village forests. Where collective action is waning, privately owned forestry could be an alternative.

Decentralization also implies respect for diversity and the recognition of *adat* rights of indigenous communities. To ensure that weaker groups will have secure livelihoods and the time needed to adjust, they should have access to areas of forests that are secure from encroachment by others. For the Punan, who currently cannot compete with other groups, designating special purpose forests (*Kawasan dengan Tujuan Istimewa*) might be the most equitable option. Law 41, 1999, specifically allows this designation and would provide the Punan with secure rights, as well as the freedom to choose their own way of life.

Notes

1 Special thanks are given to Peter Frost for his insightful comments and editing.

2 Formal 'decentralization' refers to the transfer of administration to a lower level of government. Although this is often distinguished from 'devolution', where rights and decision-making powers are devolved to the lower levels, in practice, decentralization is often understood by districts as including devolution. See also Chapter 1 of this book and CIFOR (2004).

3 Presidential Decree No 34, 2003, on National Policy in Regulating Land Affairs. As mandated by Decree IX of the Consultative Assembly in 2001, this decree calls for the renewal of the agrarian law, the establishment of a database using GIS and satellite data. It also lays out the tasks of the district related to land affairs: permits, land appropriation for state purposes; resolving conflicts and regulating *adat* rights and land-use planning. Taxation is still retained at the centre.

4 The nine aspects are issuing permits to use a location (for development by a third party); land appropriation for development purposes; resolving conflict over agricultural land; organizing compensation for land appropriated for development; determining the subject and object of land redistribution; arranging compensation for land exceeding maximum limits and absentee ownership; designation and solving matters of *tanah ulayat* (land under customary law); issuing permits for land clearing (but also for *ladang*); and planning land use at the district and municipal levels.

5 Interview with the head of BPN in Malinau and the head of the administrative section of the Land Office in Malinau, 23 October 2003.

6 Interview with the head of BPN in Malinau and the head of the administrative section of the Land Office in Malinau, 23 October 2003.

7 TAP MPR IX, 2001, on Agrarian Reform and Natural Resources Management, stating the need for review of all laws on natural resources management and the formal recognition of customary rights.

8 'New Order' was the term used by former President Soeharto to distinguish his regime from that of his overthrown predecessor, President Sukarno, characterized as the 'Old Order'.

9 Law 41, 1999, the Second Amendment to the Constitution (Articles 18B and 28I) and Decree No 5, 1999, of the minister of the interior, on *Guidelines to Resolve Problems of Communal Customary Rights to Land* (*hak ulayat*).

10 According to Sellato, the petty kingdom of Tidung Sesayap was probably an offshoot of the kingdom based in Tarakan. While it is said that the people descend from the long forgotten *suku Kepatal*, a Dayak grouping, for a long time they identified more with Muslim Malays rather than the interior Dayak.

11 Dt Iman Suramenggala, pers comm (2003), grandson of the last Sultan of Bulungan, founder of Yayasan Pioner and now staff of *Dinas Kehutanan* in Bulungan.

12 *Desa* is defined as settlement centres legitimized by government decree and having the required structure. Law 5, 1979, decreed that all traditional settlements had to follow a uniform structure and, as *desa*, become the lowest tier in the government hierarchy. Because the decree also set certain criteria, many traditional settlements were merged or split, which had implications on their traditional territory.

13 GR No 76 on Village Government.

14 Originally, a settlement of one main ethnic group was defined as a formal and legal village by decree of the *Bupati*. The government resettlement programme often moved whole villages, which remained legal villages separate from the other group with whom they settled.

15 This was replaced in 2007 by Law 26, 2007, on Spatial Planning.

16 PP 69, 1996, and Permendagri 9, 1998, on Procedures for Citizen Participation in the Spatial Planning Process.

17 This was based on the final revision of the land-use map as agreed upon by the governor of East Kalimantan and all the district heads, and proposed to the central government for approval (Kresno Dwi Santosa, pers comm, 2004, manager of the Malinau Research Forest).

References

ACM-CIFOR (2001) *Dokumentasi Proses Pendampingan Masyarakat Dalam Menggali Masalah dan Visi Masyarakat tentang Pengakuan Wilayah dan Pemanfaatan Lahan*, Internal report, CIFOR, Bogor, Indonesia

Alcorn, J. B. (2000) 'An introduction to the linkages between ecological resilience and governance', in Alcorn, J. B. and Royo, A. G. (eds) *Indigenous Social Movements and Ecological Resilience: Lessons from the Dayak of Indonesia*, Biodiversity Support Program, Washington, DC

Anau, N., van Heist, M., Iwan, R., Limberg, G., Sudana, M. and Wollenberg, E. (2001) *Pemetaan Desa Partisipatif dan Penyelesaian Konflik Batas: Studi Kasus di Desa-Desa*

Daerah Aliran Sungai Malinau, Januari s/d Juli (2000) Laporan Pengelolaan Hutan Bersama, CIFOR, Bogor, Indonesia

Anau, N., Iwan, R., van Heist, M., Limberg, G., Sudana, M. and Wollenberg, E. (2002) 'Negotiating more than boundaries: Conflict, power and agreement building in the demarcation of village borders in Malinau', in CIFOR (ed) *Forest, Science and Sustainability: The Bulungan Model Forest. Technical Report Phase I 1997–2001, ITTO Project PD 12/97 Rev 1 (F)*, CIFOR, Bogor, Indonesia

Andrianto, A. (2006) *The Role of District Government in Poverty Alleviation: Case Studies in Malinau and West Kutai Districts, East Kalimantan, Indonesia*, Governance Programme Report, CIFOR, Bogor, Indonesia

Barr, C., Wollenberg, E., Limberg, G., Anau, N., Iwan, R., Sudana, I. M., Moeliono, M. and Djogo, T. (2001) *The Impacts of Decentralization on Forests and Forest-Dependent Communities in Malinau District, East Kalimantan, Case Studies on Decentralisation and Forests in Indonesia*, CIFOR, Bogor, Indonesia

Césard, N. (2001) *Four Ethnic Groups (Punan, Kenyah, Merap, Lun Dayeh) Faced with Changes along the Malinau River (Kalimantan Timur)*, Forest Product and People Programme, CIFOR, Bogor, Indonesia

CIFOR (2004) 'Decentralization, deconcentration and devolution: What do they mean?', Paper presented at the Interlaken Workshop on Decentralization, 27–30 April 2004, Interlaken, Switzerland, www.cifor.cgiar.org/publications/pdf_files/interlaken/Compilation.pdf

Colchester, M. (2001) *Participatory Investigation of the Prospects for Securing Land Rights and Self-Governance among Masyarakat Adat in Indonesia*, Report of ICRAF/AMAN Collaborative Project, Forest People Programme, UK

De Soto, H. (2000) *The Mystery of Capital: Why Capitalism Triumphs in the West and Fails Everywhere Else*, Basic Books, New York, NY

Dove, M. R. (1993) 'A revisionist view of tropical deforestation and development', *Environmental Conservation*, vol 20, no 1, pp17–24

Drake, C. (1989) *National Integration in Indonesia: Patterns and Policies*, University of Hawaii Press, Hawaii

Erb, M. (1997) 'Contested time and place: Constructions of history in Todo, Manggarai (Western Flores, Indonesia)', *Journal of Southeast Asian Studies*, vol 28, no 1, pp47–77

Fay, C. and Sirait, M. T. (2002) 'Reforming the reformists in post-Soeharto Indonesia', in Colfer, C. J. P. and Resosudarmo, I. A. P. (eds) *Which Way Forward? People, Forests and Policymaking in Indonesia*, Resources for the Future (RFF)/CIFOR, Washington, DC

Gelora Mahardhika (2002) *Dibentuknya Dinas Pertanahan Kabupaten Membingungkan Masyarakat?* Laporan Utama, 1–15 November 2002

Gönner, C. (2002) *A Forest Tribe of Borneo*, Man and Forest Series No 3 D.K, Printworld (P) Ltd, New Delhi

Government of the Republic of Indonesia (1992) *Law 24 of 1992 on Spatial Planning* [*Undang-Undang Nomor 24 Tahun 1992 Tentang Penataan Ruang*]

Government of the Republic of Indonesia (1999a) *Law 25 of 1999 on Fiscal Balancing between the Central Government and the Regional Governments* [*Undang-Undang Nomor 25 Tahun 1999 Tentang Perimbangan Keuangan antara Pemerintah Pusat dan Daerah*]

Government of the Republic of Indonesia (1999b) *Law 41 of 1999 on Forestry* [*Undang-Undang Nomor 41 Tahun 1999 Tentang Kehutanan*]

Government of the Republic of Indonesia (2001a) *Government Regulation 76 of 2001 on General Guidelines for Administering Villages* [*Peraturan Pemerintah Nomor 76 Tahun 2001 Tentang Pedoman Umum Pengaturan Desa/Kelurahan*]

Government of the Republic of Indonesia (2001b) *Presidential Decree 62 of 2001 on Governmental Institutions [Keputusan Presiden Nomor 62 Tahun 2001 Tentang Perubahan atas Keputusan Presiden Nomor 166 Tahun 2000 Tentang Kedudukan, Tugas, Fungsi, Kewenangan, Susunan Organisasi, dan Tata Kerja Lembaga Pemerintah Non Departemen sebagaimana telah beberapa kali diubah terakhir dengan Keputusan Presiden Nomor 42 Tahun 2001*]

Government of the Republic of Indonesia (2002) *Government Regulation 34 of 2002 on Forestry [Peraturan Pemerintah Nomor 34 Tahun 2002 Tentang Tata Hutan dan Penyusunan Rencana Pengelolaan Hutan, Pemanfaatan Hutan dan Penggunaan Kawasan Hutan*]

Government of the Republic of Indonesia (2003) *Presidential Decree 34 of 2003 on National Policy in Regulating Land Affairs [Keputusan Presiden Nomor 34 Tahun 2003 Tentang Kebijakan Nasional di Bidang Pertanahan*]

Government of the Republic of Indonesia (2004) *Law 32 of 2004 on Regional Government [Undang-Undang Nomor 32 Tahun 2004 Tentang Pemerintahan Daerah*]

Government of the Republic of Indonesia (2007) *Law 26 of 2007 on Spatial Planning [Undang-Undang Nomor 26 Tahun 2007 Tentang Penataan Ruang*]

Li, T. M. (1999) 'Marginality, power and production: Analysing upland transformation', in Li, T. M. (ed) *Transforming the Indonesian Uplands: Marginality, Power and Production*, Harwood Academic Publishers, London

Moeliono, M. (2000) *The Drums of Rura: Land Tenure and the Making of Place in Manggarai, West Flores, Indonesia*, PhD thesis, University of Hawaii, Honolulu, HI

Moeliono, M. (2006) 'Wilderness and order: Forest conservation in Malinau District, East Kalimantan', Presented at Survival of the Commons: Mounting Challenges and New Realities, the Eleventh Conference of the International Association for the Study of Common Property, Bali, Indonesia, 19–23 June 2006

Moniaga, S. (1993) 'Towards community-based forestry and recognition of *adat* property rights in the outer islands of Indonesia', in Fox, J. et al (eds) *Legal Frameworks for Forestry Management in Asia*, Case Studies of Community–State Relations, East West Center Occasional Paper No 16, East West Center, Honolulu, HI

Onghokham (2003) *The Thugs, the Curtain Thief, and the Sugar Lord: Power, Politics and Culture in Colonial Java*, Metafor Publishing, Jakarta

Patlis, J., Moeliono, M., Djogo, T. and Iwan, R. (2001) *Analysis of Four Draft Regulations of the Kabupaten Malinau*, East Kalimantan, CIFOR internal report, Bogor, Indonesia

Scott, J. (1998) *Seeing Like a State*, Yale University Press, New Haven, CT

Sellato, B. (2001) *Forest, Resources and People in Bulungan: Elements for a History of Settlement, Trade and Social Dynamics in Borneo, 1880–2000*, CIFOR, Bogor, Indonesia

Thiesenhusen, W. C. (2003) Difficulties and Tensions in Institutionalizing Grassroots and other Civil Society Organizations During the Transformation and Post-Reform Periods, www.fao.org/docrep/005/Y8999T/y8999t03.htm

Toyamah, N. (2002) *Land Administration Authority: Power Struggle between Central and Regional Government*, Kewenangan Bidang Pertanahan, Tarik Menarik antara Pusat dan Daerah, www.smeru.or.id/newslet/2002/ed04/200204focus.htm

Wignjosoebroto, S. (2005) *Pasang Surut Otonomi Daerah: Sketsa Perjalanan 100 Tahun*, Institute for Local Development Jakarta, Indonesia

Wollenberg, E. K. (2003) 'Boundary keeping and access to *gaharu* among Kenyah forest users', *Environment and Planning A*, vol 35, no 6, pp1007–1023

12

The *Realpolitik* of Village Representation and Participation

Eva Wollenberg

Introduction

In the period of reform towards regional autonomy, rural households had more opportunity to influence government decisions. How did this change local policies, especially in relation to forests and forest land? To what extent did genuine local politics develop (Antlöv, 2003)? This chapter examines citizen participation and representation in Malinau to better understand how local political systems and decentralization shaped local people's opportunities to have voice in forest and land-use decisions. It gives special attention to small-scale, informal and village-level processes that have a major influence on local people's lives and district government decisions, yet are often overlooked in favour of more visible, larger-scale formal processes. The articulation of the formal and the informal processes of decision-making and influence at the village level constitute the *realpolitik* of local governance.

From 1998 to 2005, research was carried out in 27 villages of the upper Malinau watershed, and in depth in the 4 villages of Long Loreh, Malinau District. Most villages, especially since 2000, were involved in negotiations with other villages, companies or the district government over boundaries, land use or compensation. The research included nine months of participant observation (1998 to 2000), in-depth interviewing with key informants regarding village meetings and representation, several surveys on principles of good governance, and action research in participatory mapping and land-use planning.

The chapter begins with an overview of the limits and tensions inherent in the representation of citizens' interests in public decision-making. It then reviews policy provisions associated with decentralization and land-use planning aimed at enhancing representation. Next, findings are presented on how representation and participation occurred at the village and district levels. The chapter will show how decentralization has spurred villagers' interest in participation and how political organizations are emerging in parallel with the government to act as advocates and lobby on behalf of different ethnic groups. Real changes in representation and participation have occurred but there were still many hurdles for people to have a say in public decision-making in Malinau.

The contradictions of representation and participation

In democratic models of public decision-making, citizens have a say by directly participating in governance decisions – based on the presumption that people best know their own interests – or through representation by a well-informed elite accountable to them and acting on their behalf. Neither is ideal. Direct participation and representation differ – in their efficiency, potential for debate and informed decision-making, and partiality of representation (Haraway, 1991). Both share the problem of how to aggregate diverse interests to produce a single decision. No approach to identifying and accommodating interests guarantees that all individuals' interests will be met or that any one choice will be in the public's best interest. Finding the suitable combination of direct participation and representation is the art of the political decision-maker, subject to norms and expectations of local political culture. Prescriptions are not possible.

Representation has its own special contradictions. Representatives can act as a symbol ('standing for' others) or do something for other people who are absent or who lack the capacity ('act on behalf of' others) (Bogdanor, 1987, p531). Confusion between these two roles can result in the wrong expectations and misrepresentation of citizens. In 'standing for' others, a representative shares some defining characteristics of their constituency. The ideal is to be 'an exact portrait, in miniature, of the people at large, as it should think, feel, reason and act like them' (John Adams, cited in McLean, 1996, p428). Symbolic representation is often an end in itself, rather than a means to an end (Rosener, 1978). Public consultations, which have become more common in the reform period in Indonesia, often involve inviting symbolic representatives of different interest groups.

While this form of public involvement allows expression, it obfuscates how public preferences are determined and what kind of weight is given to them in decisions (Steelman and Asher, 1997, p74). The identity represented can also be problematic as groups have multiple, fluid identities that can vary in different contexts. A representative acting as a symbol of a group is unlikely to fully capture the complexity, tensions and dynamism of a group's identity. Symbolic representation is the most appropriate for 'nonbinding direct involvement' meetings (Steelman and Ascher, 1997, p71), where an individual contributes his or her personal input to a deliberative process, but is not beholden to a larger constituency and does not participate in decisions that affect others.

In contrast, when 'acting on behalf of' others, the representative functions as an agent for the interests of a specific individual or group. A smaller group of representatives presumably acts more efficiently and knowledgeably than an entire constituency (Lee, 1993). Members of political assemblies and official leaders are agent representatives. Decentralization in Indonesia encourages agent representation that is more closely linked to people by requiring village- and district-level assemblies.

Agent representation raises at least three tensions. First, the representative has leeway to act independently from the principal individual or group whom he or she represents (Rao, 1998). Second, the representative needs to aggregate different points of view by, for example, voting with majority rule, building consensus or working with only one element of the constituency (Haraway, 1991). Representing only part of the population is one means of coping with the aggregation problem, but can alienate the rest of the constituency and result in instability. A third tension is the degree to which there are checks and balances on the representative's accountability to his or her constituency (i.e. how much power representatives wield and if they use it in the interest of the constituency) (Ribot, 2001).

Representation issues have become increasingly important with participatory politics (Peters, 1996; Lewis, 2000). A distinction can be made between voluntary participation and participation by invitation. Voluntary participation occurs when the general public is encouraged to make their views known through letters, visiting officials, attending meetings or public protest. Participation by invitation occurs when a convener calls different individuals to a meeting, usually as representatives of larger groups.

Participation by invitation can be more problematic because participants often do not know until the meeting, if at all, whether they are acting on behalf of others or standing for others – and for whom. This makes it difficult for them to undertake prior actions of accountability to their constituency – for example, to find out their constituency's position on an issue. Where invitees act as symbolic representatives, the group whom they represent is likely to be defined in the context of the meeting and other participants. 'Such and such a group did not attend, could you represent them, please?' is all too common a phrase in some meetings. There is a natural tendency to focus more on identity vis-à-vis other members of the stakeholder group, rather than their identity as a representative of their community of origin (Bickford, 1999). There is also a risk when representatives are assigned constituencies that the group itself may not wish to be represented by that person or in that forum. Some groups have strong feelings about whether they want to be represented by a member ('the poor should represent the poor'; see Rao, 1998) or by an advocate. Groups can demand representation.

Conveners of meetings tend to focus more on symbolic representation (Rao, 1998) and their own internal dynamic than on the relationship of the representative to their constituency. Conveners seek an optimum composition of symbolic identities and people who contribute to a productive and congenial work environment. Participants are often told to act in their personal capacities, yet it is not clear what this means in terms of responsibilities and expectations that they may carry personally to their group. Personal perks can tempt representatives of groups to maintain their personal position in negotiations, even at the cost of failing to achieve benefits for their constituents.

We suggest that, in practice, voluntary participation is often limited to certain representatives who feel more at ease with providing their views because of language, status, disposition or familiarity. Both invited and voluntary participants become part of a specialized class of people that is not necessarily representative of the larger populace (Melucci, 1996). They may receive perks for their attendance that further set them apart and create incentives to maintain relationships with the conveners rather than their constituents. Villagers complain that their representatives have been co-opted. Participation by invitation is also usually biased insofar as conveners usually select participants for certain traits. Some traits are desirable for representation, such as a willingness to speak up and understanding of the range of views in a group. Other traits can create biases, such as selecting people based on acquaintance, loyalty to the convener, or merely presence and availability. Because of these biases in representation, input from public participation is always partial and needs to be situated (Haraway, 1991).

Accepting that representation is always situated and problematic, we concur with Bickford (1999, p94) that 'the act of representation determines what is represented'. To guide our observations in Malinau, we relied on the following principles on acts of representation:

- Acting on behalf of others promotes democratic representation more than standing for them.
- Representatives who act on behalf of others need to be accountable to them (Ribot, 1999). Accountability reflects the degree of citizen control over their representative (Schumaker, 1991) or a person whom citizens can trust to act in their interest because of their integrity (Lee, 1993). A constituency should have the power to select and de-select their representatives (Strachan, 1997).
- Representatives who act on behalf of others need to be responsive to them. Responsiveness carries the double responsibility of knowing what the constituency wants and exercising a voice, influence and authority to make decisions accordingly (Schumaker, 1991; Rao, 1998). Mechanisms for aggregation should be clear, and communication is two way.
- Representatives acting on behalf of others and facilitators of public input need to acknowledge pluralism in the constituency and work to represent and facilitate principles for agreement among different identity and interest groups (Rescher, 1993; Rossi, 1997; Bickford, 1999; Edmunds and Wollenberg, 2001). Interest groups should be allowed to define themselves and resist categorization (Bickford, 1999).

These principles should apply at all levels, from village to district and national decision-making. In villages, and even at the district level, however, representatives have multiple identities and links with members of the constituency that affect how they play their role. Acting on behalf of the group may be compromised by special interests or obligations to family, friends or neighbours. Representatives often come from elite *adat* (customary) social classes that hold special privileges of decision-making and appropriation of benefits. They are not necessarily expected to consult broadly with lower social strata. Some representatives are employed with the government or company with whom negotiations are being held. Others are outsiders who cannot 'stand for' local people, but adopt the identity of local people and presumably act on their behalf. There is little or no oversight to check transgressions of roles. Understanding this multi-layered social context is essential to unravelling the impacts of reforms on local politics.

Decentralization policies

With decentralization's formal implementation in January 2001, the district government provided improved channels for local people to express their views and exercise influence, which included:

- Representatives in the district assembly – *Dewan Perwakilan Rakyat Daerah* (DPRD) – and village assembly – *Badan Perwakilan Desa* (BPD). By electing these officials and having better geographic access to them, there are new opportunities for local people to make government officials accountable and represent their interests.
- Technical requirements in policies for government consultations with local communities – for example, providing input to land-use plans.
- Legal provisions through district regulations (*Peraturan Daerah*, or *Perda*) and decrees (*Surat Keputusan*, or SK). In Malinau, *Perdas* provide for the registration and recognition of customary organizations and in village governance for the creation of the BPD.
- Improved geographic accessibility of district government officials compared to provincial or national officials.

With the enhanced role of local government, organizations such as WWF Indonesia–Kayan Mentarang National Park Project and CIFOR shifted from dealing with policy-makers in the provincial capital in Samarinda and Jakarta to those in Malinau. Both organizations organized multi-stakeholder meetings in Malinau for the first time.

As people realized that much representation in official events was by invitation, they recognized the importance of establishing representative organizations. These organizations were modelled after the existing government structure consisting of institutions at the district level, sub-district level and, sometimes, village level. These organizations are effective since, in general, they are invited to public meetings; however, their links to their constituency are often weak (see also Chapter 3).

Representation and participation in practice

Given these opportunities for local people to influence government decision-making, what occurred after reforms began in 1998 to 1999?

Representation in and among villages

Open meetings

Within villages, most aspects of representation and participation remained unchanged with decentralization. Representation within villages occurred mostly through direct participation, even in the largest villages of about 200 households. Open-invitation, village-wide meetings were still the most common way for most people to participate in the affairs of their community. The village leader (usually an elected *kepala desa*, sometimes the *kepala adat*, if there was one) or his representative arranged a meeting by sending out word to key individuals who were then expected to inform others. There were also frequently announcements in church. In Kenyah villages, a caller walked around the settlement announcing the meeting (*pekesat*). Announcements were commonly made on the same day or night before the meeting, but could be made up to a week in advance. Open meetings were usually held in public buildings such as meeting halls, churches and schools to organize village-wide events, share information, or introduce projects by the government or by other groups.

Although these meetings provided a seemingly ideal setting for participatory politics, participation varied substantially. Participants typically represented some factions of the village better than others, particularly the inner circle of more powerful individuals that villagers called the *tokoh masyarakat* (village elders[1] or leaders). For the Merap and Kenyah, these individuals were drawn from the aristocratic class and were often close relatives of the current *adat* leader or village head. They also sometimes included teachers or outsiders with more education or experience with government or companies. Teachers and the top elite of the village were more likely to be in the village at any particular time than those busy in the field or in the forest. One individual noted that he rarely attended meetings since he often slept in his field. It was usual for the leaders' networks to be well represented, while factions representing the opposing view to the village leader were poorly informed about meetings or underrepresented. Information about meetings clearly favoured the leader's closest networks. Factions sometimes boycotted meetings.

In any meeting, usually only a small proportion of the participants were active in discussions and decision-making. In Long Loreh meetings, typically only three to five individuals from each of the villages of Sengayan, Pelancau and Bila

Bekayuk, and six to ten individuals from Loreh were active in meetings. The same people tended to participate in meetings.

The village leader, or in his absence the *adat* leader or secretary, facilitated the meeting and took final decisions. Occasionally, others with strong grievances or opposing views spoke out. Often, smaller informal exchanges took place before the meeting to build support for a consensus. In all villages, a much lower proportion of women than men attended general meetings on village affairs. Women said that they 'were too busy', 'don't know how to speak well and would be shy and embarrassed', 'do not know anything about the topic' or that the meeting was 'men's business'. Women who did attend were almost always silent observers. The wife of the village leader was more likely to attend meetings than other women. According to the wife of the village leader in Sengayan, women were more likely to exchange information and make collective decisions during visits to their rice fields (*ladang*) or during cooperative work (*gotong royong*) sessions. Women also made group decisions during preparations for events or visits from other villages. A number of women said they acquired information about village meetings from their husbands. Even though certain government programmes specifically targeted women – such as *Pembinaan Kesejahteraan Keluarga* (PKK), a programme aimed at family well-being and health – women noted that these were not important forums for them.

Meetings by invitation

In addition to these open public meetings, village leaders hosted smaller meetings by invitation. The smaller meetings were used to settle differences among households and to manage specific tasks (such as prepare for the visit of a government official; negotiate with outsiders; and prepare or follow up on open meetings) in order to make decisions quickly. They were usually held in the village leader or *adat* leader's home. The village or *adat* leader facilitated the meeting and took final decisions. Participants ranged from 3 to about 30 individuals, depending upon the village and issue at hand. Negotiations on money were limited to only a handful of individuals.

Smaller closed meetings were more frequent than large open meetings. During busy periods, such as the negotiation of an agreement or preparation for an event, there could be meetings every evening lasting long into the morning hours. There was an unspoken understanding that individuals closest to the village leader were invited to and attended these meetings. Sometimes, these individuals held positions as village officials, but not necessarily. Other individuals were invited according to need, based on the issue or task at hand.

As venues for settling conflicts, smaller meetings did involve representation to the extent that villagers representing different sides of the conflict needed to attend. Nevertheless, groups outside the leaders' network sometimes complained about the lack of representation of their interests in meetings. For the most part, representation *within* villages was not a guiding principle underpinning the village's public decisions.

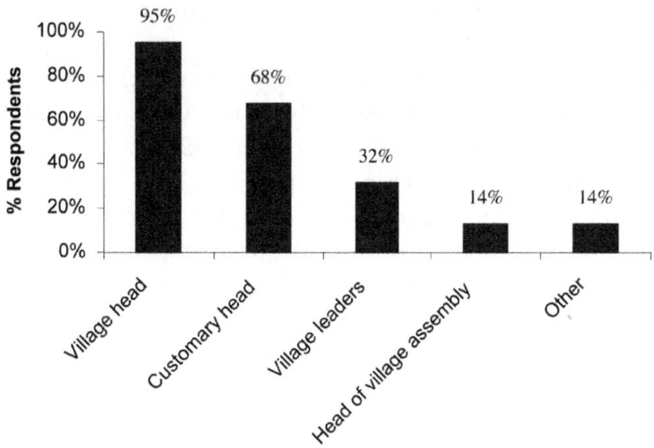

Source: Wollenberg et al, 2006, p11

Figure 12.1 *Who usually represents communities? A survey of 95 individual villagers in 19 villages, 2004–2005*

Meetings among villages

Among villages, representation took on more importance. The village leader, or *kepala desa*, was the nominal representative of most villages and, according to 95 per cent of people surveyed, was the usual representative (see Figure 12.1). But the customary head was also highly likely to represent the village, as was the *sekretaris desa* (village secretary) or other members of the *aparat desa* (village administration) or elders. Weaker villages sometimes appointed outsiders to positions in the village administration, with the result that an individual from Sulawesi represented a Kenyah village, an individual from Flores represented a Punan village, and a Kenyah represented a Punan village. In these cases, the outsiders were married to women in the village, and villagers viewed them as having superior public speaking and negotiation skills, as well as more worldly knowledge.

When asked about the aspects of a good village representative, all of the people interviewed invariably responded that the person should at least be '*pintar bicara*', or able to speak well. Furthermore, the representative should not be a *pemalu* (shy person) in front of many people, should have *kewibawaan* (authority) in the community, should be able to make presentations, and should be able to summarize and communicate lessons and information from elsewhere to the community (e.g. in workshops). Villagers also valued reading and writing skills in their representatives. One respondent explained that, in general, villagers who worked with the government had experience that enabled them to become good representatives. The same person noted, however, that many *wakil* (representatives) did not report back to the community. Bad representatives were those who were immoral or received money from other parties to make decisions counter to the village's interest. Sometimes, villagers trusted the person second in command because they believed he received fewer payments from outsiders. Villagers assumed that the representative should act on behalf of the village and that his or her behaviour reflected upon the community.

Meetings among villages fell into two categories: those sharing a single settlement area and those not. Where villages shared a settlement area, such as in Long Loreh, Paya Seturan or Setarap, decisions were based on a mix of direct participation and representation.

Participants from all villages were free to attend open meetings, but felt hesitant to make decisions if their village leader was not present. Certain villages tended to be better represented than others in these meetings. One village almost always took the lead in organizing meetings where several villages shared a location. In the four villages of Long Loreh, for example, the larger and more dominant village of Long Loreh usually organized meetings and managed the village meeting hall, which was in their section of the settlement. There were always more participants from Long Loreh than from the other villages. Sometimes, there were no participants from one or both of the two Punan villages. Punan villagers said they did not always receive invitations or participate in the meeting because they were frequently in the forest collecting *gaharu* (*Aquilaria* spp, or eaglewood) and other products. Punan villagers also frequently felt their attendance made little difference to outcomes and often preferred not to attend.

Weaker villages often felt out of place, unwanted or intimidated in meetings with more powerful villages, both within and across settlements. The participatory mapping activities and negotiations facilitated by CIFOR were often dominated by the stronger villages and resulted in unstable agreements. Stronger villagers tended to host negotiations, which further disadvantaged the villagers travelling from far away. It is interesting to note that Pelancau, a Punan village located in Long Loreh, arranged its own separate meeting hall, enabling the people of Pelancau to have more independence in organizing meetings.

Among villages, the single most important factor affecting the level and quality of participation and representation was the location of the meeting. The distance and cost of travel among villages were substantial in the upper Malinau watershed. Travellers required up to three days to go between Long Jalan, the village furthest upstream, and Sentaban, the village furthest downstream. We observed at least ten meetings where host villages had 20 or more people present and where visiting villages had no more than two or three representatives each, or a boatful of people. Agreements resulting from such meetings frequently failed when factions back at home challenged the decisions that their leaders made in a distant location. Larger numbers from villages gathered only for annual celebrations, such as harvest festivals or ethnic group meetings.

Women, especially those with children, found it even more difficult to attend meetings where they had to travel. In five inter-village meetings, the CIFOR team always specified that at least one of the three representatives should be a woman. In 2003, we even requested two women from each village. Yet, only two to five women ever attended in a single year. Women who did join tended to be still in their teens, wives of village heads, teachers or, in one case, a preacher who was also the village secretary. The younger women endured much taunting during these meetings, making it harder for them to participate, even with facilitation by a third party.

Meetings with companies and government in and across villages

One thing that changed with the reform era, partly as a result of the district government's interest in generating new income, was that villagers met more often among themselves as well as with government and company officials to negotiate land use, access to forests and boundaries. This was particularly the case between 2000 and 2003, when villagers arranged small-scale timber harvesting (IPPKs, or *Izin Pemungutan dan Pemanfaatan Kayu*) in each village. Villages with active timber concessions or coal mining also entered into more demanding negotiations with these parties. They felt more empowered to make demands during the reform period as there was little intimidation by the military and they had the backing of local government. In 2002 and 2003, two timber companies, PT Inhutani II and PT BDMS, stopped operations in the area, in part due to these negotiations.

Village participation in company-related negotiations in the pre-reform years was restricted. Initial negotiations were limited to the village head and, occasionally, a select few of the village elite. Informal meetings also took place outside the village in the offices of the company or the government. Where a company had an office in the village, such as the coal mining company in Loreh, the village head or elite often dropped by to '*ngobrol saja*' (chat), probably to maintain good relations and exchange news.

The dominant village in clustered settlements negotiated with companies or government on behalf of all the villages in the settlement. Officials from companies and local government said they preferred to deal only with the dominant group in each settlement. In Loreh and Langap, the Punan villages were almost completely ignored by the companies. PT Inhutani II gave wheelbarrows, hoes and seedlings from its village development programmes exclusively to Loreh and Langap villages. In May 1998, three Punan village leaders said they had not yet attended a meeting with the coal company. There had been an invitation to one meeting, but it was late, and there was no transportation; only two village officials attended. As one Punan leader said: 'They don't really want to approach us.'

Village representatives occasionally held consultations before or after such meetings. For example, when PT BDMS arranged to pay compensation to the ten families whose fallowed fields were to be excavated in 1998, the village head consulted with each family on how much they requested in compensation. One villager who had participated in the first round of compensation payments of about 500,000 rupiah said that he did not have a choice the first time, but now he felt he could request more money.[2] In April 1998, the village head, together with the *wakil ketua adat* (vice *adat* head) and the subsequent village head reportedly negotiated these payments alone with company officials in the PT BDMS office near Loreh.

It was in the companies' interest to make these negotiations private and limited to a small number of people. It was not, however, always in the interest of the weaker village leaders. One Punan village leader said he was woken up at 1.00 am to sign documents granting rights to a palm oil company by another village leader accompanied by a soldier.

Conflicts became more frequent as the reform period started. Delegations of villagers involved in meetings about a conflict could be large, up to 40 people. More male youths were likely to be involved in these meetings. For example, in August 1998, some 30 representatives from Loreh, 5 representatives from Bila Bekayuk[3] and 3 representatives from Sengayan went to a meeting in the coal mining company's office. The meeting was attended by the sub-district leader's staff, the police, the sub-district military branch and four staff from PT BDMS. The communities had requested the sub-district leader to organize the meeting on their behalf to demand the coal company to fulfil earlier promises to provide clear drinking water, 25 million rupiah to establish a co-operative, a car, electricity, employment for villagers and transport of villagers to their fields.

IPPK negotiations ushered in a new approach to community–company negotiations. Companies were more inclusive in their efforts to negotiate access with villages, including weaker and Punan villages. Village heads held more consultations with their communities to discuss potential benefits and how they were to be shared. This reflected a new concern with distributing monetary benefits in ways that recognized shared claims and avoided later conflict. Later, in 2004 to 2005, when asked which decisions required the participation of the whole community (see Figure 12.2), most people answered negotiations with timber companies (55 per cent) and village development project proposals (45 per cent).

The signing of official documents and their content, however, were not open or transparent. Most community members never saw the agreements that their leaders signed with companies.

Another impact of the reform period was that weaker villages, especially those that had been subsumed under the more dominant village's administration in clustered settlements, began to engage in independent negotiations and have more direct contact with local government and companies. As their participation in

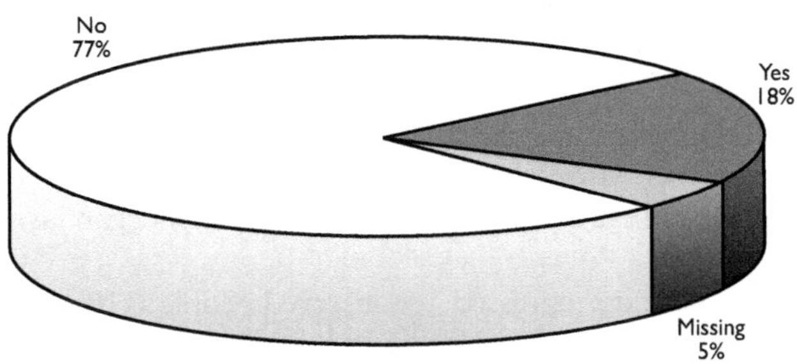

Source: field data, interviews conducted in 2004 and 2005

Figure 12.2 *Does the district assembly represent communities? Survey of individual villagers*

negotiations increased, so did their access to related benefits. They were more willing to speak out against more dominant villages. This may have been reinforced by the Punan's increased cash income from a boom in *gaharu* trade since the early 1990s and the decline of a generation that supported traditional patron–client relationships. In 1994, the Punan of Malinau also held their first *adat* gathering and elected a collective *adat* leader.

In all negotiations, it appeared that leaders acted on behalf of a constituency as the person with official decision-making authority (or their own self-interest), not as a representative of diverse village interests. Village leaders had few incentives under these circumstances to represent others or to divulge information to them. The pressures on village leaders during this time were immense.

Content of meetings and negotiations

As noted above, the reform period created new issues on which villages needed to make decisions, particularly concerning IPPKs and village boundaries. From January to June 2000, we tried to collect information on the content of all meetings concerning forests and land use and involving citizen representatives (i.e. not just village officials) among the four villages of Long Loreh. Our sample of 36 meetings is only partial as some meetings were inevitably confidential.

What were the meetings about? Forests were the subject of 36 per cent of the meetings, warranting an average of about five meetings a month (see Table 12.1). Boundary and compensation issues, which included forests issues, were the subject of nearly half of the meetings. All issues initiated from the pre-reform period: meetings were held with PT Inhutani II to request development assistance that more closely matched their needs; with PT John Holland, the subcontractor to the coal mining company, to arrange employment of villagers; with the government to complain about the impacts of the coal mine and with the coal mine to request piped clean water; with an oil palm company requesting access to the village; with the neighbouring village of Langap to negotiate boundaries; and with CIFOR to conduct mapping of those boundaries. Villagers also met among themselves to address the damage from flooding of the Malinau River; plan demonstrations to demand clean water from the coal company; prevent outsiders from cutting timber in their area; and prevent fishing with use of poison or electrical shock.

This transition period was unusually taxing in terms of the number of issues and meetings in which village leaders participated. The numbers above do not reflect any of the meetings associated with IPPKs, which began just a few months later.

As the reform period progressed, we observed people's changing expectations of their government. In the beginning of the reform era, they had high expectations of gaining new rights in new arenas, including forestry and forest land. There were unrealistic expectations that politics would become more 'bottom up' and transparent. People tested the limits of the amount of land they could claim, compensation for which they could negotiate, and relations they could develop

Table 12.1 *Subject of meetings held January to June 2000 by four villages of Long Loreh (sample = 36)*

Meeting subject	*Percentage of total*
The forest	36
Estate crops and plantations	17
Company relations	47
Boundaries or compensation	44
Government programmes	3
Meeting with government	3
Village events	3
CIFOR	2

Source: field data collected between January and July 2000 (collected by Asung Uluk, CIFOR consultant)

with the *Bupati* (district head) and his staff. In time, the local government established itself and consolidated its authority. Old patterns and power relations largely resurfaced. Citizens' expectations lessened and much went back to business as usual.

Discussion: Representation and participation in and among villages

In practice, relatively few elite were active in village deliberations and representation. Reforms did little to change these historical patterns. There were no provisions to ensure representation of the larger population. To the extent that there was democracy in villages, it was due to villagers' direct participation and access to, and influence on, their leader. Village leaders were at least morally accountable to their members,[4] who could choose to remove them and elect a new leader. Villagers rarely did this, however, out of respect for the aristocracy, lack of alternative candidates or fear of personal repercussions.

The practice of companies and government in dealing with a small circle of the elite in closed meetings, often outside of the village, encouraged village leaders to act in their own self-interest and collude with the company or government. These practices only exacerbated the tendency for a small number of elite to dominate decision-making, especially with the advent of IPPKs and involvement of more communities in negotiations. Community members were aware of the nature of these negotiations and publicly tolerated them, but in private often expressed their disappointment and suspicions. Such closed negotiations therefore undermine people's trust in the integrity of their representative, as well as in the capacity of the latter to act on their behalf or to behave consistently with their desired values.

The people who did represent the community were usually the most educated and confident members of the community, with reading, writing and speaking skills that enabled them to communicate effectively as well as experience the world outside the village, particularly with government and companies. Many were aristocrats. For the hierarchically oriented Kenyah and Merap, it was difficult for the common person to have a voice in village decisions. While disadvantaged groups such as women and the Punan should, in theory, be able to speak for themselves in village politics, they often lacked the skills to do so effectively, let alone the social status. Most women simply did not have the time. Village democracy in the upper Malinau watershed did not consist of an ideal form of popular participation. Certain groups had little or no voice in village affairs.

Decentralization sought to address this lack of more comprehensive representation in villages and create checks on the authority of the village leader by requiring a village assembly – *Badan Perwakilan Desa* (BPD). Unfortunately, most candidates did not meet the educational requirements stipulated by the law and few villagers were prepared to pay for an assembly. Few understood the function of the BPD and even requested the district government to pay for it and village leaders to appoint members. Interest in forming BPDs was low, and village leaders had no incentive to promote them. As a consequence, in the upper Malinau, there were no functional BPDs by 2004. Later, regulations reduced the BPD's authority and restored its role – dating back to the time before decentralization – as largely that of rubber-stamping village decisions.

The definition of a village and its elite became key determinants of whose interests were represented. As another consequence of reforms, districts gained the authority to define villages according to their own criteria. Malinau redefined clusters of villages as single villages. Stronger villages and ethnicities will almost certainly dominate weaker villages and ethnicities. In all cases, the stronger groups had no major objections to village mergers. The weaker groups were against the mergers, mainly due to their different ethnic backgrounds and fear of losing *hutan adat* (customary forest).

At the district level

How did decentralization to district governments and reforms affect citizen participation? To what extent did local people have a say in forest or land-use decisions?

In contrast to village-level politics, more significant changes seem to have occurred at the district level. In 2001, government officials talked, at least initially, of a genuine interest in 'bottom-up politics'. Many villagers were able to bring their concerns directly to the *Bupati* in the first year of decentralization.

More local people felt that they had improved access to government officials because of the geographical proximity of the new district in Malinau and the new priority of hiring only local people (*putra daerah*) to district government positions. More local people had ethnic, village or family relations with officials than ever before. More could speak the same local language as some of their local officials. By 2004, the local government employed 1565 people, most of who were from Malinau District. Local government employees constitute about 7 per cent of the labour force in Malinau (16.5 per cent of households).

Face-to-face interaction between villagers and district officials increased. The *Bupati* attended large community gatherings more frequently than pre-reform district leaders had. Other officials visited more often as well. In turn, local people from remote communities visited the district capital more frequently to conduct business and therefore had more contact, either intentionally or not, with district

Box 12.1 The difficulties of scheduling meetings

Early Tuesday morning, on 2 July 2002, several villagers from Tanjung Nanga suddenly arrived in Long Loreh to attend a meeting informing villagers about new government regulations. The village head of Tanjung Nanga had received a radio message informing him of the meeting. The village head of Long Loreh, who had not yet been informed about this meeting, tried to get in touch with the village head of Tanjung Nanga via CIFOR radio to ask for further information. However, he did not succeed. Later in the day, around noontime, the Long Loreh village head managed to contact staff at the office of the sub-district head, who informed him that the meeting would not be held until the next day, around noon. They advised him to contact the sub-district office again the next morning.

Much to the surprise of the Long Loreh village head, later in the afternoon, the group of government officials who were going to conduct the meeting arrived. While busy organizing accommodation and meals, the Long Loreh village head was asked whether or not any notice of the meeting had been distributed to the neighbouring villages. After he had told the officials that none of the 24 villages that were to attend the meeting had been informed, he was kindly asked if he could arrange something so that early the next morning all the invitations could be distributed, picking up participants of the neighbouring villages at the same time.

Accordingly, at 6.00 am the next morning, two boats went to inform the neighbouring villages and pick up participants. The meeting was scheduled to start at 9.00 am, but only started at around 11.00 am. Due to the short notice, representatives of only six other villages attended the meeting.

officials. Village leaders who arranged IPPKs interacted especially intensively with government officials during 2000 to 2002.

The number of public hearings, consultations and efforts to promote government programmes (*sosialisasi*) increased noticeably. However, most public hearings took place in the town of Malinau, where the seats for community groups were often empty. In a sample of six meetings, the *Laporan Pertangungjawaban*, or district leader's accountability speech, had the highest attendance, with about 50 community members. Three other meetings only had an attendance of 10 to 30 community members, and two meetings had an attendance of one or no community members. The community members who did attend were all, with one exception, from villages in the immediate vicinity of Malinau. None were from the upper Malinau. In another example, officials conducted a public consultation of six draft district regulations in the village of Long Loreh (two hours by car from Malinau) in May 2001 and the promotion of these regulations one year later; but to our knowledge there were no such meetings in the district's more remote locations.

Public announcements and invitations to hearings were usually circulated only the day before the event was to take place (in a district with very remote villages where such announcements can take up to a month to reach their destination) (see Box 12.1). From the government's perspective, last minute notice was necessary because the schedule of important top officials was rarely known in advance. Top officials frequently travelled outside the district.

For some consultations, it appeared that written invitations were not even sent. In one case, where the meeting was to be held in Long Loreh, district officials gave the notices to the Long Loreh village head only, who then distributed them to other nearby villages. For most meetings, upstream villages complained of never seeing invitations until after the event had taken place. Even where villagers knew about meetings, they usually did not attend due to lack of information about the process or the low priority of the topic.

Requirements to solicit public input at the beginning of a process were implemented, instead, as efforts to promote decisions that they had already made. Even these efforts, however, were not adequately planned. When district officials held a consultation in Long Loreh with 18 villages to review draft regulations, they came prepared with only 5 photocopies of the 6 regulations, 1 set of which was given to CIFOR. Villagers were given only five days to provide input to officials in Malinau.

The local consultative assembly – *Dewan Perwakilan Rakyat Desa* (DPRD) – had the mandate to represent the citizens of Malinau in the district government. However, 77 per cent of the villagers felt that the assembly did not represent them (see Figure 12.2).

DPRD members did little to solicit input or report back to community members. They appeared to view their accountability as linked to their success in election only. When challenged about a decision by community members, it is common to hear an assembly member reply: '*Siapa yang suruh memilih saya*?' ('Who forced you to choose me?), implying that the community should trust the DPRD member since they chose him in the first place. Local people complained that their representatives rarely visited villages and that local people needed to go to Malinau should they have to conduct business; even then, it is often not easy to meet them.

More than half of the people interviewed said they felt that the DPRD members could do a better job by meeting with the community directly; having more discussions with villagers; and visiting the village and making direct observations in the village. Others suggested that the DPRD could improve their role by facilitating communities' proposals to the government; inviting communities to deliver their requests to the DPRD in Malinau; sharing with communities the results of government discussions and regulations; and serving as a channel of information to and from communities.

Although the changes have been significant, it has been hard to meet the ambitious expectations of the early days of reform. In 2002, after one year of experience with decentralization, and one and a half years of experience with the IPPKs, local government could be compared to a rider on a bucking horse. Increasing conflict and fierce struggles for control among ethnic groups created an unstable political environment in which the district government's first priority was to establish its authority and political survival. One district official's comment summarizes the downgraded spirit of democracy at the time: 'We'll get their [communities'] input, then we will make the decisions.' In 2002, another district official made it clear that the district government would set village boundaries without the local people's involvement. That same year, in a government–community dialogue, government officials were more concerned about being the target of attack than with the opportunity of interacting with community members. The role of the sub-district leader (*camat*), which had been abolished with reforms, was reinstituted as that of liaison for villagers with the government. The *camat* was put in charge of managing village-related conflict. Direct meetings with the *Bupati* became more infrequent as he became busier and other officials were hired in the district. By 2004, the district government was well established, and using the new decentralization law as guidance, retreated to the newly built government complex.

Despite reforms, the villagers who met with district officials were the same elite who represented and dominated interests at the village level. Again, although access to government officials was, in theory, open to anyone, only certain individuals felt *berani* (confident) to approach government officials or had the authority to conduct business with them. Village heads or their representatives also had official business that they regularly conducted in Malinau, including a collection of village development funds (*dana Bangdes*) and distribution of rice to poor people, among other activities.

Where local people took the initiative to request appointments directly with government officials, we observed that local people who were more confident were more likely to succeed. One Punan man reportedly waited in a district office for three days to discuss a project, while another man, who was Kenyah from Long Loreh and worked for the local coal mine, walked right into the *Bupati*'s office. The Punan man admitted that he felt *kurang berani* (not brave) and did not persist in asking the receptionist about his meeting. He allowed others to slip in front of him. Most meetings concerned business deals with logging companies, public works projects or requests for government mediation of conflicts. During the last two years, we did not observe a single woman waiting for an appointment with a high-ranking district official.

At times, a local person's appointment with high-ranking officials was facilitated by a prior series of communications and actions. In the case of Pelancau village, the Punan community delivered complaints about a local logging company verbally to lower-level officials. When this proved ineffective, they sent a formal letter to the district leader. Finally, they demonstrated. After the demonstration, the district secretary and a DPRD member visited the village. When Pelancau's village leader subsequently went to see the *Bupati*, he was seen immediately.

The government sometimes considered these meetings annoying because they perceived demands made by communities as excessive or ignorant. We also observed that communities frequently viewed their new rights to participation as an opportunity to make more demands on the government for compensation or projects, often without any basis. One community made exorbitant and continually changing requests for compensation for damage done to birds' nest caves near a logging and mining area. In a training session on preparing a proposal to government officials, most community members were unaware of the different offices in the district government and their functions, as well as how to prepare and justify a proposal. Most local people were used to addressing their concerns directly to the *Bupati* only. The *Bupati* of Malinau quickly reinstated the role of the *camat* as the representative of the district government to delegate more of this work to others.

In sum, government officials were making efforts to be more public and transparent about their decisions, but there was still much scope for improving community participation. District officials lacked the capacity to be transparent and involve local people in a meaningful way. Ingrained habits and attitudes from the New Order continued to shape how government interacted with communities. The difficulties of coordination within the district government, and the huge transportation and communication costs of sharing information with remote villages, made it difficult for the district to fulfil even minimal legal requirements of public involvement. The intensity of struggle for control over forest resources among different parties raises the question of the district's genuine will to share benefits or control over forests with communities.

From the communities' perspective, most villages would have liked to participate more in district decisions, but lacked information about how to do so. Many village leaders took a proactive approach to meeting with district officials; but most villagers lacked the confidence to do so. Local leaders viewed increased access to district officials as an opportunity to negotiate projects or money for themselves and their village, but did not have the legal background to justify their demands.

The biggest gains in community influence may have occurred through increased casual contact and the expectations associated with local people's informal relationships with officials. These informal relations and the expectations of reciprocity or exchange in associated negotiations are necessary to understand how formal rules of participation might be implemented and an important aspect of how communities may have had real influence. As the district expanded by hiring more and more government employees, the webs of informal relationships only increased. Unfortunately, thus far, interactions have been confined to a small elite of mostly village leaders. Furthermore, these leaders have generally demonstrated little accountability to their constituents in terms of consultation or reporting.

Conclusions

Given the incipient nature of reforms, improved citizen participation in local government is a long-term prospect. For remote communities, physical obstacles and costs raise questions about how regular participation can ever genuinely occur. Based on our observations of the current government's interactions with rural households, we make the following conclusions about the potential for more responsible representation in places such as Malinau.

Efforts to strengthen opportunities for public hearings and discussion of government programmes would support government officials' capacity and confidence to communicate with local communities. Topics of shared priority among government and communities, especially economic development, are more likely to be seen as non-threatening than those such as land tenure, mapping or legal reform. The government's role in community meetings facilitated by non-governmental organizations (NGOs) or the communities themselves could be expanded beyond providing opening and closing remarks.

Given the importance of informal contacts and the confinement of these contacts to an elite of village leaders, efforts to increase the accountability of village leaders to their constituents would improve representation. This work requires understanding the micro-politics of village decision-making and the overlapping between customary and government networks of authority.

Certain highly marginalized groups, such as the Punan, women and very remote villages, require extra attention from outside parties such as NGOs to facilitate their interaction with government officials. We believe that in the short term, the prospects for these groups to interact through formal or even informal channels are low. A tradition of customary authority and elitism persists (Moore, 1973).

Responsible representation and citizen participation do not occur automatically (Diemer and Alvarez, 1995; Daniels and Walker, 1997; Rossi, 1997) and take dramatically different forms in different contexts. Under decentralization, better representation of wider interests of society requires:

- open debate and learning;
- capacity-building;
- a more informed populace, especially among the non-elite;
- interaction with personalized politics; and
- more critical discussion on the kind of participation and representation desired.

As indigenous groups develop new relationships with the state political system, the political practices of both parties will need time to evolve to a mutually satisfactory level.

Local governments in Indonesia currently have high expectations that they will help democratize natural resources. Yet, the formal capacities for doing so are low. Efforts by third parties to increase participation independently run the risk of alienating district authorities and not building institutionalized capacity, but might accomplish real empowerment of communities. Efforts by third parties to increase participation collaboratively with districts run the risk of being co-opted out of addressing issues of high priority to communities such as land rights and inefficient use of project resources, but might enhance trust with the district. We suggest that facilitating real change requires a delicate balancing act of both approaches, together with a long-term perspective.

Notes

1 The *tokoh masyarakat* are not necessarily old but are considered the people most capable of making decisions on behalf of the community.
2 The respondent made this comment on 6 May 1998, just before Soeharto's resignation two weeks later.
3 These included (unusually) four women, probably because the able men were in the forest. There were no representatives from Pelancau for this reason.
4 Village leaders are officially appointed by district head decree and their salaries are still paid by the district government.

References

Antlöv, H. (2003) 'Not enough politics! Power, participation and the new democratic polity in Indonesia', in Aspinall, E. and Fealy, G. (eds) *Local Power and Politics in Indonesia: Decentralisation and Democratisation*, Institute of Southeast Asian Studies, Singapore

Bickford, S. (1999) 'Reconfiguring pluralism: Identity and institutions in the inegalitarian polity', *American Journal of Political Science*, vol 43, no 1, pp86–108

Bogdanor, V. (1987) *The Blackwell Encyclopaedia of Political Institutions*, Blackwell Reference, Oxford, UK

Daniels, S. and Walker, G. (1997) 'Rethinking public participation in natural resource management: Concepts from pluralism and five emerging approaches', in *Proceedings of the International Workshop on Pluralism and Sustainable Forestry and Rural Development*, Rome, Italy, 9–12 December 1997, Food and Agriculture Organization, Rome

Diemer, J. A. and Alvarez, R. C. (1995) 'Sustainable community, sustainable forestry: A participatory model', *Journal of Forestry*, vol 93, pp10–14

Edmunds, D. and Wollenberg, E. (2001) 'A strategic approach to multistakeholder negotiations', *Development and Change*, vol 32, no 2, pp231–253

Haraway, D. (1991) *Simians, Cyborgs, and Women: The Reinvention of Nature*, Routledge, New York, NY

Lee, K. N. (1993) *Compass and Gyroscope: Integrating Science and Politics for the Environment*, Island Press, Washington, DC

McLean, I. (1996) *The Concise Oxford Dictionary of Politics*, Oxford University Press, Oxford

Melucci, A. (1996) *Challenging Codes: Collective Action in the Information Age*, Cambridge University Press, Cambridge

Moore, S. F. (1973) 'Law and social change: The semi-autonomous social field as an appropriate subject of study', *Law and Society Review*, vol 7, pp719–746

Rao, N. (1998) 'Representation in local politics: A reconsideration and some new evidence', *Political Studies*, vol 46, no 1, pp19–35

Rescher, N. (1993) *Pluralism: Against the Demand for Consensus*, Oxford University Press, Oxford

Ribot, J. (1999) 'Decentralization, participation and accountability in Sahelian forestry: Legal instruments of political-administrative control', *Africa*, vol 69, no 1, pp23–65

Ribot, J. C. (2001) 'Integral local development: 'Accommodating multiple interests' through entrustment and accountable representation', *International Journal of Agricultural Resources, Governance and Ecology*, vol 1, no 3/4, pp327–350

Rosener, J. B. (1978) 'Citizen participation: Can we measure its effectiveness?', *Public Administration Review*, vol 38, no 5, pp457–463

Rossi, J. (1997) 'Participation run amok: The costs of mass participation for deliberative agency decision making', *Northwestern University Law Review*, vol 92, no 1, pp173–247

Schumaker, P. (1991) Critical Pluralism, Democratic Performance, and Community Power, University Press of Kansas, Kansas, US

Steelman, T. A. and Ascher, W. (1997) 'Public involvement methods in natural resource policy making: advantages, disadvantages and trade-offs', *Policy Sciences*, vol 30, pp71–90

Strachan, P. A. (1997) 'Should environmental management systems be a mechanistic control system or a framework for learning?', *The Learning Organisation*, vol 4, no 1, pp10–17

Wollenberg, E., Edmunds, D. and Buck, L. (2001) 'Anticipating change: Scenarios as a tool for increasing adaptivity in multistakeholder setting', in Buck, L., Geisler, C. C., Schelhas, J. and Wollenberg, E. (eds) *Biological Diversity: Balancing Interests through Adaptive Collaborative Management*, CRC Press, Boca Raton, FL

Wollenberg, E., Lumberg, G., Iwari, R., Rahmawati, R. and Moeliono, M. (2006) *Our Forest, Our Decision: A Survey of Principles for Local Decision-Making in Malinau*, CIFOR, Bogor, Indonesia

13

New Experiments Shaping Government and Society Interactions

Moira Moeliono and Godwin Limberg

The IPPK (small-scale timber harvesting) period proved to be a short-term boom but left a lasting impression. No longer were forests merely home and reserve area for swidden or subsistence needs. Forests had become cash, and 'compensation and fees' had become part of local people's vocabulary. Government and society relationships changed forever. Decentralization brought the state to society, and IPPKs brought a cash economy. Meanwhile, the government tried to fulfil its regulatory role over a society that perceived itself to be 'free' without realizing that freedom has its price.

As the new district government established its authority and presence, much of the opportunism and fluidity of the early years of decentralization dissipated. In its place, we now find government officials, villagers and companies using more sophisticated and nuanced strategies for seeking control over forest resources. Political alliances become more entrenched and new political elites emerged. District decisions appear more deliberate and the authority of the district leader is respected by local government officials and local people. Opportunities for claiming forest benefits have become more circumscribed and access to government officials has become formalized and distant. By reinstalling sub-district leaders (*camat*) as links between villages and the district government, district officials have reduced their interaction with villagers. Local government is at ease with the rhetoric of civil society participation, but is suspicious and unsure of exactly how to make it happen. The heady early days of reform are over; but the roles of district government and local people in the forest remain as opaque as ever and the struggle for control of the forest continues.

The period dominated by the IPPKs constitutes a first phase of decentralization. Learning from this experience, local governments started a second phase, which initiated several experiments to fulfil its new role.

Local government in Malinau and incentives to address poverty

Decentralization reforms and national poverty strategy processes have given district governments the mandate to address poverty. Over the past three decades, efforts to tackle poverty have been unsuccessful and poor people remain vulnerable to changes in economic, social and political conditions and natural disasters. Past weaknesses have included centralized policies; a focus on charity and macro-economic growth; an economy-oriented viewpoint on poverty; positioning of communities as objects; and assumptions that the causes of poverty and, thus, its alleviation strategies can be uniform throughout the country (KPK, 2002).

With autonomy, *district governments have been placed at the heart of efforts to alleviate poverty*. New concepts were considered. Poverty alleviation should be specific to local conditions. Poor communities should not be the passive objects of development, but, rather, should be actively involved in the planning, implementation, supervision and evaluation from the outset of a poverty eradication programme.

Poverty

Poverty in forested districts such as Malinau is significant, with more than half of all families considered poor by national standards. Nevertheless, the current commitment, vision and means to reduce poverty at the district level remain fragmented and insufficient. Programmes still focus on meeting the basic needs of the poor, rather than addressing the underlying causes of poverty or providing opportunities for long-term asset growth and economic progress. Most are driven by national mandates and agency priorities, rather than by locally perceived needs.

As different stakeholders have different perceptions of poverty, it is hard to gauge the real situation in Malinau District. Officials in the Malinau District government have various perceptions and opinions regarding poverty that can be linked to the 'main duties and functions' that formally apply in their workplaces. Thus, according to the agricultural service, poverty is defined as the inability to fulfil a decent standard of living caused by the lack of an enabling environment to develop agro-business. The Forestry Service defines poverty as a lack of capability and capital to manage resources; the district consultative perceives poverty as caused by the lack of physical infrastructure. Officials, however, consider poverty to be caused by the 'lack of work ethic and motivation of rural communities'.

There is no official data set or statistics on the number of poor people in Malinau District. While different sources give different figures, the percentage never falls below 24 per cent (BKKBN, 2001). The *Malinau District Poverty Reduction Strategic Plan*, developed in 2003, could not state the number or distribution of poor people in Malinau District. It only mentioned that 126 of the 135 villages across 9 sub-districts (*kecamatan*) were poor. In 2005, a district decree, SK Bupati 144, 2005, stated that 58.5 per cent of Malinau's population is poor.

Members of the Malinau District Consultative Assembly were of the opinion that the poverty situation was often manipulated and used by the district government as a tool for securing attention and aid. Using isolation, backwardness and poverty, proposals for government subsidies are more effective, and more development funds are received.

The poverty situation is also used by certain community groups, which manipulate and exploit it as a means of securing aid and attention from the district government. Legislative assembly members pointed out a number of demands from community groups asking for facilities and infrastructure to be built. Both assembly and community members compete to secure projects in the villages or sub-districts in which they live. However, the proposals do not come from poor people in the village and will not necessarily help in managing poverty in their areas.

Gerbang Dema

In Malinau, the government envisioned the alleviation of poverty by creating a just society in Malinau District through a Village Self-Sufficiency Development Movement (*Gerakan Pembangunan Desa Mandiri*, or *Gerbang Dema*). In 2002, the district government claimed to have carried out 96 village development projects, spending approximately 74.6 billion rupiah (see Table 13.1).

District visions for self-sufficient villages were inspired by regional autonomy reforms and the reflections of the district leaders and their advisers. These visions have noble aims. In their implementation, however, local officials treat *Gerbang Dema* in Malinau like one-size-fits-all clothes, where all programmes are promoted as something that can be applied universally. Everything proposed by officials is presented as part of *Gerbang Dema* (Andrianto, 2006).

Officials seem to interpret the movement with their own self-interests in mind in order to implement as many projects as possible. Since projects are the typical means through which officials can earn additional money, the projects may benefit local officials more than they do the intended beneficiaries.

To improve the second phase of the programme, local government adopted the model introduced by the Community Empowerment for Rural Development (CERD),[1] where university graduates were tasked as facilitators to work with local communities at the sub-district and village level. Ideally, these facilitators would support village communities in planning and implementing the *Gerbang Dema* programmes, according to a formula of 40 per cent for infrastructure development, 35 per cent for human resource development and 25 per cent for agriculture. In practice, programmes were decided more at the district level with hardly any input from communities. The general public receives little information

Table 13.1 *The Gerbang Dema Programme, 2002 (in billions of rupiah)*

Sub-district	Production sector		Transportation sector		Social sector		Economics sector		Village infra-structure sector		Total	
	No. of projects	Value (billion rupiah)	No. of projects	Value (billion rupiah)	No. of projects	Value (billion rupiah)	No. of projects	Value (billion rupiah)	No. of projects	Value (billion rupiah)	No. of projects	Value (billion rupiah)
Kayan Hulu	2	0.8	–	–	5	3.3	1	0.2	6	5.6	14	9.9
Sungai Boh	–	–	2	1.5	3	3.0	–	–	–	–	5	4.5
Kayan Hilir	1	0.2	2	1	4	0	1	0.2	4	1.7	12	3.1
Pujungan	–	–	3	1.2	7	3.4	1	0.2	7	5.5	18	10.3
Malinau South	–	–	–	–	–	–	–	–	–	–	–	–
Malinau West	–	–	3	6.5	3	1.9	–	–	2	2.1	8	10.5
Malinau North	–	–	2	6.0	4	1.1	–	–	1	0	7	7.1
Malinau	–	–	6	9.3	8	7.1	–	–	3	2.1	17	18.5
Mentarang	2	0.6	2	4.6	4	1.9	1	0.2	6	3.4	15	10.7
Total	5	1.6	20	30.1	38	21.7	4	0.8	29	20.4	96	74.6

Source: BPS Kabupaten Malinau (2003)

on district government policies and programmes. Minimal awareness-building and a lack of transparency have meant that there is no 'path' or 'front door' for communities to participate in these movements. Government officials' behaviour contradicts their visions and leaves poor communities increasingly powerless to fend for themselves.

Despite a substantial amount being spent, many respondents felt that these programmes had not been participatory or transparent in their implementation. There were many questions from respondents from communities in South Malinau and West Malinau sub-districts. These questions showed that the principles of *Gerbang Dema* (being accountable, participatory and transparent) were not being implemented. Communities commonly felt the programme was ineffective and wasteful and inappropriate to the needs of the communities whom it involved.

The government has no common shared vision of poverty. Poverty alleviation programmes still focus only on fulfilling the basic and immediate needs of poor communities. They do little to support the growth of household economic assets, the expansion of economic opportunities, the sustainability of vital forest and water resources, or the empowerment of marginalized groups in the political process. In addition, poor communities are still considered homogeneous, and there is no distinction made among target groups with specific characteristics. The district government in Malinau has collected only limited data on poor people in the district as of mid 2005, and the analysis of these data is not yet available. Poverty is looked at primarily at the village level, resulting in field interventions that focus on physical infrastructure development in the villages, rather than on the people themselves.

Box 13.1 Steps towards village self-sufficiency

1 *Pre-self-sufficient:* a traditional village, with no outside influences, low productivity with earnings only from the primary sector.
2 *Partially self-sufficient:* a village that is one stage more advanced than a pre-self-sufficient village, where customs and traditions are undergoing transition; outside influences have begun to appear in the village causing more progressive thinking and increased employment; and villagers' incomes begin to develop from the primary to secondary sectors. Productivity increases along with improved village infrastructure.
3 *Self-sufficient:* a village that is one stage more advanced than a partially self-sufficient village, where customs involve more rational interpersonal relationships; villagers' incomes have become more diversified; tertiary sectors are emphasized; new technology is being utilized; and high productivity is balanced by sufficient infrastructure.

Source: Government of Malinau District (2002)

Although communities are highly dependent upon the forests for their livelihoods (farming land, protein, fruit, clean water, medicines, rattan, timber and cash income) and forests are essential to the food security of most of the population, the district governments still do not consider the forests important for the people. The different forest harvesting policies that allow community groups to make money by exploiting their forests appear to be encouraged only for securing larger locally generated revenue – *pendapatan asli daerah* (PAD).

This frequently leads to land conflicts among communities, damage to their safety net and reduction in the quality of their environment. After several years, it is clear that these policies have failed to increase the standard of living of poor communities living around the forests.

Poverty alleviation efforts in Malinau consist of:

- subsidies through the forestry sector;
- subsidized rice for the poor;
- subsidized flights;
- transportation subsidies; and
- provision of electricity and clean water for rural areas.

Forestry subsidies are provided through reforestation funds and are intended to provide local people with additional income through reforestation and rehabilitation activities. Rice subsidies for the poor in Malinau consisted of 105 tonnes of rice in 2002 and 45.5 tonnes in 2003. Since there are no official data regarding the recipients or conditions of recipients, actual distribution is often decided at the discretion of sub-district and village authorities. Recipients have to come to the sub-district office to collect their 20kg of rice, although the cost to collect the rice in the sub-district capital (by motorized canoe) can be higher than the subsidies. Villagers also feel that subsidized rice is only needed a few months before harvest.

Flights and transportation were subsidized by the national budget – *Anggaran pendapatan dan Belanja Negara* (APBN) – and provincial budget – *Anggaran pendapatan dan Belanja Daerah* (APBD) – until 2001. From 2002 until 2004, subsidies were provided through the provincial and district budgets. Subsidies increased from 4.6 billion rupiah in 2002 to almost 7.7 billion rupiah in 2003. This policy also helps to stabilize prices so that remote communities can manage to buy basic necessities.

More recent efforts to address poverty

To mediate the impact of the sharp hike in fuel price, the national government designed several programmes targeted at poor people and channelled through the local government. The first programme, implemented in mid 2005, was a health insurance providing free health care for poor people. It was in preparation for this programme that the Village Empowerment Agency – *Pemberdayaan Masyarakat Desa* (PMD) – developed a list of local poverty indicators, conducted a survey and established the number of poor people at 32,429 individuals (58.5 per cent). One problem is the very limited time for preparation. Senior staff in PMD said that the central government had already fixed the number of poor people at 11,000 individuals based on a figure from the district government in an earlier report. Therefore, in order to cover all 32,429 people, the district government would have to make additional funds available from the district budget.

Another problem is that such programmes are often onerous. In order to obtain free healthcare, patients have to go to the local health centre and obtain a letter referring the patient for further medical attention to the district hospital. Given the geography of Malinau, this sometimes means that people have to travel four hours upstream to obtain this letter, whereas travel from the village to the hospital takes only two hours in the opposite direction. The information is also not generally known and many people only find out when they register at the district hospital to receive urgent medical care.

A second programme was direct financial assistance to poor families, at 100,000 rupiah per family per month, disbursed once every three months. This programme started in September 2005 for a one-year period. The Central Statistics Agency was in charge of surveying poor households to compose the lists of eligible families. The cash was distributed through post offices; but in Malinau, there are only post offices at the sub-district level, and not even all have one. Again, travel costs from the village to the post office can be almost as high as the amount of money to be collected.

A more successful programme consisted of providing block grants for rural village infrastructure development. This national programme was channelled through the district Public Works Agency, providing 250 million rupiah per village to be used to construct village infrastructure by the villagers themselves, who could thus earn some income. In Malinau, 56 out of the 106 villages received this assistance. Community members were pleased with this programme due to several positive aspects: income generation (through labour or sale of local available construction material); improved village infrastructure (cemented paths, water facilities or landing places along the river); and empowerment of community members since the whole programme was designed, implemented and partly supervised by them.

In addition, all state schools received subsidies to provide additional funds enabling schools to provide good-quality education and increase poor families' access to education (SMERU, 2006). This programme started in the 2005 to 2006 school year and continued for the 2006 to 2007 school year. Unfortunately, lack of information resulted in suspicion over the use of these funds.

The government is also looking into the possibility of working with investors from Malaysia to develop palm oil plantations, coal mines and roads to connect sub-districts in Malinau District. Senior district government officials believe that this kind of development will bring prosperity to Malinau and thus result in increased welfare of the people, a technocratic view on development inherited from the New Order. Despite originating from the area, they lack the experience to steer those kinds of development activities to better fit the local conditions and needs of the communities.

While the district governments' roles in poverty alleviation are in transition, the national government may need to continue to be partially involved in order to encourage more public-minded programmes. However, these programmes need to include not just coordination structures, but also the necessary legal clout, funding and training to make the programmes operational. On the other hand, the central government needs to create space to enable districts to define poverty locally, identify their own priorities and begin to develop more integrated poverty programmes across district agencies.

In the short term, given current perceptions and incentives, poverty alleviation at the district level is likely to occur mostly through efforts to improve local infrastructure and to build the district economy. While these efforts directly benefit government officials, their benefits to the poor are not guaranteed. The risk is that these programmes will increase economic disparities since some groups are able to take better advantage of them than others.

From timber harvest (IPPK) to sustainable forest management?

Forest resources remain a promising resource for income generation, both for the government and rural village people. Malinau had discontinued IPPK after the ban by the national government and after considering its negative impact. In its place, the local government adapted the national concession model on a smaller scale. With a district regulation on business permits for the use of timber (Perda 6, 2001) and through a Ministry of Forestry decree,[2] the local government assumed the authority to issue concessions of up to 50,000ha.

Although lacking clear information on the implications, villagers have expressed eagerness for implementing these small logging concessions, known as

'HPH mini'. People dreamed about the compensation and fees that they would receive since the IPPK experience had shown the cash value of forests.

Will the logging system under HPH mini be different from IPPK?

The rules applying to the new logging licences are identical to the guidelines that applied to logging concessions during the New Order. The maximum area is 50,000ha and the licence is valid for 20 years. The company has to apply selective felling of trees of 50cm in diameter or more. The district regulation also provides that licences grant rights to exploit an area of forest in a certain location with a defined volume and species composition with the responsibility to plan, log, transport and use the timber logged, as well as to replant and safeguard the forest and to monitor the environment (there is no clear specification on what to monitor).

When a company applies for a licence, it must attach a map of the location and approximate size. The district forestry service provides technical advice and must check the location (e.g. size of production and protected forest in the requested area). The district head issues the permit that specifies exact size and location. Co-operatives, state companies and private companies can apply for a licence. In Malinau, the district government has issued 11 of these licences.

Although it is not mentioned in the regulations, villagers expect that companies applying for a licence will have to reach an agreement with the village(s) in whose territory they will operate. Local brokers have visited several villages to discuss possibilities for logging in the village territory. The discussions with the brokers are conducted on a village-by-village basis. As stated above, since the experience with the IPPK, villagers now consider it their right to receive cash payments from forest exploitation in their village territory, irrespective of the lack of clarity on their legal rights and/or entitlements to the area.

Concurrently with this change in system and in efforts to increase sustainability of forest management, the Ministry of Forestry is trying to enforce a quota system to assign maximum allowable annual cutting volumes per province. The province then divides its quota among the districts. The district forestry service stated that the annual volume will be considerably lower than previous levels of logging. Furthermore, they indicated that the set level will not be profitable to companies.

Initial experience with the new system

Although the licences were issued early in 2002, by 2004 only five companies were active (see Table 13.2). The experience with PT Gunung Sidi Sukses Makmur, which started its operations in April 2004 in Laban Nyarit, shows that some of the problems during the IPPK are repeated. Within a week of the official launching of its operations, villagers demonstrated against the company. They demanded more employment for local people, compensation for new claims and fair distribution of benefits. Despite suspicion of unfair distribution, money paid

Table 13.2 *Eight of 11 HPH mini-licences issued by Malinau District and their status*

Name	*Date of permit*	*Size (ha)*	*Location*	*Status*
PT Gunung Sidi Sukses Makmur	15 February 2002	38,000	Laban Nyarit	Active
PT Wana Adi Prima Mandiri	2002	32,000	Warot–Sembuak–S. Gita Mentarang	Active
PT Rajawali Sakti Perkasa	14 December 2001		S. Semendurut	Active
PT Bumi Anugrah Lestari	14 December 2001		S. Gong Solok	Active
PT Lestari Rimba Raya	14 December 2002	14,000	S. Sembuak and S. Luso	Active
PT Glory Sejahtera Mandiri		12,000	S. Semendurut–S. Sembuak	Active
PT Gunung Makmur Perkasa			S. Hong–Laban Nyarit	Non-active
PT Rimba Makmur Sentosa				Non-active

Source: Discussion with the Malinau Forestry and Plantation Service (*Dinas Kehutanan dan Perkebunan Malinau*), 2003

during the launching was divided among the villages where logging will eventually be, even though in some villages logging would only start a few years later.

Will 'HPH mini' guarantee sustainable forest management?

One reason for changing the logging system from IPPK to HPH mini is that exploitation by IPPK was not sustainable and it is assumed that 20-year licences will better ensure sustainability. However, many issues remain unresolved.

A main concern is that the size of the concession is not sufficient for profitable exploitation given the system that logging companies use. For example, PT Glory Sejahtera Mandiri will exploit 12,000ha in a 20-year cycle, which means a maximum of 600ha per year. Given the scale of operations of logging companies, it is unlikely that this will be economically interesting for the companies.

The ongoing struggle for authority over forestry affairs between the central government and district governments will cause confusion and, possibly, a myriad of fees, levees and royalties to be paid to the various government levels with no control of actual logging operations in the forests.

Community expectations, in terms of financial benefits, remain unchanged, while according to the district forestry service officials, it is unlikely that HPH mini will be able to afford to offer benefits at the IPPK level. In addition, the lack of transparency as well as overlapping and unclear forest claims remain unresolved, a potential source of conflicts within and between villages.

Conservation district

With over 1 million hectares of national park and over 600,000ha of protection forest, the government of Malinau had to find some innovative approaches to development. It chose to exploit its environmental constraints, and in 2003 declared itself a conservation district, pledging to apply principles of conservation development. In this initiative, Malinau was supported formally by the ministers of forestry and of the environment, representatives from the National Planning Board, several universities, and international organizations such as CIFOR,[3] WWF and Tropenbos.

In a national seminar entitled 'Malinau Regency Heading towards a Conservation Regency', held in Malinau on 5 July 2005, a general set of criteria was developed to define a conservation district in Malinau (Santosa et al, 2005). These criteria include, among others, the presence of large areas allocated for conservation, high dependence of people upon these areas and a commitment to conservation-based development. Malinau's vision of the development of self-reliant communities was also mentioned.

In declaring Malinau a conservation district, the *Bupati* (district head) gave the following rationale:

> *The concept of conservation district as development model is driven by the consciousness of the local government and local communities of the importance of Malinau as steward of the rich biodiversity, lungs of the world and source of water. This concept is the result of in-depth discussions, debates and reflections on the meaning of these future assets which have to be protected and passed on to the next generation.* (Translated from the Proceedings of the National Seminar, 4–5 July 2005)

In order to gain support, the *Bupati* often resorts to the rhetoric on traditional harmony between Dayak and nature (Li, 2000; Billa, 2005). However, during the same seminar, the few community representatives who were present clearly expressed their concern that conservation would mean no possibilities for community development, including exploiting forest resources.

The district government of Malinau, however, expects to gain financially from its role as protector of the 'lungs of the world' in the interest of the local, national and international public (Rudiwidjaja, 2004). Indeed, from the start of regional autonomy, Malinau already insisted that it deserves compensation payments for not logging the over 1 million hectares of the Kayan Mentarang National Park. Referring to Malinau's rich resources and primary forest, the *Bupati* was quoted in *Kompas* (Anon, 2004):

> *It is no surprise that with this condition, many parties insist that the district of Malinau become a conservation area. The virgin forest should not be logged. However, we have not received anything from the central government, not even from the non-governmental organizations that pressured us to conserve.*

In the national seminar, a conservation district would expect to access various funding opportunities, such as:

- compensation payments for environmental services within the province, based on upstream–downstream agreements;
- a higher proportion of shared revenues from the national and provincial governments;
- special allocation funds for conservation districts as an incentive to conserve natural resources;
- direct payment for environmental services;
- conservation concessions;
- funds disbursed for the Clean Development Mechanism;
- Debt for Nature swaps;
- grants from international donors;
- partnership with international and national institutions.

There is, therefore, a clear expectation that being a conservation district is a means to obtaining direct cash benefits and higher revenue. It is not obvious, however, that Malinau District government understands that is requires a clear conceptual framework, hard work and visible signs of commitment in order to receive them.

In an effort to develop incentives for conservation, a system for conservation payments has been proposed where conservation districts were to receive additional funds to finance conservation efforts and to compensate for income unearned. Unfortunately, lack of understanding and concern by the decision-makers in Jakarta resulted in conservation funds being distributed equally among all districts, a sum of about 200 million rupiah in the total district budget of 3 billion to 4 billion rupiah (F. Agung Prasetyo, pers comm, 2004).[4] To date, the national government has not responded very enthusiastically and the effort to provide additional funds earmarked for conservation has met with failure. Reforestation funds are paid per cubic metres logged; but most of the funds are disbursed to the province rather than to the district (Oka and William, 2004).

Within the current system of administration, there is not much encouragement for conservation. With regional autonomy, district governments are required to generate sufficient revenue for its administration and development. Malinau, like many other districts, still greatly depends upon shared revenue from natural resources; it receives 32 per cent of the royalties from logging and mining in the district, and a share of royalties raised in the province (Cahyat, 2005). Clearly, it is more to the advantage of districts to allow high rates of extraction. No incentives for conservation are built in the system. On the contrary, the way in which reforestation funds are distributed is a clear example of a disincentive for conservation. According to the revised budgeting law (Law 33, 2004), 40 per cent of the collected reforestation funds are returned to the 'contributing' region, where 'contributing region' means the region where the forest is logged and funds are collected, and 60 per cent is to be used for national level rehabilitation projects. The issue is over the definition of 'region', where log production is often based on the area of a district, but revenues are returned to the province and distributed to all districts in the province.

On the other hand, there is much discussion and debate internationally on the different incentives for conservation and mechanisms of implementation, including direct and indirect payments (Ostrom, 1999; Ferraro and Kiss, 2002; Pagiola and Platais, 2002; Hayes and Ostrom, 2005; Wunder et al, 2008), and no effective mechanism to pay for these has yet been created. As a result, interest in conservation districts is waning and districts are looking for other options, such as the development of plantations. The private sector had already expressed interest to invest in oil palm despite the objections of local communities.

In the early discussions about the conservation district, it was envisaged that this concept could be linked to compensation payments through the carbon trade. One senior government official heavily involved in the discussion on conservation districts regularly requested information on carbon trade and the implementing mechanisms. A senior scientist on this topic at CIFOR clearly explained that carbon trade is rigidly regulated and a burden on the receiving party. He also explained that at that time, in 2003, under the Kyoto Protocol, carbon trade was also not applicable to Malinau District since it did not include conservation of primary rainforests. The scientist pointed out that there are many other, more flexible, mechanisms to receive compensation payments and/or subsidies for conserving forests and its biodiversity.

In 2005, in the midst of ongoing discussions on conservation districts, Malinau gained the attention of a Swiss-based NGO, the Borneo Tropical Rainforest Foundation (BTRF), which offered to support Malinau through the voluntary carbon market. BTRF then brokered a carbon trade agreement between Global Eco Rescue (GER), an environmental service firm, and the government of Malinau. In return for 1 Euro per hectare per year, Malinau has agreed to protect 325,000ha of protection forest (Anon, 2007; Sinyal and Warouw, 2007; Williamson, 2007).

GER claims to promote 'the transition to an ecologically and energy constrained global economy through developing, funding and managing a large diversified portfolio of natural resource, renewable energy and environmental projects that will be run on a best practice, sustainable and profitable basis'. In the voluntary market, GER has – in close collaboration with its sister organization BTRF – created a cutting-edge forestry business model, which rewards conservation and re-growth of tropical forests (BTRF–GER, 2007).

The project is still in its infancy and actual implementation is, as yet, unclear. The *Bupati* is quoted as saying that the actual value involved has not been determined and still needs to be assessed. In fact, at least some of this money will be used to identify the amount of carbon sequestered (Anon, 2007). The location of the 325,000ha also remains unclear, with one source putting it in three separate areas of protection forest (i.e. Pasilan Tabah Hilir Sungai Sembakung, Long Ketrok and Gunung Laung–Gunung Belayan) (Dewi, 2007).

The amount of money is relatively small, but still amounts to a significant sum for Malinau. In order to administer this money and ensure its use for development, the government is planning to establish a public service agency, Badan

Layanan Umum (BLU), run by several independent individuals. BLU is to work as a grant-making agency, receiving and assessing proposals submitted by rural communities and disbursing funds to approved proposals.

A scheme to protect large areas of forest cannot work without firm agreement and collaboration with the people living in or near those areas. However, most people in Malinau are still in the dark with no information made public on where these forests are located and who will actually benefit. Effective protection and sustainable community development requires trust between all stakeholders. The most effective way is to design the project from the bottom up and to ensure that communities receive immediate benefits and not only in the long run (Kusumaatmadja and Thoumi, 2007), a lesson not quite internalized by most government staff of Malinau.

A further concern is that often, especially at the community level, agreements are fluid and change when situations change, as discussed below.

Agreements

The wave of changes in governance in the post-Suharto period and decentralization has specifically fostered rapid social transformation across the country. Social change has been particularly acute in regions with rich natural resources and historically marginalized people. This is illustrated by the case of Malinau, where many opportunities were missed due to the inability to enforce agreements.

Villagers note that in earlier times, promises and/or agreements were kept, although they were all verbal. With decentralization and the associated changes, villagers now lament that agreements are all written, but rarely 'stick'. This section discusses why this is the case and the implications for sustainable resource management.

It is uncertain whether people always fulfilled agreements and promises in earlier times, and also highly unlikely. Accordingly, important agreements were sealed with a vow. Breaking such a vow, it was believed, would inevitably lead to divine punishment, so villagers rarely transgressed such vows lightly. With Dayak conversion to Christianity and, to a lesser extent, to Islam, these vows lost their 'social power' because they were seen as artefacts of animism and primitivism, indeed paganism. Christianity and Islam, however, have not succeeded in inculcating an equivalent for the previously held belief of 'divine retribution'.

Moreover, with decentralization and opportunities for entrepreneurship at the local level, the Indonesian state proved to be incapable and/or unwilling to enforce written agreements. The IPPKs are illustrative – the notary deeds that were made gave the impression of very strong agreements; but communities were not greatly concerned or knowledgeable about how these documents could assist them in enforcing contracts with a company. Furthermore, little action was taken against those who violated these contracts. Communities or their leaders often made agreements on village boundaries or resource use that were seen, at best, as temporary, and sometimes disputed after only a few months. Examples include disputes over the boundary between Sengayan and Nunuk Tanah Kibang, or the IPPK contract between Langap and CV Hanura.

Logging and mining companies hold a similar attitude. Of the initial agreements between companies and communities, none were kept without action from the communities to force the companies to fulfil their promises. Some communities initially tried to ensure that companies fulfilled agreements by sending letters both to the company and the district government. As this had little impact, communities resorted to road blockades and confiscation of equipment to draw attention to the problem. Since these types of activities resulted in quick reactions from either the company or the government, communities have come to believe that demonstrations are the only means to enforce an agreement. However, they also learned that broken promises are replaced by other promises that will most likely be partially broken again unless more action is taken to enforce the agreement.

If stable agreements cannot be reached between communities, companies and the government, sustainable resource management will remain illusive. What needs to change? Perhaps the most urgent need is for the various levels of government to take a more active role in brokering agreements, and articulating and enforcing clear sanctions. For example, villages have had little experience with business agreements and lack the capacity to broker equitable ones. Recognizing this, district governments could set a fair level of compensation for communities, and while this has been requested by villagers, it has yet to be acted upon. With decentralization, the role and authority of the sub-district government is entirely unclear. Although according to the law the sub-district has almost no authority, the district government seems keen that the *camat* take the lead in settling disputes over village boundaries, resource use and disputes between companies and communities.

Conclusions

This chapter described several responses to the rapid change brought by reforms and decentralization in Malinau. Furthermore, the chapter also illustrates a learning process of both the district government and communities to adjust to change and to try to make the most of new opportunities. After the IPPK boom – which brought some significant benefits to local communities, but also increased conflict and visible environmental problems – the government tried to regulate logging in a more rational manner. It introduced a new logging system, taking the environment into consideration (e.g. long-term concessions). However, some other aspects fundamental to sustainable forest management have not yet been dealt with. The matter of rights for communities has not been clarified and most likely will continue to be a source of conflict among villages, and between villages and companies. This will severely decrease chances for more sustainable resource management. The issue of equitable distribution of benefits from forest exploitation has also not been settled. This may well result in an increased burden to companies who will then look for other ways to maintain profit margins.

The three experiments described above were good initiatives; but implementation has been difficult. National government regulations, which shifted decision-making on forests back towards the centre, have prevented the implementation of the district regulation on HPH-mini. National disincentives to conservation have not been conducive to developing conservation districts.

The learning process has now gone through at least three cycles. The first years of decentralization brought the state (i.e. the government) closer to society. After the first years, the government retreated and sub-district heads were set up in order to mediate between citizens and district administration. In this third cycle, a new understanding has developed between the government and its citizens. The government is learning to collaborate with local people, who are beginning to understand their own and the government's roles and responsibilities.

Notes

1 CERD is a programme funded by the Asian Development Bank (ADB) through the Ministry of Home Affairs, and emphasizes improving linkages between rural communities with urban markets. The programme focused on poor communities with development potential that are not too far from urban centres.

2 The Ministry of Forestry No 16/Kpts-II/2003 Decree on work plans, five-year and annual plans and the organizational chart for use of timber from natural forests (*Surat Keputusan Menteri Kehutanan nomor 16/Kpts-II/2003 tentang rencana kerja, rencana kerja lima tahun, rencana kerja tahunan dan bagan kerja usaha pemanfaatan hasil hutan kayu pada hutan alam*).

3 CIFOR has been doing research in Malinau since 1998 and is a major stakeholder. Similarly, WWF has been working in the Kayan Mentarang National Park since 1990.

4 F. Agung Prasetyo is a CIFOR scientist involved in the Forest Partnership (WWF, Tropenbos and CIFOR).

References

Andrianto, A. (2006) *The Role of District Government in Poverty Alleviation: Case Studies in Malinau and West Kutai Districts, East Kalimantan, Indonesia*, Governance Programme Report, CIFOR, Bogor, Indonesia

Anon. (2004) 'Daerah konservasi mestinya dapat kompensasi', *Kompas*, 15 December, www.kompas.com/kompas-cetak/0412/15/otonomi/1440097.htm

Anon. (2007) 'Kabupaten Malinau masuki pasar karbon dunia', *Kompas*, 9 November, www.kompas.co.id/kompas-cetak/0711/09/humaniora/3980513.htm

Billa, M. (2005) *Alam Lestari dan Kearifan Budaya Dayak Kenyah*, Pustaka Sinar Harapan, Jakarta

BKKBN (2001) *Field Data on Poor Households per Subdistrict*, BKKBN Malinau, Malinau, Indonesia

Boedhihartono, A. K., Gunarso, P., Levang, P. and Sayer. J. (2007) 'The principles of conservation and development: Do they apply in Malinau?', *Ecology and Society*, vol 12, no 2, www.ecologyandsociety.org/vol12/iss2/art2/

BPS (Badan Pusat Statistik) Kabupaten Malinau (2003) *Kabupaten Malinau Dalam Angka Tahun 2002*, BPS, Malinau, Indonesia
BTRF–GER (2007) *Brochure*, www.greenrenaissance.org/files/brochure/btrfger_brochureweb.pdf
Cahyat, A. (2005) *Perubahan Perundangan Desentralisasi:Apa yang berubah?, Bagaimana dampaknya pada upaya penanggulangan kemiskinan?; Dan, Apa yang perlu dilakukan?* [*Changes in Decentralization Laws:What Has Changed? What Is the Impact on Poverty Alleviation? and What Shall We Do?*], Governance Brief No 22, CIFOR, Bogor, Indonesia, www.cifor.cgiar.org/publications/pdf_files/govbrief/GovBrief0610.pdf
Dewi, S. U. (2007) *Perdagangan Karbon Sukarela: Suatu Alternatif Kreatif*, http://sudewi.blog.com/2276188/, accessed 3 December 2007
Ferraro, P. J. and Kiss, A. (2002) 'Direct payments to conserve biodiversity', *Science*, vol 298, pp1718–1719
Government of Malinau District (2002) *Gerbang Dema Strategy General Guidelines*, East Kalimantan, Indonesia
Harijono, T. (2007) 'Ironis, daerah yang membabat hutan malah diberi kompensasi ratusan miliar rupiah', *Kompas*, 7 November
Hariwibowo (2008) *Pilot Project Skema REDD Gagal, Pemerintah Indonesia akan Sangat Malu Besar*, 22 January, www.modus.or.id/parlemen/redd.html
Hartiningsih, M. and Laksmi, B. I. (2007) 'Perdagangan karbon libatkan semua pihak', *Kompas*, 10 December
Hartono, B. (2008) 'Marthin Billa, penggugah dunia dari pedalaman', *Tribun Kaltim*, 10 January
Hayes, T. M. and Ostrom, E. (2005) 'Conserving the world's forests: Are protected areas the only way?', *Indiana Law Review*, vol 38, no 3, pp595–617
Hudiono, U. (2007) 'Indonesia could net US$2 billion from forest conservation', *The Jakarta Post*, 30 November 2007
Kolesnikov-Jessop, S. (2006) 'Fighting to save Borneo's vital last rain forests', *International Herald Tribune*, 19 September 2006, www.iht.com/bin/print.php?id=2852374
KPK (Komite Penanggulangan Kemiskinan) (2002) *Dokumen Interim Strategi Penanggulangan Kemiskinan*, KPK, Jakarta, Indonesia
KPK (2005) *Strategi Nasional Penanggulangan Kemiskinan*, KPK, Jakarta, Indonesia
Kusumaatmadja, R. and Thoumi, G. (2007) 'Land dictates the rules and rural communities are the gatekeepers', *The Jakarta Post*, 4 December
Li, T. M. (2000) 'Articulating indigenous identity in Indonesia: Resource politics and the tribal slot', *Comparative Study of Society and History*, vol 42, no 1, pp149–179
Oka, N. P. and William, D. (2004) *The Policy Dilemma for Balancing Reforestation Funds*, Decentralisation Brief no 1, CIFOR, Bogor, Indonesia
Ostrom, E. (1999) *Self-Governance and Forest Resources*, Occasional Paper no 20, CIFOR, Bogor, Indonesia
Pagiola, S. and Platais, G. (2002) *Payments for Environmental Services*, Environment Strategy Notes no 3, World Bank, Washington, DC
Proceedings of the Seminar Nasional (2005) *Menuju Kabupaten Malinau Sebagai Kabupaten Konservasi*, Malinau, 4–5 July 2005
Rudiwidjaja, R. (2004) 'Ekowisata alat unggulan penyelenggaraan kabupaten konservasi', *Sinar Harapan*, 6 May, www.sinarharapan.co.id/feature/wisata/2004/0506/wis02.html
Santosa, K. D., Gunarso, P. and Priyadi, H. (2004) *Kabupaten Konservasi sebagai Bentuk Model Forest: Studi Kasus Kabupaten Malinau*, Makalah Disajikan Dalam Workshop Model Forest, 6–10 December 2004, di Madiun, Indonesia

Sinyal, D. and Warouw, W. (2007) 'Kabupaten Malinau rintis pasar karbon dunia', *Sinar Harapan*, 25 October, www.sinarharapan.co.id/berita/0710/25/kesra02.html, accessed 8 February 2008

SMERU (2006) *Kajian Cepat Pkps-Bbm Bidang Pendidikan Bantuan Operasional Sekolah (Bos) 2005*, www.bappenas.go.id/.../&view=443/SMERU_ BOS_PKPS_BBM.pdf, accessed 9 February 2008

Syaifullah, M. (2003) 'Kapuas Hulu mestinya jangan hanya basa-basi', *Kompas*, 7 November

Williamson, L. (2007) 'Pricing the Indonesian forests', *BBC News*, Kalimantan, 12 December, http://news.bbc.co.uk/2/hi/asia-pacific/7136345.stm

Wunder, S., Campbell, B., Iwan, R., Sayer, J. A. and Wollenberg, E. (2008) 'When donors get cold feet: The community conservation concession in Setulang (Kalimantan, Indonesia) that never happened', *Ecology and Society*, vol 13, no 1, pp12

14

Riding the Rapids: Synthesis and Conclusion

Eva Wollenberg, Moira Moeliono and Godwin Limberg

Malinau's story highlights the mixed potential of global governance reforms to more localized and democratic government. Decentralization and democracy reforms in post-Soeharto Indonesia have increased state control over forests at the district level, while dramatically weakening control of the centre. As a result, local social forces have gained more influence in localized governance, which in turn structure and limit the control of the district government. In Malinau, and we suspect many forest areas, these social forces have been and continue to be strongly influenced by local ethnic affiliations. Local political orders are emerging, reinforced and organized by power relations among local ethnic groups. Many of these local political orders grew out of early networks of influential people and their descendants. Decentralization and reforms gave these orders new visibility and voice.

Yet, without strong democratic measures in place or the oversight of central government, the potential for collusion among local ethnic elite, district officials and entrepreneurs to further their own interests continues and might even be strengthened. Malinau's forests have been spoils up for grabs, especially during the uncertainty of the early transition years. The benefits from forest resources in the current era have accrued to those who, for the most part, were already locally influential or better organized. Weaker groups lost, negotiated poorly and faced threats to their claims to forest, village territories and benefits from timber companies. Conflicts among groups and with government and companies increased.

In the early years of decentralization, Malinau's forest has been degraded at rates previously unknown, albeit to a limited geographic extent. Unclear policies and authorities, as well as uncertain land-use designations and management rights for most of the district, leave Malinau's forests dangerously in limbo. The uncertainty and lack of clarity give powerful groups room to manoeuvre informally to meet their own interests. Although local people feel they have better access to local officials, government shows little responsiveness to local people's interest in demarcating land or their priorities for forest use and conservation. Timber cutting and forest conversion driven by commercial interests appear likely to continue under these circumstances without regard for environmental or local people's priorities, especially since attempts to gain financial support for conservation have not been very successful to date.

The rapid pace of changes has been as important as the changes themselves. The swiftness of reforms influenced perceptions of uncertainty and made it difficult for institutions and people to adapt quickly enough to address problems and opportunities. At the same time, this perceived uncertainty allowed rent-seeking and opportunism to develop unhindered. Although the initial scramble is over, the lingering indeterminate state of forests partly reflects decision-makers' need for more time to develop appropriate institutions. Similarly, local people's political organizations are still growing as the state system evolves. The pace of forest regeneration will make it difficult to ever regain what has been lost.

Malinau's recent history thus captures the transformation in social and state control that has occurred in the effort to establish more localized governance. In this book we tried to explain these changes by answering three questions:

1 How are new roles of state and society leading to new local political orders?
2 How have these changes affected marginalized groups?
3 How have these changes affected forests, how they are managed and their long-term benefits?

Emerging local political orders

The manifesto: Decentralization and democracy

Decentralization and democracy reforms in Indonesia set in motion typhoon-force winds of change. Although these changes reflected and built on longer trends and tensions between the Indonesian state and its diverse societies, the last five years have produced a surge of rapid developments. How far has the recent burst of reform gone? Can we say that Indonesia has achieved decentralization and democracy in places such as Malinau?

In Malinau, we see a blend of trajectories of change, adjustment, contradiction and reversal. Governance reforms established the structure for local government to operate. Expectations, rhetoric and the desire and openness to change were high in the first two to three years of reforms as this structure was being established. There were new initiatives, alliances and voices, especially of weaker groups such as the Punan or groups in remote areas such as the Bahau River and Apo Kayan, who began to meet more regularly with government officials. Officials in local government were familiar with traditional values and lifestyles, even if they did not support them. Society became bolder in their critique of government and people had more rights to express conflict and dissent. There was talk of allocating land rights and building new economic opportunities. Local groups had more access to and influence on local government than ever before.

As the pieces were increasingly in place, however, the complications of reforms became apparent. Within the frame established by reforms, the substance of governance – as measured by responsiveness, efficient allocation of resources, transparency, accountability and justice – has been weak on two accounts. In the formal realm, the capacity of most officials and departments of government has been low. Infrastructure and economic power has remained concentrated in economic capitals, such as Tarakan, Surabaya and Tawao.[1] Reliance on funds from Jakarta and the need for lobbying in the centre has remained high. Enforcement of laws and regulations has been poor. The division of roles among the district, province and centre has been murky. Development of unified enforceable institutions for defining and distributing property, managing natural resources, handling conflict or promoting participation has been uneven, unclear and slow. Agreements have been weak and reversible. The exigencies of change and accompanying local economic boom in the early years left no time for coherent thinking and action (de Soto, 2000). Most officials have not had adequate exposure, training or experience in democratic local government. Indeed, most have been educated and/or trained in the centres run by the Department of Home Affairs. Thus, although they find the concepts of democracy and autonomy appealing, they lack the knowledge and confidence to use them. Some expressed fears that reforms could unleash more chaos and conflict. New Order-driven habits and cultural norms have remained ingrained. Attitudes about trying to make traditional people modern (Dove, 1988) have persisted. Suspicion about non-governmental organizations (NGOs) and community organizing has remained high. The period of adaptation has been brief.

If engaged citizens are the 'core of the democratic process' (Moyer, 2001, p2), society has also been a handicap to local government fulfilling its role. Local people have remained relatively unorganized and uninformed despite a strong initial interest among the majority in democratic change at the district level. Ironically, only certain factions have been strongly supportive of village reforms, usually the opposition to the village head. Both officials and local people were excited at first about the possibilities for increased participation in government affairs, but soon became disillusioned by the slow pace of reform. They also recognized that it was costly, both in material terms for transport and accommodation, as well as in time. Participation returned to the old networks and their informal contacts with the bureaucracy.

Officials found many local people's demands exorbitant or difficult to respond to. Apart from the systematic dissection and prevention of local organization during the Soeharto era, there has also been a continuing belief that government should be the patron or *bapak* that takes care of its populace. People's demands on government have been more for material gain than to promote self-determination and improvement of collective institutions. The government has been unable to impose obligations that should accompany rights, implying that new benefits would be subject to taxation.

Aside from the lack of capacity at this formal level, there has been an unofficial level of inwardly focused activity that has mitigated against substantive reforms. The handing over of power to local officials has accelerated the possibilities for collusion, corruption and the promotion of self-interest (Smith and Obidzinski, 2003). Often, political power is misused to obtain financial gain, which is then invested in people and strengthening networks to increase and sustain their power. As the financial and political stakes have risen, so have the internal deal-making and haggling. Higher-level officials have pursued lucrative profits, including those from timber and mining, justified under the cover of producing income from the district. They have tended to use, ignore or adapt policies in ways that benefit them most, the IPPK story being a prominent example. Lower officials have concerned themselves with their own cut from projects, opportunities to travel with its extra allowances, as well as trivial power struggles. This excessive preoccupation with personal gain and petty politicking within the bureaucracy has provided strong incentives against developing substantive reform. It has also siphoned off scarce time and attention.

The central government has reacted to what it sees as local government's excessive self-interest by trying to re-establish control, especially in forestry. A new law in 2004 reframed decentralization, primarily by reinstating the province with authority to supervise districts. In the forestry sector, a series of new policies progressively decreased formal authority over forests by the district level. Adjustments in the balance of power between the centre and districts are still ongoing, although most agree that it would be difficult for the centre to reassert its authority fully unless it resorts to coercive force. Many saw the election of President Susilo Bambang Yudhoyono in 2004 as a reflection of the popular desire for a return to such authority.

The new structures of local government thus provided a vast potential for change; but the factors described above have slowed the progress of substantive decentralization and democratic reforms. Despite the increased knowledge and experience gained through the processes, both officials and citizenry seem to have set their standards lower as high expectations slipped by unmet in the short run. People may also be retrenching into comfort zones according to what feels familiar from the past. Local as well as central government would need to feel more security before they could promote more democratic measures. In building this security, however, local decision-makers seem to be retreating back to old patterns of control rather than establishing stronger foundations for more democratic local governance. Many outside of the government feel that while numerous elements of governance have changed, the results remain disappointing.

State and community

Before reforms, 'government' was a distant, unknown, inaccessible entity to most residents of Malinau and other remote forest areas. People were citizens of the country of Indonesia, but without influence or voice in formal decision-making. The governance with which they were familiar was the local unofficial politics of customary leaders, ethnic relationships, local bosses and the odd official passing through to provide health services, police potential conflicts or attend festivals. Reforms have now brought the government closer to the people, and in doing so, merged the two domains of state and society. The government is more familiar because government offices are physically more accessible and more local people staff them. About 3 per cent of the population in 2004 *was* the local government. Nearly everybody has a relative in government. The informal, personal and ethnic sphere of politics is intermingling with the formal, impersonal and anonymous state. Brought together, the two systems have catalysed an emergent new political order in Malinau, if not for most of rural Indonesia.

Politics in these emerging orders has required balancing adherence to the informal claims to resources, reciprocity and personalized relationships inherent in traditional community life with the bureaucratic, patronizing money-politics culture of the state. The most powerful politicians and local people were those who maintained influence in both domains and used that influence to bridge domains in specific initiatives. These are also the people with the knowledge and ability to misuse informal institutions in a legitimate manner. Thus, the IPPKs were a partnership of the *Bupati*'s office, the District Forest Service, local timber bosses and the heads of specific *adat* communities. The plans to develop oil palm have built upon similar alliances. In contrast, Inhutani II withdrew from the area when it failed to build local alliances after losing central government protection.

Building influence across these two domains has shifted people's attention to exercising power through local alliances, rather than through top-down decrees or alliances upward in the hierarchy of the state. In doing so, no one entity has held the ultimate base of power, although the *Bupati* is still perceived as being the highest authority.[2]

Entities within the government have been more interdependent, as has been district government with local communities. As attention has shifted to these local alliances, initiatives have taken place through a mix of formal and informal agreements that have allowed district officials to dodge the attention of central authorities and be less than fully transparent to communities. The resulting collusive behaviour has created huge benefits for some select elite, usually those also favoured in the past. The old hierarchies regained their status since only those from the nobility and early elites had the education and, thus, the capability to perform in the new government.

Ethnic groups, local communities and family have selectively been used as the common social foundation to the informal and formal political spheres. Reforms have therefore enabled ethnicity to take an elevated importance far beyond Soeharto-era cultural exhibitions. Ethnic groupings have been the basis for claims to land and forest resources. They have also been the basic building blocks of social capital and political alliances. Local people use ethnic identity to 'place' others as ally, rival or unknown outsider, and indicate social position. In response to the wider scope for civil society action in the post-Soeharto period, people have organized themselves according to ethnic associations, such as the Lembaga Adat Punan, Lembaga Adat Tidung and Lembaga Adat Lundaye, invoking the informal authority of customary institutions, albeit newly invented. Ethnic groups have been the basis for alliances above and below the district level. The Persekutuan Dayak Kalimantan Timur reflects efforts to build ethnically based alliances at the provincial level. The Lundaye have both sub-district- and district-level organizations.

For government officials, their ethnicity or affiliation with other ethnic groups by marriage or adoption has reflected their loyalties and social obligations, as well as, frequently, their constituency. District government has tried to maintain its support and legitimacy through balanced representation of powerful ethnic groups in key positions, such as the Kenyah, the Lundaye and the Tidung. They have made increased use of, and appeal to, cultural symbols and events such as the forest and harvest festivals in order to gain popular support. They have taken the new ethnic associations seriously, inviting their representatives to public district events and participating in association events. To unite the diverse groups, the *Bupati* tried to assume the status of the chief *adat* leader (*kepala adat besar*), expecting that this status would help him to increase his influence, making governing easier. However, the people, supposedly traditional, have not unanimously accepted this role. Due to reform, more people increasingly hope and demand that their government is more transparent and accountable. The role of a feudal chief *adat* leader does not fit in this picture anymore. Yet, few real mechanisms for genuine consultation and accountability are in place.

Malinau's experiences of the last five years suggest that the power of civil society will lie in the degree to which local ethnic groups act in coordination or in competition, both within their groups and among themselves, to form coalitions. Groups with ethnic, community or family associations to local government officials will have the most opportunities for building alliances. Where there is only one ethnic group or a set of allied ethnic groups, and they are well organized, they are likely to be more influential in district governance and to reap more benefits from the new order. Where there are many ethnic groups who compete with each other, they will be weaker and benefit less.

Localized governance therefore gives more space to the current trend of increasing ethnic associations in the face of a weaker nation state. The local state recognizes its interdependency on these groups and has sought to work with them, both within and outside government.

This emerging political order remains incoherent and unstable because of tensions in local officials' instincts to work in the realm of local community, as well as in the realm of the state. If the central state allows districts to maintain their authority and downward accountability to local people, we can expect pendulum-like alternation in emphasis between these two realms as the tensions play themselves out in different contexts and election years pass. The swing of the pendulum will correspond to officials' willingness to take the extra effort and risk in developing messy democratic models of governance, or a preference to ease back into comfortable old authoritarian patterns of money politics that allow them to reap large financial rewards. Based on historical contractions and expansions of state power, we expect periods of increased conflict and social upheaval that will encourage officials to revert to greater authoritarianism, as appears to be the case at present. Whether democracy or authoritarian rule, either will be built on the foundation of ethnic politics and stronger local alliances that are more independent from Jakarta.

Impacts upon marginalized groups

The changes described above have had important implications for marginalized groups. The implications for both the population of Malinau and marginalized groups within Malinau are considered below. The main points of the previous chapters are also summarized.

As a whole, Malinau has benefited from decentralization and democratization reforms.[3] The district government has been able not only to tap into (Li 1999a, b), but also to usurp, previously centrally controlled power over forests and other resources. Equally importantly, because of its status as a district, Malinau received direct allocations of funds from Jakarta, resulting in more financial resources and infrastructure for what were remote regions relative to the Bulungan District capital. Financial benefits flowing to Malinau have increased significantly: more than 30 per cent between 2002 and 2003 (Andrianto, 2006).[4] Health and education, infrastructure and economic activity have improved in most parts of the district.

The district government has also operated semi-autonomously from the centre and the province. With more funds and travel opportunities, district officials have been able to develop more extensive networks and alliances. More local people have had more influential positions in a more powerful local government. For the reasons discussed in the previous section, local society has a much closer relationship with the state and control over it. Although Malinau may be marginal relative to Samarinda and Jakarta in geographic and political economic terms, the district has cleverly taken advantage of its proximity to Malaysia and China to develop new forest-related trade opportunities and relationships. It has also shrewdly turned its remoteness and underdevelopment into an international asset, positioning itself as a conservation district of global importance that wishes to sell environmental services, while using these as arguments to request additional central government funding.

Communities in Malinau have experienced some corresponding benefits, but for the most part, these have been short term, less than their potential and captured by the elite. In the case of villages studied by Anderson and Kamelarczyk, 50 to 72 per cent of the households in each village reported better health conditions, better roads, better transportation, better education and higher general income. Similarly, in our survey of 14 villages, 52 per cent reported that their prosperity had increased despite greater expenses.

IPPKs have had the most visible impact on the local economy. The cash benefits from three years of IPPK fees averaged about US$1000[5] per household in seven villages of the Malinau watershed cooperating with IPPKs. The variation among villages was high, however, ranging from US$61 to US$2235. Households valued IPPK incomes for the large amounts of cash they generated; but compared to other valued incomes such as wage income and forest products, IPPK fees were irregular and uncertain. Most households used the extra income at least partly for daily living expenses, but there was also substantial investment in health, education and homes that enhanced human capital. Less than 30 per cent reported saving part of the fee. Benefits provided to the villages, such as roads and public buildings, were significant, although it remains to be seen how they will be maintained. A landslide cut off the road to one village for almost a year.

The IPPK experience showed that villagers in Malinau lost potential benefits because of poor negotiation processes. Acting on behalf of Malaysian companies, brokers coordinated IPPK negotiations with communities and the district. There appears to have been little competition among brokers: they acted similar to a cartel in which the terms had already been rigged. Brokers often brought a map prepared by the district of the area to be logged. They also rarely had direct contact with the community as a whole, except after a decision had been taken. They usually negotiated terms with only a few select individuals from the community – leaders of the *adat* institution (*lembaga adat*), a *pewaris* (inheritor of rights to birds' nests) or the village head. In one case, other villages even negotiated on behalf of Punan villages without the latter being aware of it. Village decision-makers met the broker away from the village and alone, sometimes even without the village head. Government officials frequently accompanied negotiations, usually sitting with the broker, and arriving and leaving with him.

Most village decision-makers did not challenge the terms offered to them in the first round of negotiations. Some villages received only 5000 rupiah (US$0.59) per cubic metre, believing that they had reached a good deal as timber concessions had paid compensation rates of 3000 rupiah (US$0.39) per cubic metre. The average in Malinau was 27,000 rupiah per cubic metre and the highest was 50,000 rupiah per cubic metre (US$5.90). Communities learned over time to negotiate fees closer to this latter figure. The opportunity and willingness to renegotiate was a key factor influencing the level of benefits obtained in Malinau. Yet, these sums were still vastly below that which villagers obtained in some other districts. In Kutai Barat, there was competition among companies, but more importantly, communities were better informed. Villages coordinated permits themselves and met directly with logging contractors in the village. Fees in Kutai Barat ranged from 30,000 rupiah (US$3.50) to 150,000 rupiah per cubic metre (US$17.65), averaging 80,000 rupiah per cubic metre (US$9.40).

As the comparison with Kutai Barat suggests, villagers in Malinau negotiated far below the potential value of the timber. We estimate that villagers received only

1.4 per cent of the gross value of the timber harvested, based on timber prices in the town of Malinau. Common villagers benefited less compared to decision-makers, who sometimes received an individual fee of about 10,000 rupiah per cubic metre more than was paid to the village, in addition to possibly other less transparent payments. Punan villages were the biggest losers. The Punan were the least involved in negotiations, often represented by other villages, the most uninformed and vulnerable in dealing with brokers, and the least aware of the amounts logged. The villages that negotiated the best benefits were those that acted collectively with other villages (Palmer, 2003). Where there was strong internal social capital and transparency, households within a village acquired higher levels of benefits.

Most significantly, the cash benefits for local people from IPPKs were short term. Although people invested some of this income in assets with longer-term benefits, the forest, one of the most important local assets for local people's food security, became more fragmented and degraded. The brokers and government officials who received the largest sums of money were those who benefited most. The balance of profits and associated power over forest benefits remains highly skewed at the local level.

Further skewing this balance of power is the ongoing fact that the two most important assets for the Malinau population remain formally unsecured – land and forest. Disconnections between national and district policy, and between policy and practices on the ground, make it unclear who is responsible for land. The district government has become the default decision-maker but has little legal backing or capacity. The belief of the national- and district-level offices in Malinau that they can offer certificates to people on state forest land appears to be due to a basic lack of understanding of national policy and decentralization. Only ethnic groups' efforts to organize village territories and traditional communities' claims to land have a current legal foundation, although the process for doing so is still unclear.

The increased possibility for compensation payments and fees based on rights to the forest has complicated the claim process by increasing competition for land, especially in the early transition years. Claims to land increased from an estimated 1.3 cases per year during the New Order to at least 50 per year from 1998 to 1999. Yet, the district has not taken a stand on tackling the issue of rights over land and resources. It may be in the district's interest to maintain multiple levels of ambiguity to allocate land in ways that they judge most important and that meet their own interests. Any legalization of rights to others would detract from the district's opportunities for control. Since formalization would result in winners and losers, the district is wisely wary of the possibilities of increased ethnic strife and a decline in political support from key groups.

Unfortunately, such ambiguity encourages a short-term use of resources and puts weaker groups at a disadvantage, such as the Punan, who have to struggle with more powerful groups to retain rights over their land. Without legal security and clarity, land and forests have formally become 'dead capital' in the new policy setting (de Soto, 2000, p32). According to de Soto, where assets remain in the extra-legal sector, development opportunities are limited because assets are not described or organized; their unfixed and uncertain nature results in misunderstanding and unstable agreements; transactions are not protected; and assets are not convertible or easily traded. Security and clarification of Malinau's forest and land assets would encourage more transparent use, more accurate valuation, investment and the development of a longer-term perspective on Malinau's natural resources.

The increasing forest values, competition, legal ambiguity and resulting opportunism of the transition period have resulted in dramatic increases in the expression of conflict. Although the numbers are difficult to estimate precisely, villagers reported at least 86 cases of conflict during 1997 to 2002, compared to 8 cases that villagers could recall in the New Order period. Conflicts have shifted from those in the New Order years concerning rights to agricultural land and non-timber forest products, such as *gaharu*, to reform era conflicts over village boundaries, rights to timber, sharing of IPPKs and government project benefits. Many of these conflicts were linked to historical tensions among ethnic groups and their claims to land. As noted elsewhere (Wulan et al, 2004), the reform era enabled freer expression of conflict that has unleashed tensions that had been built up and were latent during the Soeharto period.

The IPPKs catalysed more conflict than any phenomenon in the reform period. Of the 22 villages in Malinau that had IPPKs, 19 mentioned that they had encountered problems with the broker, including unpaid fees, no development of promised facilities, logging in the wrong places or no planting of horticultural trees in the cleared areas. Of these 19 villages, 11 took action against the IPPK, usually reporting them to the district. At the household level, 52 per cent of the households sampled in Adiu, Tanjung Nanga and Sengayan reported IPPK conflicts. Conflicts among households concerned the amount of money received per household, accusations of bribery and lack of transparency about the distribution of the fees.

The willingness of communities to protest company or government actions reflects an unmistakable shift in grassroots power. Since the reforms, district officials have talked more about bottom-up politics. Public meetings in Malinau have increased, although most people attending the meetings live in the immediate vicinity of the town. Visits of officials to villages have increased and government officials in Malinau report increases in letters and visits from villagers. The numbers are still small, however, and most proposals made by villagers go unanswered, unless coinciding with government plans or concerning financial ventures. Villagers are seeking more self-determination and feel freer to convey their views to government officials, especially on matters concerning their own villages. Setulang village's initiative to decline offers from timber companies and to organize the protection of part of its forest on its own reflects this grassroots power. Similarly, Laban Nyarit and Pelancau demanded and obtained fees for the timber cut on land cleared to build a public road for which they had not been consulted.

Yet, as described above, democracy in Malinau, as elsewhere in Indonesia, is still a project in process, with both the government and the people unsure of how to proceed. The geographic distances and poor infrastructure between the government offices and villages make timely invitations difficult and visits costly, as they would in any remote forested district. The most remote villages are still the most politically marginal. Development and political opportunities have been concentrated in the immediate vicinity of the town of Malinau. Since the establishment of the district capital in Malinau, there has been a steady in-migration to the town, both from more powerful villagers and from other places in Indonesia resulting in increasing pressure on resources.

Villages remain the unit of political organization for most people and little has changed in village-level politics. An elite of village leaders (*tokoh masyarakat*) still controls decisions and most common people are still reluctant to speak out. Leaders are valued more for their leadership qualities than their accountability to a constituency. *Adat* leaders have not regained the power they lost during the Soeharto period. Women have very little voice in politics and the Punan are often disadvantaged by more dominant groups in negotiations among villages of different ethnic groups. Petty jealousies limit many communities' capacity to act coherently.

District officials still treat villagers in forest areas as *terbelakang*, (backwards) (Li, 1999a, b). Shifting cultivation, particularly the nomadism of the Punan, is viewed as an unfortunate holdover that prevents modernization. Officials see it as their duty to help people to modernize. There is little acknowledgement of locally important species or parts of the landscape that local people value. These prejudices are longstanding and common throughout Indonesia (Dove, 1988). They reinforce the marginality of those who choose a traditional lifestyle in the eyes of the government. It is odd that they have not changed with the empowerment of officials who themselves are at least familiar with shifting cultivation and local cultural values. The values inculcated by New Order education and experiences remain strong in this generation. Officials also represent a class of villagers who chose to depart from their traditional livelihoods and may lack empathy with those who choose to maintain these livelihoods. These attitudes remind us that just because the state comprises local people, the values of local society do not automatically become part of the state.

The most politically and economically marginal of all groups in Malinau remain the Punan, who, for a large part, continue to live in the least accessible areas, especially in the Malinau and Tubu watersheds. Although our survey of 14 poor villages in the district shows the majority of the Punan reporting that their lives have improved with decentralization (58 per cent of the Punan compared with 36 per cent of the Kenyah, 2 per cent of the Merap and 65 per cent of the Lundaye),[6] in most aspects of their lives and in relation to other groups they remain discriminated against or taken advantage of. On the one hand, their cash income and political influence have vastly increased compared to a decade ago; on the other, they have also lost out to other groups as competition for the forest has increased. In the effort to keep up with this competition, they have made important changes to try to maintain their stake in land and the forest, including relocating their settlements, recording their history in Malinau to establish prior claims and accepting fixed village boundaries. The Punan remain constrained by their lower levels of education, small village sizes, frequent dispersal of village members, absenteeism from the village when members enter the forest, weaker village organizations, higher frequencies of disease and historical patron–client relations with certain ethnic groups.

Nevertheless, their status as a distinct and marginal group has also acted in some ways to their advantage. Because of their marginality, they have attracted the attention of development-oriented groups, such as a missionary of the Catholic church, the International Fund for Agricultural Development (IFAD), CIFOR and Indonesian NGOs. A number of capable and experienced leaders have developed relations with outside groups. As with other ethnic groups, the Punan have been increasingly willing to speak out on their interests with district officials, often arguing on the basis of their disadvantaged status. In the survey of poor villages, 53 per cent of Punan households reported that government officials have given their demands more attention compared to the pre-reform years, in contrast to the Merap (50 per cent), Lundaye (37 per cent) and Kenyah (96 per cent) households.[7] The sharp contrast with the Kenyah reinforces the observation that ethnic ties encourage close relationships between the local government (where the district leader is Kenyah) and villagers.

Perhaps most importantly, the Punan have developed an ethnically based institution (the Lembaga Adat Punan) that has held well-organized meetings and drawn on substantial support from international organizations. The Punan are also the only ethnic group who early on, in 1994, established their own NGO, Yayasan Adat Punan, with a development agenda. The question is whether these types of support will be sufficient to enable the Punan to act in more politically powerful ways, or whether existing prejudices and competition will continue to marginalize them.

What, then, are the prospects for marginalized groups in Malinau? In the excitement of the early years of reforms there was a sense that the common people would finally get their due: formal recognition of their village lands; cash benefits from the timber in those lands; compensation for damaged forest and loss of resources from companies; and a government composed of people whom they knew and who were willing to listen to them and fight for their interests. There was also a sense that Malinau had vast economic potential that was only waiting to be developed so that all could share her riches. Those were the hopes. People in Malinau were potentially non-marginal in a potentially non-marginal place.

Such hopes have not been realized, of course. Instead, individuals and groups who previously held influential positions have tended to become more powerful, while marginalized groups like many of the Punan have become weaker in terms of land tenure, benefits from forest exploitation and representation. Competition and conflict have increased, with local government preferentially serving the elite and powerful. Whereas communities have traditionally helped to manage the production and distribution of economic benefits across households in a village based on an ethical obligation of sharing, there is now an increasing value on individual entrepreneurship and acceptance of accumulation.

Decentralization and democratization reforms have generally enhanced life for people in Malinau, especially in the short run, even if initial high expectations have not been met. A more competitive environment, however, has disproportionately benefited the previous elite and stronger ethnic groups. The elite have significantly influenced local government, both formally and informally. Others remain, for the time being, constrained by village politics or deeper historical relations of marginality. For most, the economic gains of the last five years have been short lived. The lack of clarity about how forest and land assets are distributed and managed remains the biggest constraint to addressing equity and economic development in the district. As long as the district maintains ambiguity, they will maintain control. Local government has itself become a new elite of local society, a part of society entitled to its own rights and privileges. Marginality has moved to lower social levels among and within villages. The gap has become narrower but deeper.

Impacts upon forest management and the forest

How has the role of the state and society changed in the use, conservation and control over forests – Malinau's most valuable asset? How is that control organized? What are its implications?

Decentralization enabled dominant local social groups to tap into state systems for exploiting forest resources. Local officials, in turn, relied heavily on personal networks and alliances, especially between companies and villages, to organize alliances for forest exploitation. Nearly everyone, including those in local government, used the transition to test the limits of what they could gain in the face of the disorganization of the state.

Logging in the early reform years was guided by entrepreneurship and opportunism rather than by public institutions concerned about managing the forest as a resource. The people involved in forest exploitation sought immediate profits rather than sustainability. The conversion value of the forest became higher than the benefits of protecting and maintaining it. The local government actively promoted this entrepreneurship in the name of generating district income. They made no effort to impose selective cutting or other harvesting regulations, to enforce protected land designations, to guide road construction, to prohibit activity near waterways, to ensure that protected species were not cut, or to enforce regeneration and plantation requirements. There was no effort to manage forests across districts. No one sought to make local government accountable for these actions either. Almost everyone was more interested in the money to be made.

The result was that forests were significantly more damaged under the IPPK system than they had been under relatively un-enforced selective cutting regulations – *Tebang Pilih Tanam Indonesia* (TPTI). In a comparison of the impacts of two selective logging concessionaires with three IPPK permit holders, Iskandar et al (2006) found that the IPPK holders harvested more intensively, removing four times as many trees per unit area, including a larger proportion – more than 25 per cent of trees under the 50cm diameter at breast height (dbh) size class compared to 3 per cent by concessionaires. The larger proportion of small trees may reflect the trend of IPPK elsewhere to sell timber to local mid-size and smaller sawmills rather than to larger more regulated plywood factories outside of the district. It may also reflect the fact that IPPKs were expected to clear cut and convert the land to plantations.

IPPKs also resulted in significantly more damage to residual trees (38 versus 20 trees per 100m in primary forest) and larger canopy openings (0.3ha versus 0.1ha per hectare), resulting in a more degraded and fragmented forest. This damage further threatened the population of future trees and potential for regeneration in addition to damaging habitat and corridors for wildlife. The future options for harvesting from areas logged by IPPKs are, thus, low compared to areas held by concessions, although it might be ecologically better to clear cut once than repeated selective cutting. The IPPKs were an inefficient and ineffective use of the forest. They did not serve the purpose of higher economic value conversion to plantations or of sustainable production forestry. They degraded the value of Malinau's most valuable resource. Given the low future potential for sustainable logging, the most likely use will eventually be conversion to harvest of the last remaining trees.

Only those with a long-term view and high dependence upon the forest can be expected to want to use it carefully and to conserve it (Wollenberg et al, 2006). The state has played this role in most nations, as have traditional communities in some places. The case study of Setulang, where the community protected 5000ha of their forest despite offers from IPPK brokers of US$300,000, indicates the possibilities for community-based conservation. The Setulang experience also indicates the difficulties of community-based conservation: tenure is insecure; boundaries are disputed and difficult to enforce; funding for conservation support is difficult to acquire for small amounts of land; neighbouring villages show jealousy over the extra attention given; and internal factions in a village threaten the longevity of long-term commitments.

Most villages in the upper Malinau watershed have consistently wanted to set aside protected forest and forests for shifting cultivation in their areas. In 1999 and 2000, Sheil et al (2002) found that the perceived future value of forests was higher than its past value, with people responding that they took the forest for granted before. Indeed, the increasing scarcity of forest resources awakened some people to the need to protect what forest remains, even before IPPKs operated. In the survey by Sheil et al (2002), people mentioned that forest was their single most important land use. People valued forests primarily for the food they provided, with bearded pig and sambar deer rating highest. Timber trees were the most valued plant species, especially for construction, followed by rattan and the *Arenga undulata* palm. Forests are also considered a safety need when harvests fail.

Despite local importance assigned to the forest, it would seem premature to expect communities to assume responsibility for protecting their forests in the near term, especially given the lack of a strong formal institutional basis for enforcing agreements and conservation practices. The IPPK experience showed that few communities took a long-term perspective on their forests when lured by significant cash benefits. Changing economic options in the area, insecure tenure, competition for forests and the changing importance of forests in local livelihoods has transformed local people's needs. Sheil et al (2002) found in their survey that communities showed an increasing interest in forests for timber (corresponding to the beginning of the IPPK period), saleable items and recreation, and a decreasing importance of forests for medicinal use, firewood and light construction, although the potential of the forest for some of these uses was still deemed important. Community forestry has not been, and will not be, possible in Malinau in the context of entrepreneurial logging run by powerful brokers with district government backing. For the same reason, district governments with five-year mandates and driven by money politics should *not* be solely responsible for forests.

Any responsibility for long-term use or conservation of Malinau's forest plainly needs to be shared across different levels of governance. Decentralization and reforms offer the opportunity to create this division. Communities living in the vicinity of the forest could be more involved in planning and making decisions about how it can be managed, especially in ways that yield benefits that they most value – food, timber, rattan, *gaharu* and clean water. Their proximity and knowledge of the forest should be used to good advantage. The district government could take a primary role in acknowledging and supporting different strategies among different ethnic groups, coordinating efforts among communities, enhancing economic opportunities and enforcing national regulations. The province and central government could take the lead in providing technical information, sharing lessons learned, and overseeing and creating incentives for enforcing national regulations.

In this alternative vision, the role of the state in forestry is restored, but in partnership. Key to this partnership is that in forest matters, local government remains accountable to both central government and the citizenry, rather than local entrepreneurs. Instead, what we have seen is an increase in state control over forests at the district level, with little control at the centre or community level. Because the central forestry department resisted decentralization from the start and has rejected, rather than tried to work with, local government, it forfeited its opportunity to maintain control over forests under decentralization, except by coercion. The central government, meanwhile, has been left trying to distance itself from any responsibility for the current spate of logging, with corresponding campaigns against 'illegal logging' and accusations against Malaysia for fuelling demand for Indonesia's timber.

These experiences are in contrast to devolution elsewhere, where central forestry departments devolved forestry activities to local groups but maintained – with differing degrees of success – hierarchical control over the terms of forest management through user organizations, sharing of benefits, harvesting quotas and taxes (Shackleton and Campbell, 2001; Contreras, 2003; Liu and Edmunds, 2003; Sarin et al, 2003). While we cannot recommend these models, being too bureaucratic and centrally controlled, we do think that the centre needs to play a real role in overseeing and assisting districts. Until roles and responsibilities are better defined across these levels, inefficient and unchecked use of forests and the gridlock of uncertainty will most certainly continue.

Looking forward

Decentralization and reforms in post-Soeharto Indonesia have shifted *de facto* control over forests to the district, raised conflicts, increased short-term benefits and degraded forests at faster rates. To understand the significance of these changes for the future, we have to understand them in the context of the remarkable speed with which reforms occurred. The first years of reform released people from 32 years of extreme central government control, severe limits on political and economic expression, exploitation and expropriation of resources, and systematic denigration and disempowerment of Dayak ethnic identity and organization. We suggest that much of the increase in conflict, competition and short-term benefits is related to this unleashing of opportunity in an unstable institutional environment. Seen in a broader historical context, much of what is happening now is driven by the uncertainty of change itself.

While the intensity of conflict, competition and short-term benefits should decline, we envision local political orders continuing to strengthen. These orders will result in struggles between social forces and the state at the local level, and between the centre and the districts. While these tensions will provide a healthy balance of control across different levels of society and the state, excessive struggles on the part of any one participant may also perpetuate ambiguity and prevent consolidation of formal institutions. The most important of these institutions for the management of forests and local people's well-being are those dealing with property, conflict, popular representation, and forest use and conservation. The institutions need to be simple, transparent and understandable across society and the state (de Soto, 2000), with clear incentives. To the extent that these institutions remain ambiguous, the powerful will co-opt the benefits from forests and the less powerful will make only marginal gains. Indonesia's challenge is to determine how to develop these key institutions in the context of the emerging new local political orders.

Should we feel optimistic or pessimistic about the future? The next phase of governance in Indonesia will be one of adjusting to these first years of transition and developing institutions appropriate to current circumstance. We have witnessed five years of remarkable learning and adaptation to the opportunities of change. We have every reason to believe that such learning and adaptation will continue. Local governments should not be alone in the exciting challenges that lie ahead. Collaboration will be necessary at all levels to develop the institutions necessary to achieve stable agreements and long-term benefits from forests. Riding the rapids is exhilarating as long as the greatest dangers are avoided and one can find calm waters at the end.

Notes

1 Yvan Biot of the UK Department for International Development (DFID) is thanked for this observation.
2 From the stakeholder analysis during a Workshop on Poverty and Decentralization, Malinau, October 2003.
3 Note that Sheil et al (2002) found in 1999 to 2000 that 87 per cent of respondents already saw their lives as better than five to ten years earlier, suggesting that perceptions of improvements may be based on developments that preceded the reform era. These might include the development of roads, increased availability of outboard motors and the high prices of *gaharu*.
4 The exact amount is impossible to calculate as allocations before decentralization reforms were made by Bulungan District, and we do not know how much funding reached the different parts of Malinau.
5 The quantity paid was 8.2 million rupiah, or US$965, using an exchange rate of US$1 = 8500 rupiah.
6 The high proportion of Lundaye reporting improved livelihoods may reflect the fact that the villages sampled were very close to Malinau town, where development benefits have been highest. The differences among ethnic groups are significant (P = 0.011, using the Pearson chi-square test).
7 Differences among ethnic groups were highly significant (P = 0.000, using the Pearson chi-square test).

References

Andrianto, A. (2006) *The Role of District Government in Poverty Alleviation: Case Studies in Malinau and West Kutai Districts, East Kalimantan, Indonesia*, Governance Programme Report, CIFOR, Bogor, Indonesia

Contreras, A. (2003) 'Creating space for local forest management: The case of the Philippines', in Edmunds, D. and Wollenberg, E. (eds) *Local Forest Management: The Impacts of Devolution Policies*, Earthscan, London

de Soto, H. (2000) *The Mystery of Capital: Why Capitalism Triumphs in the West and Fails Everywhere Else*, Basic Books, NY

Dove, M. R. (1988) 'The ecology of intoxication among the Kantu' of West Kalimantan', in Dove, M. R. (ed) *The Real and Imagined Role of Culture in Development: Case Studies from Indonesia*, University of Hawaii Press, Honolulu, HI

Iskandar, H., Snook, L. K., Toma, T., MacDicken, K. G. and Kanninen, M. (2006) 'A comparison of damage due to logging under different forms of resource access in East Kalimantan, Indonesia', *Forest Ecology and Management*, vol 237, no 1–3, pp83–93

Kaimowitz, D., Faune, A. and Mendoza, R. (2003) 'Your biosphere is my backyard: The story of Bosawas in Nicaragua', *Policy Matters*, vol 12, pp6–15

Li, T. M. (1999a) 'Compromising power: Development, culture and rule in Indonesia', *Cultural Anthropology*, vol 14, no 3, pp295–322

Li, T. M. (1999b) 'Marginality, power and production: Analysing upland transformation', in Li, T. M. (ed) *Transforming the Indonesian Uplands: Marginality, Power and Production*, Harwood Academic Publishers, London

Liu, D. and Edmunds, D. (2003) 'The promises and limitations of devolution and local forest management in China', in Edmunds, D. and Wollenberg, E. (eds) *Local Forest Management: The Impacts of Devolution Policies*, Earthscan, London

Moyer, B. (2001) *Doing Democracy: The MAP Model for Organizing Social Movements*, New Society, Gabriola Island, BC

Palmer, C. (2004) *The Role of Collective Action in Determining the Benefits from IPPK Logging Concessions: A Case Study from Sekatak*, East Kalimantan, CIFOR Working Paper, CIFOR, Bogor, Indonesia

Sarin, M., Singh, N., Sundar, N. and Bhogal, R. (2003) 'Devolution as a threat to democratic decision-making in forestry? Findings from three states in India', in Edmunds, D. and Wollenberg, E. (eds) *Local Forest Management: The Impacts of Devolution Policies*, Earthscan, London

Shackleton, S. E. and Campbell, B. M. (2001) *Devolution in Natural Resource Management: Institutional Arrangements and Power Shifts: A Synthesis of Case Studies from Southern Africa*, SADC Wildlife Sector Natural Resource Management Programme, Lilongwe, Malawi and WWF (Southern Africa), Harare

Sheil, D., Puri, R. K., Basuki, I., van Heist, M., Syaefuddin, Rukmiyati, Sardjono, M. A., Samsoedin, I., Sidiyasa, K., Chrisandini, Permana, E., Angi, E. M., Gatzweiler, F. and Wijaya, A. (2002) *Exploring Biological Diversity, Environment and Local People's Perspectives in Forest Landscapes*, CIFOR, Bogor, Indonesia

Smith, J., Obidzinski, K., Subarudi, I. and Suramenggala, I. (2003) 'Illegal logging, collusive corruption and fragmented governments in Kalimantan, Indonesia', *International Forestry Review*, vol 5, pp293–302

Wollenberg, E., Moeliono, M., Limberg, G., Iwan, R., Rhee, S. and Sudana, I. M. (2006) 'Between state and society: Local governance of forests in Malinau, Indonesia', *Forest Policy and Economics*, vol 8, pp421–433

Wulan, Y. C., Yasmi, Y., Purba, C. and Wollenberg, E. (2004) *Analisa Konflik: Sektor Kehutanan di Indonesia 1997–2003*, CIFOR, Bogor, Indonesia

Index

For Product Safety Concerns and Information please contact our EU representative GPSR@taylorandfrancis.com Taylor & Francis Verlag GmbH, Kaufingerstraße 24, 80331 München, Germany

Batch number: 10725973

Printed by Printforce, the Netherlands